Springer Texts in Business and Economics

Springer Texts in Business and Economics (STBE) delivers high-quality instructional content for undergraduates and graduates in all areas of Business/Management Science and Economics. The series is comprised of self-contained books with a broad and comprehensive coverage that are suitable for class as well as for individual self-study. All texts are authored by established experts in their fields and offer a solid methodological background, often accompanied by problems and exercises.

Gerhard Larcher

The Art of Quantitative Finance Vol. 3

Risk, Optimal Portfolios, and Case Studies

Springer

Gerhard Larcher
Institute for Financial Mathematics and
Applied Number Theory
Johannes Kepler University of Linz
Linz, Austria

ISSN 2192-4333 ISSN 2192-4341 (electronic)
Springer Texts in Business and Economics
ISBN 978-3-031-23866-6 ISBN 978-3-031-23867-3 (eBook)
https://doi.org/10.1007/978-3-031-23867-3

This Springer imprint is published by the registered company Springer Nature Switzerland AG
The registered company address is: Gewerbestrasse 11, 6330 Cham, Switzerland

Preface to Volume III

In Volume III we will deal with different aspects of risk management, as for example VAR and CVAR and with the credit-risk systems Credit Metrics and Credit Risk$^+$. Then we give an introduction to portfolio-selection theory, as well as to the basic techniques of optimal stochastic control theory and its application to the classical optimal consumption and investment problem. Finally, we give an analysis of 14 concrete real-world case studies where we apply our knowledge obtained in the first 9 chapters of these books.

About This Book

This book was translated by Nina Sattler-Hovdar, based on the highly successful German book "Quantitative Finance: Strategien, Investments, Analysen" by Springer Gabler (2020).

Contents

1 Risk Measurement and Credit Risk Management 1

1.1 Simple Risk Measures and Basic Features of Credit Risk Management 1
1.2 Value at Risk 2
1.3 Calculating VaR for Simple Portfolio Examples: Example 1 (an Index) 4
1.4 Excursus: Distribution of the Minimum of a Brownian Motion and Calculation of the Adapted VaR 11
1.5 Calculating VaR for Simple Portfolio Examples: Example 2 (a Stock Index in Foreign Currency) and Stress Testing 16
1.6 Example 3: VaR of a Portfolio of Two Stocks as a Function of Their Correlation 20
1.7 Example 4: VaR Estimation for a Simple Options Strategy 21
1.8 Conditional VaR 28
1.9 CVaR Estimation: Some Examples 30
1.10 Credit Risk Management: Introduction 33
1.11 Credit Metrics, Part I: Basic Approach 38
1.12 Credit Metrics, Part II: Ratings and Fair Value of Bonds at Risk of Default 39
1.13 Credit Metrics, Part III: Rating Transition Probabilities and Expected Value of the Loan Portfolio in 1 Year 42
1.14 Credit Metrics, Part IV: Variance of a Credit Portfolio's Value in 1 Year 44
1.15 Credit Metrics, Part V: Variance of a Credit Portfolio's Value in 1 Year 46
1.16 Credit Metrics, Part VI: The Asset Value Model for Determining Joint Credit Rating Transition Probabilities 47
1.17 Credit Metrics, Part VII: Determining the Percentile of A_1 Using Monte Carlo Simulation 52
1.18 Credit Metrics, Part VIII: Detailed Example with Variants 57
1.19 Excursus: Generating Positive Definite Random Correlation Matrices 70
1.20 Credit Risk+, Part I: Basic Concept and Expected Number of Defaults Assuming Independent Loans 75

1.21 Credit Risk+, Part II: Expected Loss Amount Assuming Independent Loans 80
1.22 Credit Risk+, Part III: Allocating Loans to Mutually Independent Sectors 83
1.23 Credit Risk+, Part IV: Allocation of Loans to Mutually Independent Sectors and Determination of Required Capital Ratio 86
1.24 Credit Risk+, Part V: An Illustrative Example 88
References 93

2 Optimal Investment Problems 95
2.1 Classic Portfolio Optimization According to Markowitz, Part 1: Fundamental Principles 96
2.2 Classic Portfolio Optimization According to Markowitz, Part 2: Two Financial Assets 100
2.3 Classic Portfolio Optimization According to Markowitz, Part 3: Two Assets, Efficient Border 106
2.4 Classic Portfolio Optimization According to Markowitz, Part 4: Two Assets, with Short Selling 110
2.5 Classic Portfolio Optimization According to Markowitz, Part 5: Two Risky Assets and One Risk-Free Asset. Portfolios with Maximal Sharpe Ratio 113
2.6 Classic Portfolio Optimization According to Markowitz, Part 6: Two Risky Assets and One Risk-Free Asset: Practical Example 119
2.7 Classic Portfolio Optimization According to Markowitz, Part 7: Sketch of a Sensitivity Test 121
2.8 Classic Portfolio Optimization According to Markowitz, Part 8: Arbitrary Number of Assets, Principal Form of Opportunity Sets 122
2.9 Classic Portfolio Optimization According to Markowitz, Part 9: Arbitrary Number of Assets, Explicit Calculation of Efficient Border, and Portfolio with Maximal Sharpe Ratio: Planning the Procedure 127
2.10 Classic Portfolio Optimization According to Markowitz, Part 10: Arbitrary Number of Assets, Explicit Calculation of Efficient Border, and Portfolio with Maximal Sharpe Ratio 127
2.11 Classic Portfolio Optimization According to Markowitz, Part 11: Arbitrary Number of Assets, Explicit Calculation of Efficient Border, and Portfolio with Maximal Sharpe Ratio: Practical Example 138
2.12 The "Market Portfolio": The Portfolio with Maximal Sharpe Ratio 141
2.13 Portfolio Selection Based on a Single-Index Model 142

2.14 The Optimal Investment and Consumption Problem: Introduction and Formulation of the Problem 149
2.15 Fundamentals of Stochastic Control Theory, the HJB Equations ... 154
2.16 An Application Example for the HJB Equation: The Linear Regulator 159
2.17 Solving the Optimal Consumption Investment Problem 163
References 168

3 Case Studies 169
3.1 Case Study I: The Fynup Ratio 171
3.1.1 The Task 171
3.1.2 Definition of Terms and General Remarks.................. 172
3.1.3 Technical Preparations 174
3.1.4 One Possible Intuitive Approach 176
3.1.5 The Principle of the Fynup-Ratio Approach: Determination of Weights – Schematic 178
3.1.6 The Principle of the Fynup-Ratio Approach: Determination of Weights – Technical Details 180
3.1.7 Optimization Using a Monte Carlo Method 182
3.1.8 Adaptation of Weight Selections 183
3.1.9 Indication for Funds with at Least 10 Years of History 185
3.1.10 Informal Indication for Funds with Less Than 10 Years of History 186
3.1.11 The Result and the Percentile Representation 186
3.1.12 Informative Value of the Indication Results – Conclusions 188
3.2 Case Study II: Expert Witness Opinion on Churning 189
3.2.1 Specifics of the Case 189
3.2.2 The Question in a More Precise Wording.................. 192
3.2.3 Modelling the Underlying Assets 193
3.2.4 Profit Probabilities for the Individual Positions When Held to Expiration 194
3.2.5 Profit Probabilities for the Individual Positions in the Case of Profit-Taking Before Expiration............... 197
3.2.6 Profit-Taking Probabilities for the Option Portfolio........ 199
3.3 Case Study III: Pricing an Interest Rate Swap and Analysing Its Suitability for Optimizing a Loan Portfolio...................... 201
3.3.1 Presentation of the Case, the Products, and the Issues 201
3.3.2 Fair Value of the Swap at Inception: Part 1 – Interest Rate Swap Component....................................... 206
3.3.3 Fair Value of the Swap at Inception: Part 2 – Option Component 208
3.3.4 Regarding the Intent of Achieving "Portfolio Optimization" by Using the Swap.......................... 212

3.4 Case Study IV: Valuation of Callable Range Accrual Swaps as Part of an Expert Opinion 219
3.4.1 The Product 219
3.4.2 Motivation for Entering into the Swap and Further Background Information 221
3.4.3 Modelling and Calibration of the Requisite Interest Rates in the Vasicek Model 223
3.4.4 Valuation of the RAS Based on the Vasicek Model *without* Considering the Termination Right 229
3.4.5 Valuation of the RAS Based on the Vasicek Model *with* Termination Right 232
3.4.6 Test of the Risk-Free Model's Calibration Quality and Sensitivity Analysis of the Results 233
3.5 Case Study V: Analysis of the Performance of Put-Write Strategies 236
3.5.1 Setup of the Tested Put-Write Strategies 238
3.5.2 Discussion and Examples of Some Parameter Choices 241
3.5.3 The Various Parameter Choices and Selection of Some Analysis Results 245
3.5.4 Variant: 2-M Strategy 247
3.5.5 Instructions for the Use of the Analysis Program on the Website 247
3.5.6 Closing Remarks on Operational Risks and Strategy Variants 251
3.6 Case Study VI: Valuation of an Asian Option on the Basis of a Foreign Currency Loan Within the Scope of an Expert Witness Report 254
3.6.1 Overall Product Structure 255
3.6.2 Fair Price of the Guarantee Bond in Asian and European Style 257
3.6.3 Preparation of an Ex Ante Risk Assessment from the Perspective of Realistic Earnings Prospects When Using Debt Financing 261
3.6.4 Actual Performance of the Product Combination 269
3.7 Case Study VII: Expert Opinion on the EUR CHF Stop-Loss Order Fiasco in January 2015 271
3.7.1 Analysis of Comparable Events: GBP DEM September 1992 272
3.7.2 The GBP DEM Case: Comparison of Stop Loss Versus Hedging with Put Options in the Period October 1990 to October 1992 277
3.7.3 The GBP DEM Case: Hedging with Put Options for 10-Year Loans as from October 1990 278
3.7.4 Alternative Hedging Method for the EUR CHF Exchange Rate Using Put Options 279

3.7.5 Hedging Using Put Options for 10-Year or 20-Year Loans as from January 2012 ... 283
3.7.6 Comparison with the Stop-Loss Approach ... 286
3.8 Case Example VIII: Analyses of the Risk of Individual Higher-Quality Bonds Versus Portfolios of Lower-Quality Bonds ... 287
3.8.1 Two Simple Hypothetical Examples for Illustration ... 288
3.8.2 Analyses Based on Realistic Data ... 290
3.9 Case Study IX: Portfolio Selection Based on Sustainability Parameters ... 296
3.9.1 The Problem ... 296
3.9.2 Monte Carlo Determination of Efficient Frontiers and Market Portfolios Subject to Sustainability Constraints and Optimal Sustainability Sharpe Parameters in the Case *without* Short Selling ... 302
3.9.3 Determination of Efficient Frontiers and Market Portfolios Subject to Sustainability Constraints and Optimal Sustainability Sharpe Parameters in the Case *with* Short Selling ... 304
3.9.4 Outlook ... 310
3.10 Case Study X: Comparison of Two Basket Derivatives ... 311
3.10.1 Detailed Description of the Two Products ... 311
3.10.2 First Brief Comparison of the Two Products ... 314
3.10.3 Comparison of the Two Product Values on 1 June 2008 ... 315
3.11 Case Study XI: Optimal Hedging with Future Contracts in Connection with Liquidity Risks and the "Metallgesellschaft" Case ... 320
3.11.1 Presentation of the Hedging Strategy and the Problem ... 321
3.11.2 Translation of the Problem into the Continuous Case and Glasserman's Strategies ... 326
3.11.3 The Solution to the Optimal Hedging Problem ... 329
3.11.4 The Proof Sketch ... 330
3.12 Case Study XII: Computing the Payout Ratios for a Financial Market Game ... 336
3.12.1 Description of the Game ... 336
3.12.2 Computation of the Payout Ratio ... 338
3.12.3 Computation of Probability P_1 ... 339
3.12.4 Computation of Probabilities P_2 and P_3 ... 340
3.12.5 Testing the Results Using Monte Carlo Simulation and Concluding Remarks ... 343
3.13 Case Study XIII: Trading Strategies Based on Hedging ... 344
3.13.1 Background ... 344
3.13.2 An Initial Idea of a Trading Strategy ... 345
3.13.2.1 The Strategy ... 345
3.13.2.2 Implementation in Reality ... 346

3.13.2.3 Estimating Realized Volatility 346
3.13.2.4 Executing the Strategy on an Example 347
3.13.2.5 Analysing the Strategy Through Backtesting ... 348
3.13.2.6 Extension of the Strategy 348
3.13.3 A Second Variant of the Strategy........................... 350
3.13.3.1 Analysing the Strategy Through Backtesting ... 353
3.14 Case Study XIV: Analysis of the Derivative Trading Strategy "Lambda +" 356
3.14.1 Introduction 356
3.14.2 Definition of the Strategy and Its Variants 357
3.14.3 The Task and the Analyses 362
References 368

Risk Measurement and Credit Risk Management

1

Keywords

Value at risk (VaR) · Conditional VaR · Diversification · Credit risk management · Rating classes · Rating transition probabilities · Default probabilities · Credit Metrics · Credit Risk+

1.1 Simple Risk Measures and Basic Features of Credit Risk Management

In Chap. 1, we will introduce –again in terms of their basic principles only– the most essential risk measures for financial products and financial portfolios, the "value at risk (VaR)" and the "conditional value at risk (CVaR)", which is also referred to as "expected shortfall".

In the course of our discussion of how to calculate these risk measures using historical simulation, we will also briefly address the topic of "backtesting" (and "stress testing") of trading strategies. These two topics will then play a major role when we get to our case studies in the last chapter.

We are going to perform the VaR calculation already in this Chap. 1, applying it to a challenging problem: namely, computation of the risk of a large complex credit portfolio. For this purpose we will introduce J.P. Morgan's "Credit Metrics" system (which is based on VaR calculation). As an alternative approach to calculating credit risk, we will also look at a second system, Credit Suisse First Boston's "Credit Risk+" method.

Determining credit risk may seem to be a bit of a special problem, yet the principal approaches used in the two systems presented here can also be used and applied –in slightly adapted ways– in many other contexts. Its application here to credit risk calculation is therefore more of a "discussion starter'" for explaining the approaches in principle.

G. Larcher, *The Art of Quantitative Finance Vol. 3*, Springer Texts in Business and Economics, https://doi.org/10.1007/978-3-031-23867-3_1

As so often in the course of this book, we reiterate: Our aim here is to convey basic principles and not to elaborate in detail on certain methods and principles.

1.2 Value at Risk

We consider a portfolio PF consisting of financial products whose value PF(t) at time t we specify in one currency (e.g. the EUR).

We assume a specific time range T (e.g. $T = 1$ year, or $T = 10$ days, or $T = N$ days).

A certain percentage X% is also given (e.g. X = 99%).

The **value at risk (VaR) of the portfolio PF for the time horizon *T* and the confidence level *X*** is then the value V, for which we have: "With probability X, the portfolio loss at time T (as from now), will be at most V."

This definition is somewhat informal from a mathematical point of view and can sometimes lead to ambiguities in the details. A mathematically correct definition of the VaR V would be:

$$V = \inf\{v \mid \textit{probability}\ (PF(0) - PF(T) < v) > X\%\}$$

That is:

- PF(0) – PF(T) is, after all, the amount of loss from today (time 0) to time T.
- We look at all possible (large) losses v that are so large that they only occur with a probability smaller than X.
- From all of these v, we choose the smallest one (the infimum).

Graphically, the concept can be illustrated as shown in Fig. 1.1:

- We assume a current portfolio value of 1,000,000 EUR in Fig. 1.1.
- Losses can be up to 1,000,000 EUR.
- Profits in this example are capped at approximately 160,000 EUR.
- The area under the graph representing the distribution density of the possible losses for the next time period of length T (negative values represent profits) to the right of the value 865,000 is 0.01, i.e. 1% of the total area under the density function.
- So there is only a 1% probability that the loss at time T will be higher than 865,000 EUR.
- The probability for the loss at time T to be smaller than 865,000 EUR is thus $X = 99\%$.
- The VaR for the time range T at confidence level 99% is thus 865,000 EUR.

VaR is often expressed not in absolute terms but as a percentage of the current portfolio value.

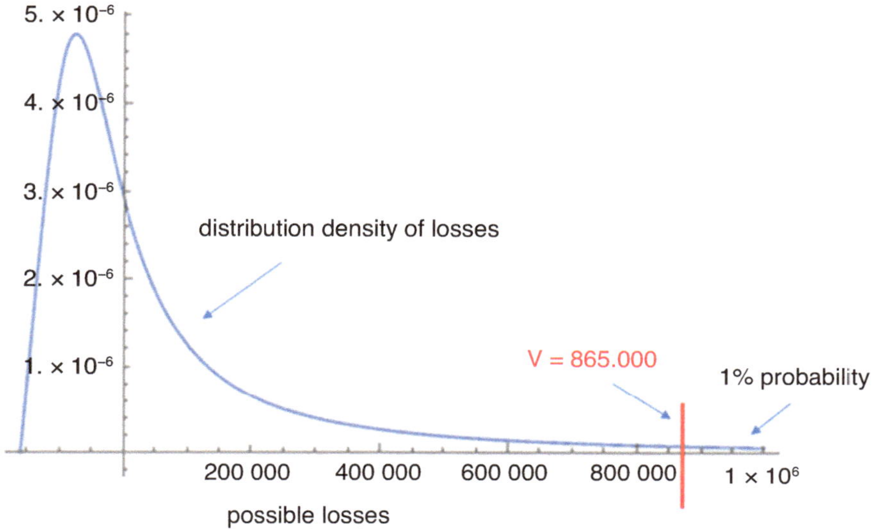

Fig. 1.1 Illustration of the concept of value at risk

By the way, VaR estimation can be applied not only to a given static portfolio but also to any trading strategy that has been clearly defined in advance for the time interval $[0, T]$. In that case, however, the usual precautions that need to be heeded also when backtesting trading strategies (and which we will discuss further below) must then be taken into account.

In many cases, the definition of VaR is somewhat inexact in terms of the risk measurement period: The VaR relates exactly to the time point T. Thus, it measures the loss of the portfolio **at** time T (from now). It is an estimate of the probability that the loss **at time *T*** will be this or that amount.

In some cases, however, it can be quite relevant to know how likely it is for losses to exceed a certain level at some point in time **up to** time T. Normally this probability is of course higher than the VaR (as we will also see in some of the following examples).

This essential distinction is missing in many definitions of VaR, or they do not make clear that VaR is about the exact point in time T and not about the time span up to time T!

VaR is essential, for example, in calculating the capital requirements of banks. The 10-day VaR at a confidence level of 99% is primarily used for this purpose.

1.3 Calculating VaR for Simple Portfolio Examples: Example 1 (an Index)

VaR is usually calculated (or, more precisely: "estimated") using historical simulation or modelling and, in most cases, subsequent Monte Carlo simulation.

Historical simulation is essentially equivalent to backtesting the portfolio strategy. In this process, a portfolio (or trading strategy) is analysed over a longer test period in the past. More specifically, one systematically takes periods of length T in the past and determines how large the losses of the portfolio (or the trading strategy) were in these periods. This yields an empirical distribution function of the loss levels. One then uses this distribution function to estimate the VaR from the current situation's perspective. (We will do this below using simple examples.) An important factor in this approach is, of course, what past period is chosen to create the empirical distribution function.

Calculation (estimation) of the **VaR based on modelling**. The approach here is to model the individual components of the portfolio in a suitable manner. Taking into account the correlations between the components, this provides a model for the portfolio's performance. If the resulting model is not too complex, then the VaR for this model may even be calculated exactly (which is then an approximation of the portfolio's "actual" VaR). Normally, however, Monte Carlo simulation is necessary to determine the model's VaR. Of course, choosing a suitable model and its parameters is of the essence for this approach.

In the following we will discuss some simple examples to illustrate VaR calculation for individual products, for a combination of products, and for a simple trading strategy. In so doing, we will always analyse both the correct VaR approach (with respect to the specific point in time T) and the adapted approach (with respect to the entire period up to the point in time T).

Example 1.1 As a warmup exercise, we consider an investment of 100,000 EUR in the DAX and are interested in finding the 10-(trading)-day VaR at a 99% confidence level from the perspective of 11 March 2019. (We will calculate the VaR for other parameters as well, which can be done with no extra effort once we have created the environment for **one** such calculation.)

First we create an **estimate using historical simulation**.

To illustrate how the results depend on the period that is selected for estimating the empirical distribution function, we choose three past periods of different lengths. Figure 1.2 shows histograms of the distributions of the 10-day DAX returns in the last 13 years, in the last 9 years, and in the last 3 years.
In each case, the data (the 10-day returns) are sorted by magnitude:
$x_1 < x_2 < x_3 < \ldots < x_{m-1} < x_m$.

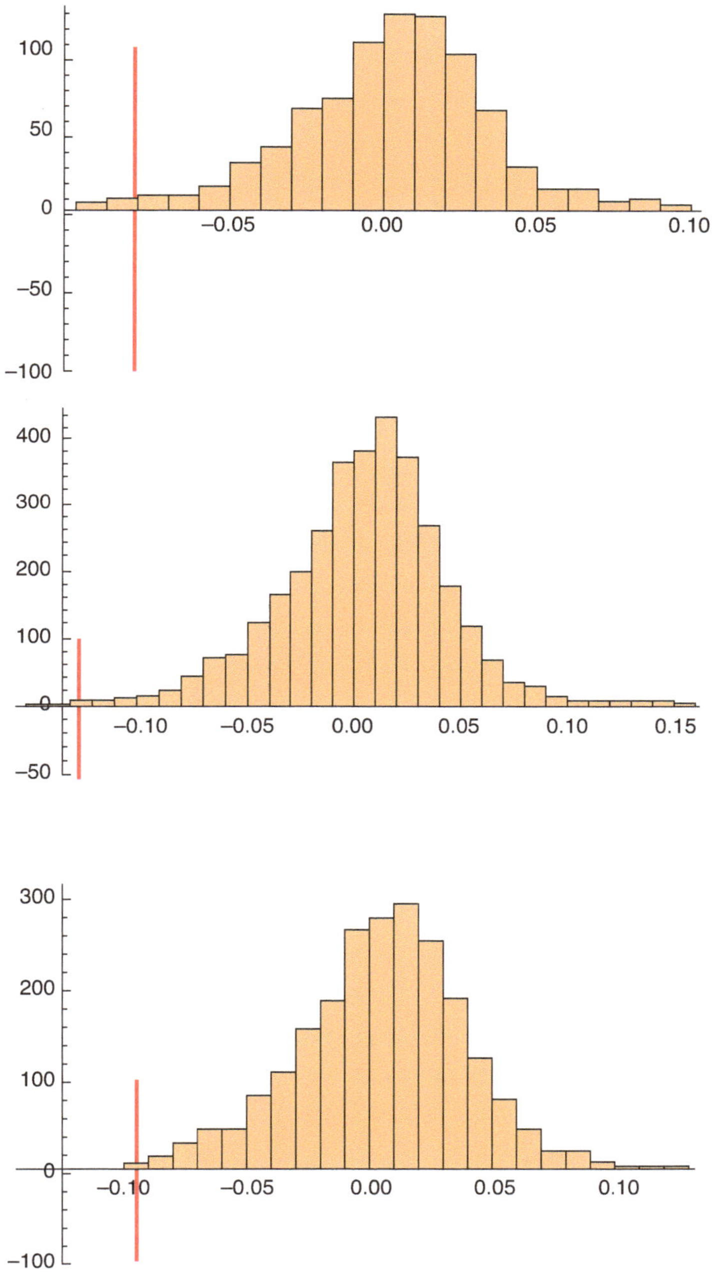

Fig. 1.2 Histograms of historical 10-day returns in the time periods March 2006–March 2019 (top), March 2010–March 2019 (middle), and March 2016–March 2019 (bottom) with plotting of the left 1% percentile

Then we determine the index, which determines the first percentile of the data. This is the index $i := \left[\frac{m}{100}\right]$. ($[x]$ in this case denotes the largest integer that is less than or equal to x.) Using x_i, the VaR is then calculated as $100,000 \cdot x_i$. In our example:

If we work with data of the last 13 years, we get $x_i = -0.1282$ $(= -12.82\%)$, and this results in a VaR of 12,820 EUR.

If we work with data of the last 9 years, we get $x_i = -0.0952$ $(= -9.52\%)$, and this results in a VaR of 9520 EUR.

If we work with data of the last 3 years, we get $x_i = -0.0817$ $(= -8.17\%)$, and this results in a VaR of 8170 EUR.

For comparison, we now calculate the 1-day VaR at a 99% confidence level. Figure 1.3 shows histograms of the distributions of the 1-day DAX returns in the last 13 years, in the last 9 years, and in the last 3 years.
Using the analogous procedure to the one described above for 10-day returns, we now get:

If we work with data of the last 13 years, we get $x_i = -0.0400$ $(= -4.00\%)$, and this results in a VaR of 4000 EUR.

If we work with data of the last 9 years, we get $x_i = -0.0336$ $(= -3.36\%)$, and this results in a VaR of 3360 EUR.

If we work with data of the last 3 years, we get $x_i = -0.0273$ $(= -2.73\%)$, and this results in a VaR of 2730 EUR.

Now let's turn to **estimating the VaR using the modelling approach**:

For this purpose, we assume that the DAX price $S(t)$ follows a Wiener model. Hence, $S(t) = S(0) \cdot e^{\mu \cdot t + \sigma \cdot \sqrt{t} \cdot w}$ with an $\mathcal{N}(0, 1)$-distributed random variable w.

We now need to estimate the parameters μ and σ. There are two ways to do this:

One possible approach is to estimate μ and σ from historical data. We already noted above how much this estimate depends on the data used for estimating the parameters.

Another approach –which is the one we will take here– is the following:

As we know from our earlier considerations in the context of derivatives pricing, the market currently sees the trend μ –assuming the EMH (efficient market hypothesis)– at a value of $r - \frac{\sigma^2}{2}$. Here, r is again the risk-free interest rate (we can agree on a short-term overnight interest rate for this, since the VaR estimate is usually for short forecast periods). For σ we could use the value of the DAX

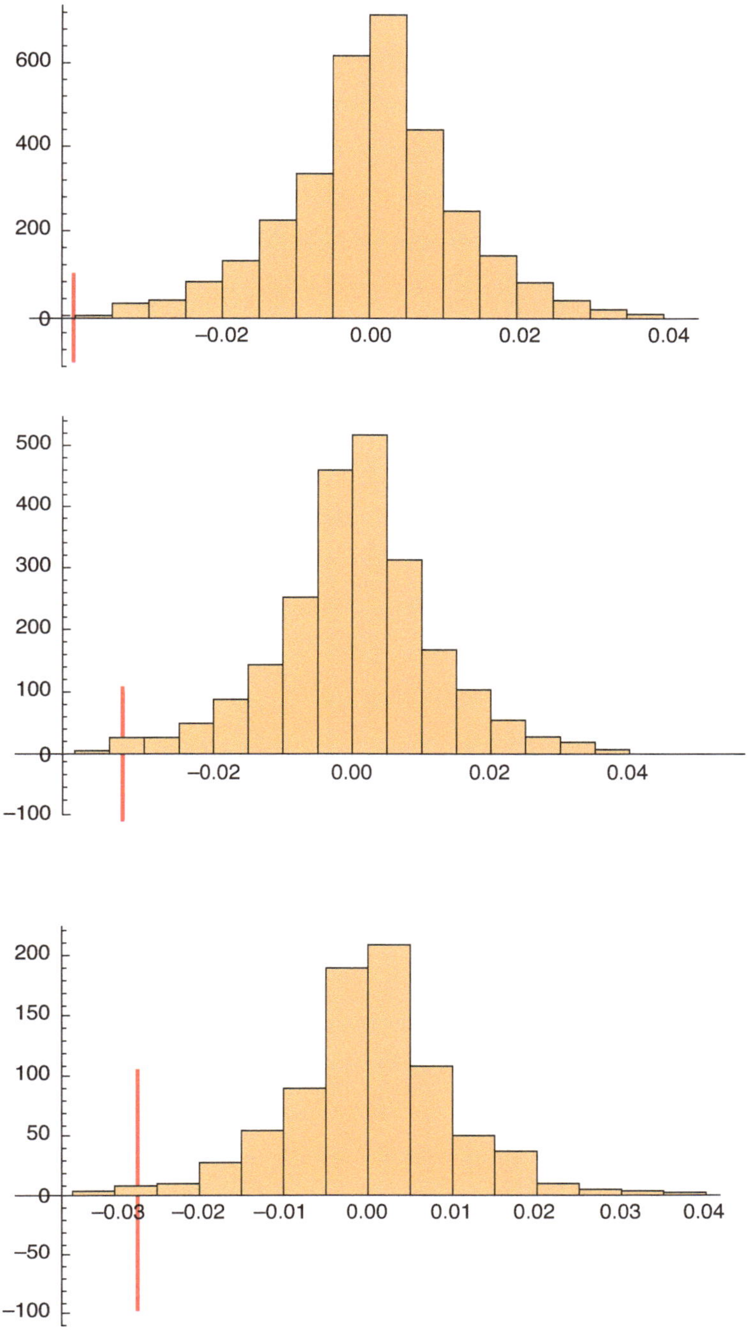

Fig. 1.3 Histograms of historical 1-day returns in the time periods March 2006–March 2019 (top), March 2010–March 2019 (middle), and March 2016–March 2019 (bottom) with plotting of the left 1% percentile

volatility index VDAX of 11 March 2019, which was 15.03% at the start of trading. This would correspond to a σ of 0.1503.

The Eonia overnight rate was −0.364% on 11 March 2019. Thus our chosen $\mu = r - \frac{\sigma^2}{2}$ would be $\mu = -0.015$, and the Wiener model would be

$$S(t) = S(0) \cdot e^{-0.015 \cdot t + 0.1503 \cdot \sqrt{t} \cdot w}.$$

Let us now consider the 10-day percentile again. Since we are referring to trading days here, the time parameter t, which is given in years, is approximately $t = \frac{10}{250}$. The Wiener model for the price moves over the next 10 trading days thus has the form

$$S(t) = 100,000 \cdot e^{-0.0006 + 0.03 \cdot w}$$

with the $\mathcal{N}(0, 1)$-distributed random variable w.

Now we only need the 1% percentile y of the standard normal distribution, that is, the value that w will fall short of with exactly 1% probability. y is obtained as the value of the inverse of the distribution function of the standard normal distribution applied to the value 0.01. This value y is approximately −2.33.

Thus, with this approach we get: The probability that the DAX on day 10 from now will be below the value of $S(t) = 100000 \cdot e^{-0.0006+0.03 \cdot (-2.33)} = 93.193$ is almost exactly 1%. The 10-day VaR with this approach is therefore 6,807 EUR.

To determine the 1-day VaR, we only need to change the above procedure in that for t in the equation $S(t) = S(0) \cdot e^{-0.015 \cdot t + 0.1503 \cdot \sqrt{t} \cdot w}$ we now substitute $t = \frac{10}{250}$ by $t = \frac{1}{250}$. For w, we again set $w = -2.33$.

And so, the probability that the DAX on day 1 from now will be below the value of $S(t) = 100000 \cdot e^{-0.00006+0.0095 \cdot (-2.33)} = 97,805$ is almost exactly 1%. The 1-day VaR with this approach is therefore 2195 EUR.

As we mentioned above, it is sometimes necessary to know not the 1% loss level **at time T** but the 1% loss level **up to time T**. We refer to this in the following as the **adapted VaR**.

A scenario where we would want to know the adapted VaR is, for instance, when we are running a marginable derivative trading strategy with expiration in T and where the margin requirements also depend on the strategy's price. To estimate the probability of a margin call occurring at any point before T, we need the adapted VaR up to time T.

We will now determine this adapted VaR for our DAX example from above and $T = 10$ days for comparison. The approach to this **calculation of the adapted VaR using historical simulation** is obvious: Instead of using historical data to calculate 10-day returns, we calculate for each 10-day period the return from that period's start time to that period's lowest level (starting with the following day). With these values $x_1, x_2, \ldots, x_m$, we now proceed in the same way as above in the case of conventional VaR and obtain the following plots Fig. 1.4.

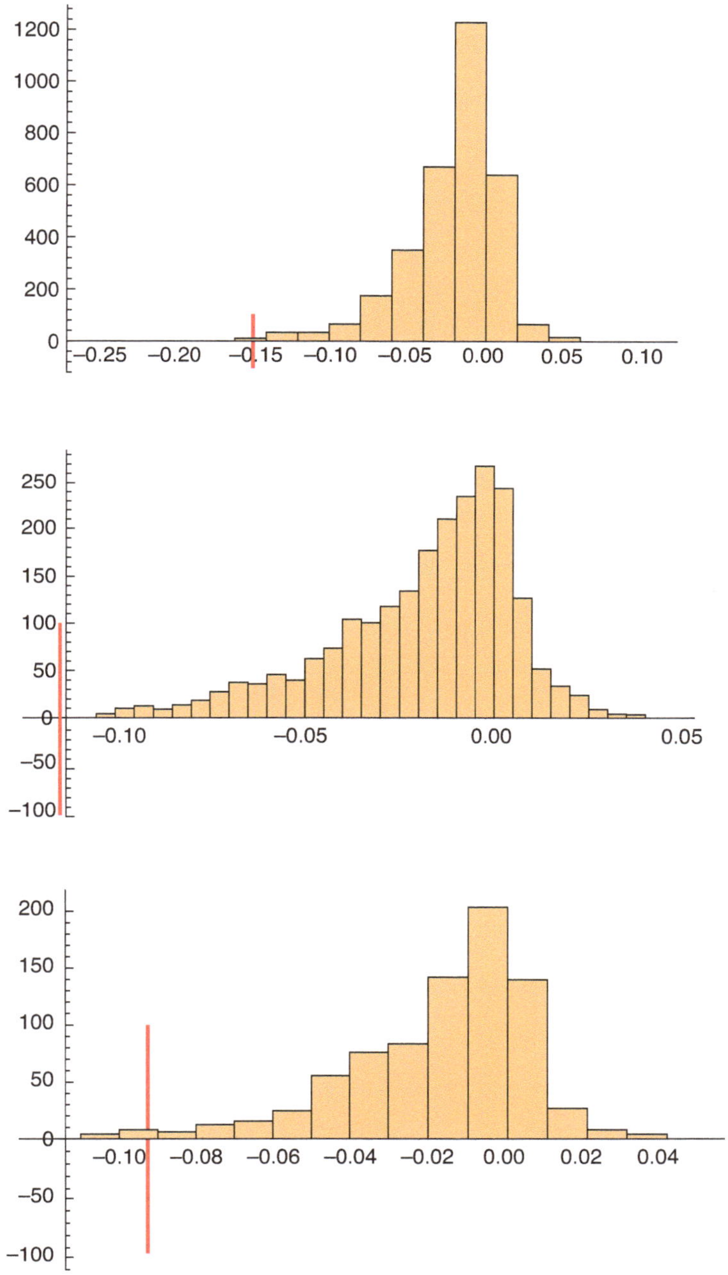

Fig. 1.4 Histograms of historical 10-day maximum losses in the time periods March 2006–March 2019 (top), March 2010–March 2019 (middle), and March 2016–March 2019 (bottom) with plotting of left 1% percentile

If we work with data of the last 13 years, we get $x_i = -0.1490$ $(= -14.90\%)$, and this results in an adapted VaR of 14,900 EUR.

If we work with data of the last 9 years, we get $x_i = -0.1149$ $(= -11.49\%)$, and this results in an adapted VaR of 11,490 EUR.

If we work with data of the last 3 years, we get $x_i = -0.0920$ $(= -9.20\%)$, and this results in an adapted VaR of 9200 EUR.

Warning Depending on the application of this adapted VaR, it may not be sufficient to consider the minimum of the daily closing prices of the next 10 trading days, that is, you may actually need to use the intraday low on each of these following 10 trading days. The calculations are then analogous to the above. Here and in the following, we will be using daily closing prices only, however.

Explicitly calculating the **adapted VaR directly from the DAX modelling** (we use the same modelling approach with the same parameters as above) is possible in principle (see the following Section 1.4) but much more intricate (especially for somewhat more complex portfolios). It is therefore easier to work **with Monte Carlo simulation** here.

So we simulate a large number N ($N = 10{,}000$ in our example) of possible price paths for our portfolio (100,000 EUR invested in the DAX) in ten steps from time 0 to time T (= 10 days) based on the selected performance model (from each time point t to the respective following trading day)

$$S\left(t + \frac{1}{250}\right) = S(t) \cdot e^{-0.015 \cdot \frac{1}{250} + 0.1503 \cdot \sqrt{\frac{1}{250}} \cdot w}$$

Figure 1.5 shows a selection of 30 of these simulated paths. For each path, the portfolio loss is determined from the starting point to the lowest point over the

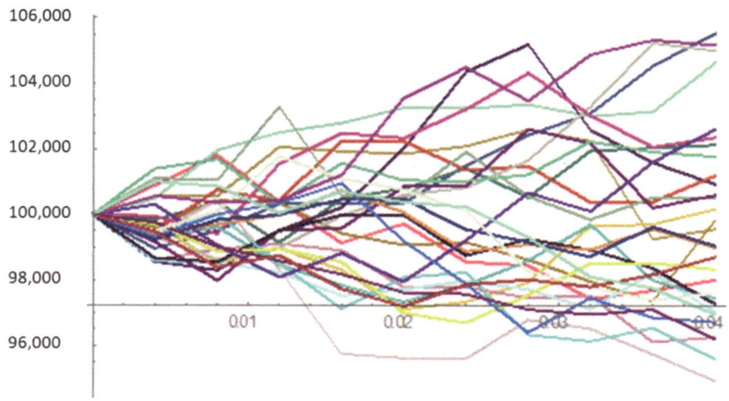

Fig. 1.5 30 simulations of the daily portfolio performance for the next 10 trading days

following 10 days. These N values are again sorted by magnitude, and then the 1% percentile is taken and used as the estimate for the adapted VaR.

The value of the adapted VaR obtained by this procedure is approximately 9300 EUR.

1.4 Excursus: Distribution of the Minimum of a Brownian Motion and Calculation of the Adapted VaR

In Sect. 1.3 we noted, in calculating the adapted VaR through portfolio modelling, that calculating the adapted VaR directly is sometimes possible in principle but is not easy even in the simplest of cases, i.e. when the portfolio moves according to a geometric Brownian motion (a Wiener model).

In this –mathematically somewhat more challenging– section we want to solve this task.

For this we will have to determine the distribution of the maximum of the Brownian motion, and that requires us to delve a bit deeper into the world of the Brownian motion and its mysteries.

We start with a further motivation regarding the question that we will deal with in the following. For this, we recall the motivation and derivation of the Brownian motion in Volume I Section 4.15. There we had assumed a fair game –for example, a fair coin toss– between two players A and B. We then put ourselves in the shoes of player A and visualized this player's wins over the course of N games using graphs as shown in Fig. 1.6.

As we know, with the length of the time axis (T) remaining constant and as the number N of games increases, the structure of such profit curves increasingly converges to a standard Brownian motion. We were interested –obviously– in finding the distribution of the profit curve after N games $G(N)$ and found that we can compute the distribution of $G(N)$ for large N using the distribution of the standard Brownian motion. For large N we have:

$$\begin{aligned} &\textit{probability}\ (G(N) < Y) \approx \\ &\approx \textit{probability}\ (B(N) < Y) = \frac{1}{\sqrt{2\pi N}} \int_{-\infty}^{Y} e^{-\frac{x^2}{2N}} dx \end{aligned}$$

Another obvious question in this context would be in which range the **maximum profit** in these N games is to be found with which probability. That is, where is:

$$G_{max}(N) := \max\left(G(n)\,|0 \leq n \leq N\right)?$$

The maximum $G_{max}(N)$ is also indicated in the above graphs of Fig. 1.6.

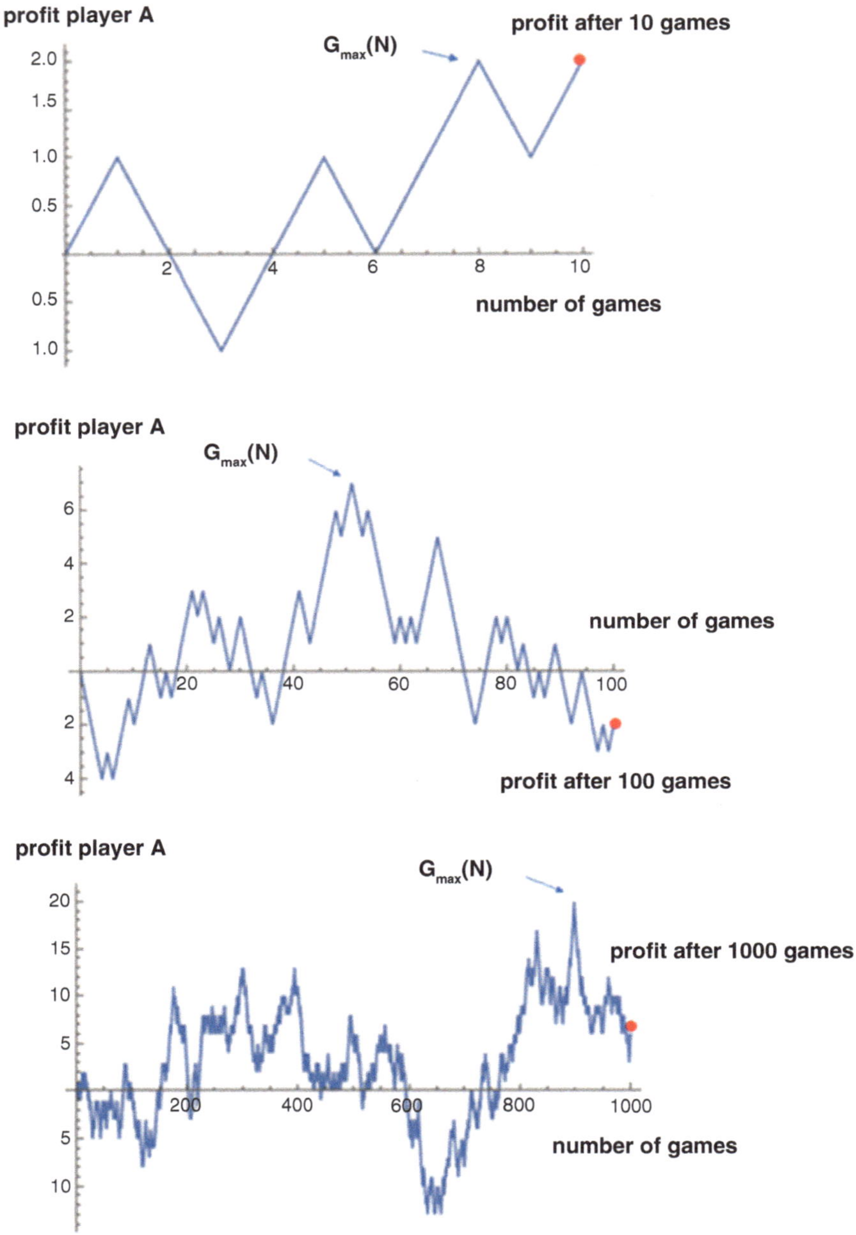

Fig. 1.6 Possible profit curves in a fair game ($N = 10$, 100, and 1000 games)

For large N the probability distribution of $G_{max}(N)$ can again be approximated by the probability distribution of the ("running") maximum $M(N)$ of the Brownian motion, where

$$M(N) := \max(B(t)\,|0 \le t \le N).$$

We are therefore interested in finding this running maximum $M(T)$ of the Brownian motion (we switch from variable N to variable T since the Brownian motion is defined for continuous arguments).

So, what is the probability distribution of the random variable $M(T) := \max(B(t)\,|0 \le t \le T)$? We will show below:

Theorem 1.1 *Let $B(t)$ for $0 \le t \le T$ denote the standard Brownian motion on $[0, T]$ and $M(T) := \max(B(t)\,|0 \le t \le T)$.*

Let $a > 0$. Then we get

$$W(M(T) \ge a) = 2 \cdot W(B(T) \ge a) = 2 \cdot \left(1 - \mathcal{N}\left(\frac{a}{\sqrt{T}}\right)\right),$$

where $\mathcal{N}$ denotes the distribution function of the standard normal distribution.

Proof The proof of the theorem is based on a very simple observation and a cleverly devised approach.

The observation is the following: We have a Brownian motion $B(t)$ on a certain interval $[0, T]$. Then we choose an arbitrary real value a. We then generate paths of this Brownian motion in the usual way, only now the procedure is as follows: As soon as the path of the Brownian motion reaches the value a for the first time, the path of the Brownian motion is mirrored from then on at the line $y = a$. That is, for the $t > a$ in the following, we now use the value $2a - B(t)$ instead of $B(t)$. The motion defined in this way is now denoted by $B_a(t)$ and is called "the Brownian motion mirrored at a". This process is illustrated in Fig. 1.7.

It is relatively obvious that the stochastic process $B_a(t)$ has exactly the same distribution properties as the original Brownian motion $B(t)$ ("mirror principle"). And here now comes the clever "trick":

For purposes of this approach, we define T_a as the first time point t at which $B(t) = a$.

We then get $B_a(t) = \begin{cases} B(t) & \text{if } t \le T_a \\ 2a - B(t) & \text{if } t > T_a \end{cases}$

(continued)

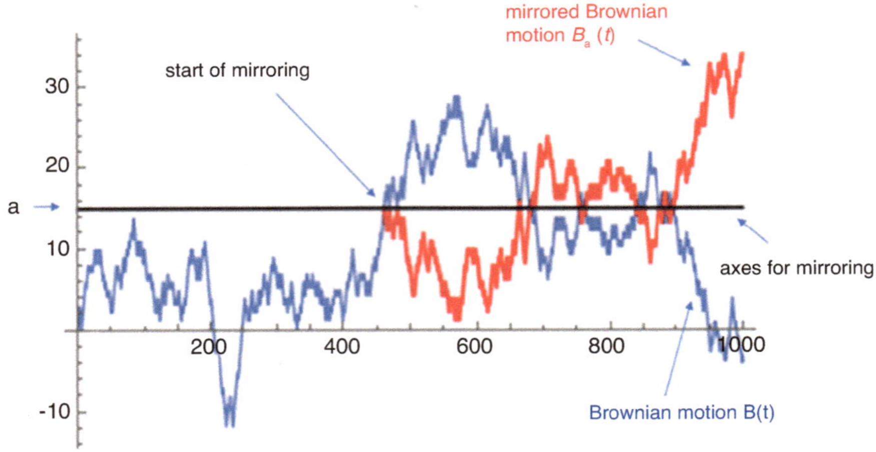

Fig. 1.7 Illustration of the mirrored Brownian motion $B_a(t)$

The following is then obviously true for any positive a:

$W(M_T \geq a) =$
$W(B(T) \geq a) + W(B(T) < a \text{ and } T_a < T)$
= (because of the mirror principle)
$W(B(T) \geq a) + W(B_a(T) < a \text{ and } T_a < T)$
= (because: if $T_a < T$, then $B_a(T) = 2a - B(T)$)
$W(B(T) \geq a) + W(2a - B(T) < a \text{ and } T_a < T) =$
$W(B(T) \geq a) + W(B(T) > a \text{ and } T_a < T)$
= (because: if $B(T) > a$, then automatically $T_a < T$)
$W(B(T) \geq a) + W(B(T) > a)$
= (because: $W(B(T) \geq a) = W(B(T) > a)$)
$2 \cdot W(B(T) > a) = 2 - 2 \cdot W(B(T) \leq a) = 2 \cdot \left(1 - \frac{1}{\sqrt{2\pi T}} \int_{-\infty}^{a} e^{-\frac{x^2}{2T}} dx\right) =$
$2 \cdot \left(1 - \mathcal{N}\left(\frac{a}{\sqrt{T}}\right)\right),$

which is what we needed to show.

□

Of course—because of the symmetry of the Brownian motion, the same relation holds analogously for the minimum ($m(T) = \min(B(t)\,|0 \leq t \leq T)$ of the Brownian motion up to time T:

Let $a < 0$. Then we get

$$W(m(T) \leq a) = 2 \cdot W(B(T) \leq a) = 2 \cdot \mathcal{N}\left(\frac{a}{\sqrt{T}}\right).$$

We will need an analogous result, however, for the Brownian motion with drift. This can be derived from the result for the standard Brownian motion, with some additional effort. Here we will only give this result without proof:

Theorem 1.2 *Let $B_{\mu,\sigma}(t) := \mu \cdot t + \sigma \cdot B_t$ for $0 \leq t \leq T$ denote a Brownian motion with drift on $[0, T]$ and $M_{\mu,\sigma}(T) := \max\left(B_{\mu,\sigma}(t)\,|0 \leq t \leq T\right)$ resp. $m_{\mu,\sigma}(T) := \min\left(B_{\mu,\sigma}(t)\,|0 \leq t \leq T\right)$*

(a) Let $a > 0$. Then we get

$$W(M_{\mu,\sigma}(T) \leq a) = \mathcal{N}\left(\frac{a - \mu \cdot T}{\sigma\sqrt{T}}\right) - e^{2 \cdot \frac{\mu \cdot a}{\sigma^2}} \cdot \mathcal{N}\left(-\frac{a - \mu \cdot T}{\sigma\sqrt{T}}\right)$$

(b) Let $a < 0$. Then we get

$$W(m_{\mu,\sigma}(T) \leq a) = \mathcal{N}\left(\frac{a - \mu \cdot T}{\sigma\sqrt{T}}\right) + e^{2 \cdot \frac{\mu \cdot a}{\sigma^2}} \cdot \mathcal{N}\left(\frac{a + \mu \cdot T}{\sigma\sqrt{T}}\right)$$

Addendum to Example 1: Calculation of the Adapted VaR by Modelling the Price and Using the Distribution of the Minimum of the Brownian Motion with Drift

Revisiting Example 1 in Sect. 1.3, we once again turn our attention to calculating the adapted VaR for the period to time T.

In that example we modelled the movements of the DAX by means of a geometric Brownian motion $S(t) = S(0) \cdot e^{\mu \cdot t + \sigma \cdot B_t}$.

We ask ourselves how to choose x so that $W(\min\{S(t)\,|0 \leq t \leq T\} \leq x) = 0.01$.

For any $x > 0$, we have (based on part b) of the above theorem)

$$W\left(\min\{S(t)\,|0 \leq t \leq T\} \leq x\right) =$$

$$= W\left(\min\left\{S(0) \cdot e^{\mu \cdot t + \sigma \cdot B_t}\,\middle|\, 0 \leq t \leq T\right\} \leq x\right) =$$

$$= W\left(\min\{\mu \cdot t + \sigma \cdot B_t\,|0 \leq t \leq T\} \leq \log\frac{x}{S(0)}\right) =$$

$$= \mathcal{N}\left(\frac{\log\frac{x}{S(0)} - \mu \cdot T}{\sigma\sqrt{T}}\right) + e^{2 \cdot \frac{\mu \cdot \log\frac{x}{S(0)}}{\sigma^2}} \cdot \mathcal{N}\left(\frac{\log\frac{x}{S(0)} + \mu \cdot T}{\sigma\sqrt{T}}\right).$$

Now all that remains to be done is to determine x such that the last expression equals 0.01. Then x is the adapted VaR at a confidence level of 99% for the period up to T.

In our example the parameters were
$S(0) = 100,000$
$T = \frac{10}{250}$
$\mu = -0.015$
$\sigma = 0.1503$

so that the equation

$$\mathcal{N}\left(\frac{\log \frac{x}{100,000} + 0.0006}{0.03}\right) + e^{2\cdot\frac{-0.015\cdot\log\cdot\frac{x}{100,000}}{0.0226}} \cdot \mathcal{N}\left(\frac{\log \frac{x}{100,000} - 0.0006}{0.03}\right) = 0.01$$

is to be solved for x. This can be done with a mathematics software (e.g. with Mathematica), giving us the solution $x = 92,457$. The adapted VaR is therefore 7,543.

1.5 Calculating VaR for Simple Portfolio Examples: Example 2 (a Stock Index in Foreign Currency) and Stress Testing

In our second example, we consider the VaR of an investment in the SPX from the perspective of a EUR investor.

There are now two risk-carrying financial products to consider in this investment, namely, both the SPX and the Euro-Dollar exchange rate.

Losses may occur in case of a falling dollar exchange rate for an investor who "thinks in euro terms" even if the S&P500 index is going up.

We will again approach the problem of estimating a VaR for this type of investment using both historical simulation and modelling for both products.

In all further examples, we will limit our calculations to a 10-day VaR at a confidence level of 99%.

To estimate the required data, we use the daily SPX closing prices and the EUR-USD exchange rates from 1 January 2000 to 1 March 2019. We denote the SPX rate (in dollars) by $S(t)$, the EUR-USD exchange rate by $X(t)$, and the SPX rate in euros by $Z(t)$. $X(t)$ denotes the price of 1 euro in dollars. We then have $Z(t) = \frac{S(t)}{X(t)}$ of course.

The evolution of these three processes is shown in Fig. 1.8.
We can already see from these pictures that due to the relatively stable EUR-USD exchange rate, the VaR of the SPX in dollars will not differ all that much from the SPX in euros.

In fact, the historical simulation based on these data yields the following values:
10-day VaR at 99% confidence level **SPX in USD**: -9.84%
10-day VaR at 99% confidence level **SPX in EUR**: -10.25%

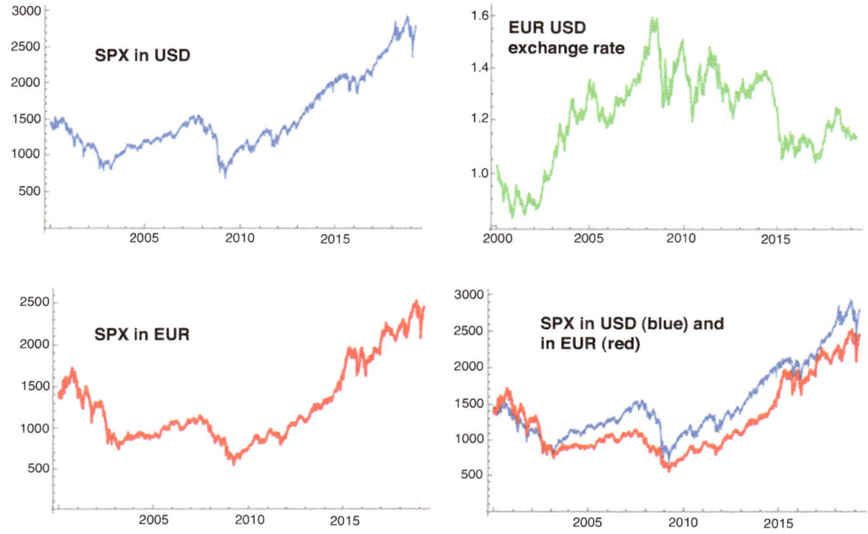

Fig. 1.8 SPX in USD, EUR-USD exchange rate, and SPX in EUR from 2000 to 2019

Adapted 10-day VaR at 99% confidence level **SPX in USD**: -12.42%
Adapted 10-day VaR at 99% confidence level **SPX in EUR**: -12.04%

Let us now calculate the VaR by modelling the rates. We could (and will) approach this task in two ways:

Method 1
We proceed directly from the price data of $Z(t) = \frac{S(t)}{X(t)}$ and use them to estimate the data needed to model $Z(t)$ as a geometric Brownian motion.

Method 2
We model $S(t)$ and $X(t)$ as two separate geometric Brownian motions based on the data from $S(t)$ and $X(t)$, respectively, and use these two models to determine a model for $Z(t) = \frac{S(t)}{X(t)}$.

Of course, in this particular case, the **first method** is a much easier and faster way to achieve our goal. However, for pedagogical reasons, so to speak, we are going to focus on the second method here in order to demonstrate how this can be done.

An estimate for the trend μ_Z and volatility σ_Z of the price $Z(t)$ is obtained from the historical daily (!) prices of $\frac{S(t)}{X(t)}$ from 2000 to March 2019, as follows:

$\mu_Z = 0.000197$ i.e. 0.04925 p.a.
$\sigma_Z = 0.01294$ i.e. 0.2046 p.a.

which leads to a model for $Z(t)$ of the form $\boldsymbol{Z(t) = Z(0) \cdot e^{0.04925 \cdot t + 0.2046 \cdot \sqrt{t} \cdot w}}$.

Regarding the application of the **second method**:

An estimate for the trend μ_S and volatility σ_S of the price $S(t)$ and for the trend μ_X and volatility σ_X of the price $X(t)$ is obtained from the historical daily prices of $S(t)$ and of $X(t)$ from 2000 to March 2019, as follows:

$\mu_S = 0.000207$ i.e. 0.05175 p.a.
$\sigma_S = 0.01204$ i.e. 0.1904 p.a.
$\mu_X = 0.000041$ i.e. 0.01025 p.a.
$\sigma_X = 0.00632$ i.e. 0.0999 p.a.

We also need the correlation between the movements of the SPX and those of the EUR-USD exchange rate (more precisely: between their returns). This correlation is slightly positive, i.e. a rising SPX has a slight tendency to occur with a strengthening euro. The correlation in the time range from 2000 to March 2019 is $\rho = 0.114$

To model $S(t)$ and $X(t)$ in parallel while taking into account this dependence between S and X, we proceed as already described in Volume II Section 2.29, using:

$$S(t) = S(0) \cdot e^{\mu_S t + \sigma_S \sqrt{t} \cdot w^{(1)}}$$
$$X(t) = X(0) \cdot e^{\mu_X t + \sigma_X \sqrt{t} \cdot \left(\rho \cdot w^{(1)} + \sqrt{1-\rho^2} \cdot w^{(2)}\right)}$$

with two mutually independent Brownian motions $w^{(1)}$ and $w^{(2)}$. For the value $Z(t) = \frac{S(t)}{X(t)}$, we then get:

$$Z(t) = Z(0) \cdot e^{\mu_S t + \sigma_S \sqrt{t} \cdot w^{(1)} - \mu_X t - \sigma_X \sqrt{t} \cdot \left(\rho \cdot w^{(1)} + \sqrt{1-\rho^2} \cdot w^{(2)}\right)} =$$
$$= Z(0) \cdot e^{(\mu_S - \mu_X) t + \sqrt{t} \cdot \left((\sigma_S - \sigma_X \rho) \cdot w^{(1)} - \sigma_X \cdot \sqrt{1-\rho^2} \cdot w^{(2)}\right)}.$$

In this exponent, the expression $\left((\sigma_S - \sigma_X \rho) \cdot w^{(1)} - \sigma_X \cdot \sqrt{1-\rho^2} \cdot w^{(2)}\right)$ is a linear combination of two independent standard normally distributed random variables and is thus normally distributed with mean 0 and with variance $\sum^2 := ((\sigma_S - \sigma_X \rho))^2 + \left(\sigma_x \cdot \sqrt{1-\rho^2}\right)^2$.

Thus we get $Z(t) = Z(0) \cdot e^{(\mu_S - \mu_X) t + \sqrt{t} \cdot \sum \cdot w}$ with a standard normally distributed random variable w. If we substitute the above values for $\mu_S, \mu_X, \sigma_S, \sigma_X$, and ρ, we get the model

$$\boldsymbol{Z(t) = Z(0) \cdot e^{0.0415 \cdot t + \sqrt{t} \cdot 0.20476 \cdot w}}.$$

To calculate the 10-day VaR at 99% confidence level, we now need to use the values $t = \frac{10}{250}$ and $w = -2.33$ again. Thus:

When calculating the **10-day VaR of the SPX in EUR at a confidence level of 99%** by modelling it directly over $Z(t)$ (**first method**), we obtain a value of **-8.91%**.

When calculating the **10-day VaR of the SPX in EUR at a confidence level of 99%** based on modelling $S(t)$ and $X(t)$ and then modelling $Z(t)$ based on those models (**second method**), we get a value of **-8.95%**.

It is clear and we have emphasized this before: The results of the risk estimates depend to a substantial degree on the time period from which historical data are taken to generate historical simulations or estimate the model parameters. Consequently, it is necessary to indicate for each result of a VaR calculation on which historical data the estimate was based.

Frequently, a **stress test** is also required for risk estimates (informally: "which results in this specific case would have to be expected under extreme market conditions" or "which result in this specific case would have to be expected in the worst case so far (in the period under review)"), in order to be able to estimate worst-case scenarios and classify any VaR result with regard to its relevance.

The information that is to be expected as part of a stress test is not clearly defined. With regard to our above example, we could contribute the following information in addition to the VaR calculation performed above for purposes of **stress testing**:

Stress test information:

Worst Case in the Observation Period

Historical daily data from the period from 1 January 2000 to 1 March 2019 for the SPX in USD and the EUR-USD exchange rate show the following strongest decline in 10 trading days for the SPX in EUR:

The strongest decline of the SPX in EUR occurred from 26 September 2008 to 10 October 2008 and was -19.25%.

The decline of the SPX in dollar terms in this period was much stronger at -25.88%, but in the same period, the EUR-USD exchange rate had fallen from 1.4615 to 1.3414, thus mitigating the drop somewhat from a euro investor's perspective.

Result of Historical Simulation by Using Observation Data Around the Period of the Financial Crisis from 1 January 2008 to 31 December 2009

Using this data for historical simulation, the 10-day VaR of the SPX in EUR at a confidence level of 99% is -15.22%.

Worst Case During the Entire Observation Period of the SPX

This worst-case value would probably occur around 19 October 1987 (which was already examined in more detail earlier in this book). However, since the euro currency did not yet exist at that time and there was no EUR-USD exchange rate therefore at that time, this value cannot be given here.

1.6 Example 3: VaR of a Portfolio of Two Stocks as a Function of Their Correlation

We will discuss VaR of large portfolios consisting of correlated financial products in more detail in later sections. In this section, for illustrative purposes, we will analyse a portfolio of two stocks that are more or less correlated and examine its VaR. For this, we are not going to use real stock price data here, but rather use artificially generated prices and Monte Carlo simulation to illustrate how the VaR changes as correlation between the two stock price movements increases (in a positive or negative direction).

First, we model the two stocks A and B using geometric Brownian motion and very simple parameter choices ($\mu = 0$ and $\sigma = 1$), thus

$A(t) = 100 \cdot e^{\sqrt{t} \cdot w^{(1)}}$

$B(t) = 100 \cdot e^{\sqrt{t} \cdot \left(\rho \cdot w^{(1)} + \sqrt{1-\rho^2} \cdot w^{(2)}\right)}$

with two mutually independent Brownian motions $w^{(1)}$ and $w^{(2)}$.

The portfolio S is of the form $\frac{1}{2}(A + B)$, and the price $S(t)$ of this portfolio is $S(t) = \frac{1}{2} \cdot (A(t) + B(t))$.

We are again interested in finding the 10-day VaR of this portfolio, thus the behaviour of

$$S\left(\frac{10}{250}\right) = S\left(\frac{1}{25}\right) = 50 \cdot \left(e^{\sqrt{\frac{1}{25}} \cdot w^{(1)}} + e^{\sqrt{\frac{1}{25}} \cdot \left(\rho \cdot w^{(1)} + \sqrt{1-\rho^2} \cdot w^{(2)}\right)}\right).$$

To determine the 1% percentile of this process using Monte Carlo, we generate 10,000 scenarios of 10-day price movements for each of various values of ρ, arrange them in ascending order, and select the one-hundredth scenario in this ranking as the estimate for the VaR. Figure 1.9 shows examples of 100 possible paths for $S(t)$ over the course of 10 days for different underlying correlations between the returns. It is immediately apparent that a lower correlation leads to significantly lower variation in the paths.

This fact is then also evident from the VaR values estimated by means of simulation (1% percentile of the terminal price values) and from the 99% percentile also estimated by means of simulation. These values are listed for different correlations of ρ in the following Figs. 1.10 and 1.11 which show how they evolve as a function of ρ.

This example is only meant to give a preliminary idea of the impact of diversifying a portfolio with low correlated stocks in a very simple case. The VaR drops significantly with a strong negative correlation of the two stocks.

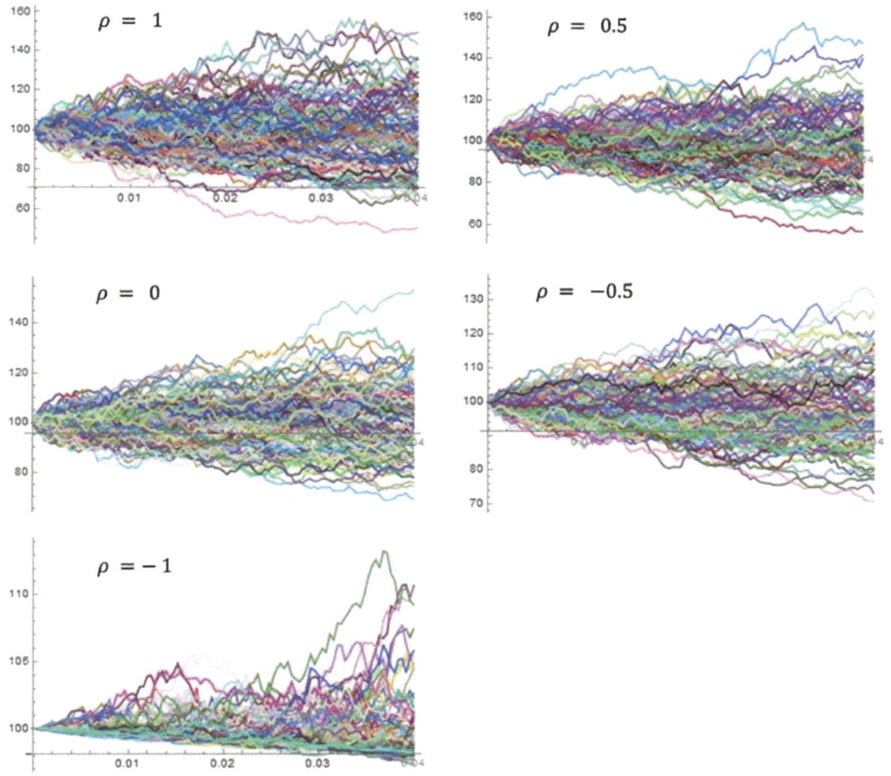

Fig. 1.9 100 simulations each of the sum of two correlated stocks with correlation ρ

1.7 Example 4: VaR Estimation for a Simple Options Strategy

We consider the following simple options strategy (which we also will analyse in more detail later):

- We assume an investment of 10,000 dollars.
- The options strategy is on the SPX (the S&P500 stock index).
- The SPX value upon opening the strategy is $S(0)$.
- We go short on one put option contract on the SPX with expiration $T = 1$ month and a strike K well below $S(0)$. At the same time, we go long on one put option contract on the SPX with expiration $T = 1$ month and strike $K - 100$.
- The margin required for this strategy will never exceed the amount of 10,000 USD because the worst-case loss is 10,000 USD. Thus, no margin call will occur during the strategy's lifetime.

ρ	99% - Perzentil	1%- Perzentil	VAR
-1	110.9	97.6	**2.4**
-0.8	117.7	85.7	**14.3**
-0.6	123.9	80.5	**19.5**
-0.4	128.5	77	**23**
-0.2	133	73.8	**26.2**
0	137.5	71.1	**28.9**
0.2	140.8	68.4	**31.6**
0.4	146.2	66.3	**33.7**
0.6	149.4	64.8	**35.2**
0.8	153.1	63.2	**36.8**
1	155.8	61.3	**38.7**

Fig. 1.10 99% percentile and 1% percentile of $S(t)$ as a function of the correlation of stock returns

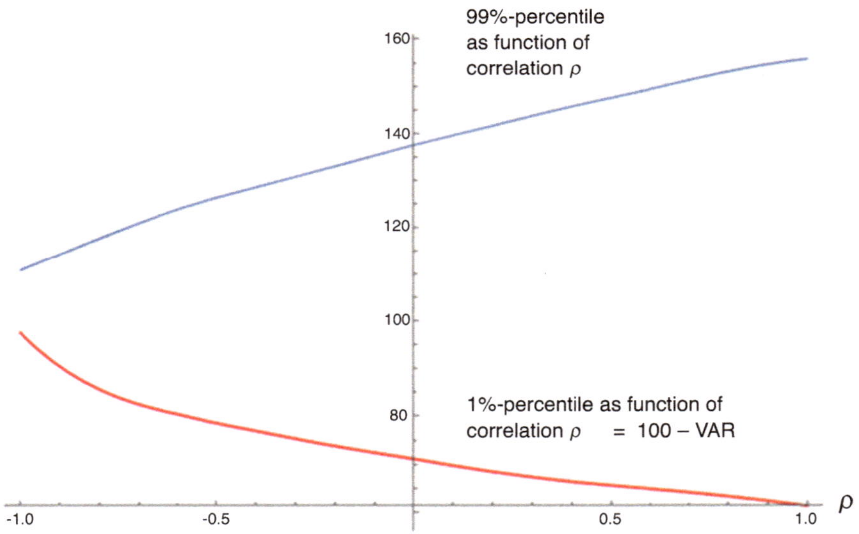

Fig. 1.11 99% percentile and 1% percentile of $S(t)$ as a function of the correlation of stock returns

We are interested in getting an estimate of the strategy's 5-day VaR at a 95% confidence level as the time left to expiration decreases and the value of the SPX changes.

For this, we **first** assume **constant volatility** $\boldsymbol{\sigma}$, a constant risk-free interest rate r, a risk-neutral trend $r - \frac{\sigma^2}{2}$, and a Wiener model for the SPX.

Additionally, to achieve illustrative results, we will use numerical values for $S(0)$ and K as well as for σ and r (of course, the procedure can then be implemented analogously for any parameter choices).

So we set $S(0) = 2,900$ points (approximate value of the SPX on 19 March 2019), $\sigma = 20\%$ (i.e. 0.2), $r = 0$ (i.e. risk-neutral trend $r - \frac{\sigma^2}{2} = -0.02$), and the strike K at 2,700. In opening the strategy, we also use the option prices for the required put options on 19 March 2019 with expiration on 18 April 2019.

The quotes for the put option with strike $K = 2700$ were 8.70×9.00.
The quotes for the put option with strike $K = 2600$ were 3.40×3.60.

We can therefore expect to sell the 2700 put at 880 USD and purchase the 2600 put at 360 USD. This results in 520 USD in premium income.

If the SPX is above 2700 points on 18 April 2019, the option contracts expire worthless, and the collected premium is the profit from the strategy at 5.2% of the investment made.

However, if the SPX is below 2600 points on 18 April 2019, the payoff to be paid is 10,000 USD, resulting in almost a total loss (except for the collected premium). The profit function is shown in Fig. 1.12.

Now, how do we best calculate or estimate, for any time t during the strategy's life and any $S(t)$, the VaR at that time t?

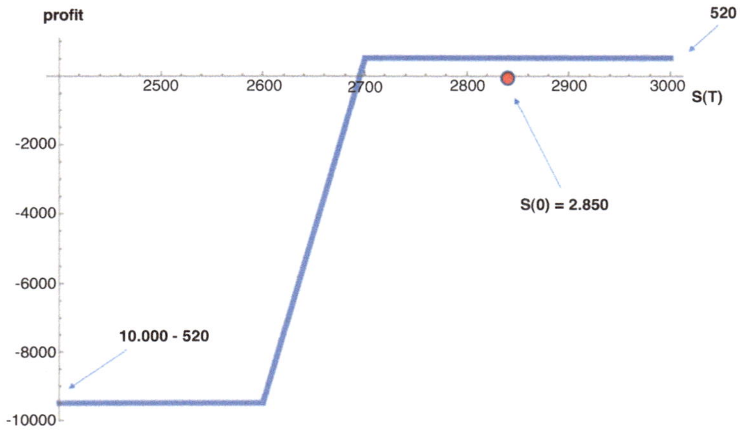

Fig. 1.12 Profit function of the put option strategy

The VaR at time t –if expressed as a percentage– refers not to an initial capital of 10,000 USD but to the total portfolio value at time t.

So, let us assume we are at time t and the SPX has a value of $S(t)$.

The portfolio value $V(t)$ is then $V(t) = 10,000 + 520 - PS(t) + PL(t)$, where $PS(t)$ denotes the price of the short put position and $PL(t)$ denotes the price of the long put position.

For each u the value $V(u)$ is monotonically increasing in $S(u)$! This fact is easily verified by calculating the delta of the combination of short put with strike K and long put with strike $K - 100$.

For the delta Δ_L of a put option with strike L, we found in Volume I Section 4.37 that $\Delta_L = -\mathcal{N}(-d_L)$ with $d_L = \frac{\log\left(\frac{S}{L}\right)+\left(r+\frac{\sigma^2}{2}\right)\cdot(T-t)}{\sigma\cdot\sqrt{T-t}}$.

The delta $\widetilde{\Delta}$ of a short put with strike K plus a long put with strike $(K - 100)$ is therefore given by $\widetilde{\Delta} = \mathcal{N}(-d_K) - \mathcal{N}(-d_{K-100}) = \frac{1}{\sqrt{2\pi}} \int_{-d_{K-100}}^{-d_K} e^{-\frac{x^2}{2}} dx$.

It is easy to see that $-d_{K-100} < -d_K$; therefore, the last integral is positive, and therefore $\widetilde{\Delta}$ is positive. The price of the combination of short put and long put is thus monotonically increasing in $S(t)$.

In the following we calculate the value $X(t)$ for which there is a probability of exactly 5% that the value S will fall below it on day 5 after t. Substituting this value $X(t)$ for $S(t + 5 \text{ days})$, we then calculate the value $V(t + 5 \text{ days})$. The portfolio will then fall below this value on day 5 after time t with a probability of exactly 5%. This value $V(t + 5 \text{ days})$ is then also the basis for calculating the VaR.

Let us now calculate $X(t)$:

We have $S\left(t + \frac{5}{250}\right) = S(t) \cdot e^{-0.02\cdot\frac{5}{250}+0.2\cdot\sqrt{\frac{5}{250}}\cdot w}$. The standard normally distributed random variable w has its 5% percentile at $w = -1.64$. Therefore, $S\left(t + \frac{5}{250}\right)$ falls below the value of $X(t) = S(t) \cdot 0.954$ with probability of approximately 5%.

We now use this value to calculate $V(t+5 \text{ days})$, i.e. 10,000 plus initial premium minus the Black-Scholes formula for the short put plus the Black-Scholes formula for the long put. The time parameter to set here is $t + \frac{5}{250}$. For T we choose $\frac{22}{250}$ (since we have 22 trading days from opening the strategy to expiration).

The value $V(t + 5 \text{ days})$ is therefore given by

$$\begin{aligned} V(t + 5\text{days}) = 10,000 + 520 - 100\,(&2700 \cdot \mathcal{N}(-d_2(2700)) + \\ &+ S(t) \cdot 0.954 \cdot \mathcal{N}(-d_1(2700))) + \\ &+ 100\,(2600 \cdot \mathcal{N}(-d_2(2600)) - \\ &- S(t) \cdot 0.954 \cdot \mathcal{N}(-d_1(2600))) \end{aligned}$$

with $d_1(L) := \frac{\log\left(\frac{S(t)\cdot 0.954}{L}\right)+0.02\cdot\left(\frac{17}{250}-t\right)}{0.2\cdot\sqrt{\frac{17}{250}-t}}$ and $d_2(L) := \frac{\log\left(\frac{S(t)\cdot 0.954}{L}\right)-0.02\cdot\left(\frac{17}{250}-t\right)}{0.2\cdot\sqrt{\frac{17}{250}-t}}$.

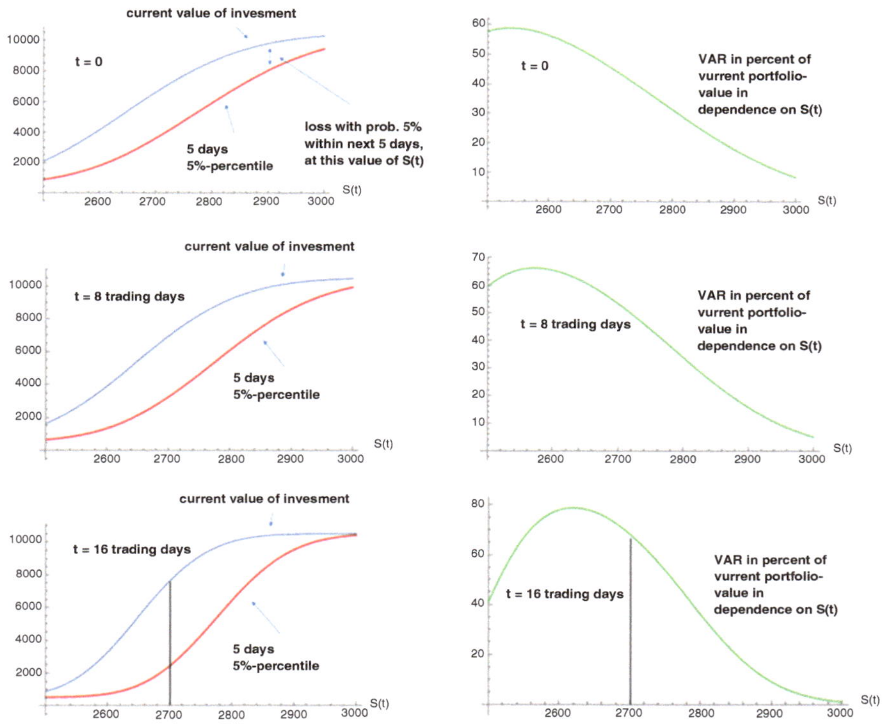

Fig. 1.13 Total value of the put option strategy (blue), the associated 5-day 5% percentile (red), and the 5-day VaR at 95% confidence level, as a percentage of the instantaneous portfolio value (green) for different time points t during the strategy's life, assuming constant volatility

We will now plot the curve of $V(t + 5 \text{ days})$ for three different values of t ($t = 0, 8, 16$ days) as a function of $S(t)$ (see Fig. 1.13, left, red line). In the same graphic, you can also see the current price of the total investment at each point in time and the price of the underlying asset (Fig. 1.13, left, blue line). To the right, you see the resulting 5-day VaR at the 5% confidence level in relation to the current value of the total investment. Subsequently, we will pick out a specific example for further explanation and discuss it in detail.

We pick a specific time t and a specific value $S(t)$ and study the situation in more detail:

We choose $t = 16$ trading days, i.e. a remaining time horizon of 6 trading days. Furthermore, we consider an SPX value of 2700 points at this time. These values are marked with a black line in the bottom row of Fig. 1.13. The VaR here is almost 70%.

The value of the short put contract with strike 2700 is −3337 USD in this setting.

The value of the long put contract with strike 2600 would be +444 USD.

This gives a total portfolio value of $10,000 + 520 - 3337 + 444 = 7627$ USD (marked by the upper end of the black line, point of contact with blue line in Fig. 1.13).

There is a 5% probability that the S&P500 will fall below $2700 \times 0.954 = 2576$ points by day 21.

On day 21 and with the SPX actually at 2576 points, we would have the following situation:

The value of the short put contract with strike 2700 would be −12, 649 USD.
The value of the long put contract with strike 2600 would be +4553 USD.

This gives a total portfolio value of $10,000 + 520 - 12,649 + 4553 = 2424$ USD (marked by the intersection of the black line with the red curve in Fig. 1.13).

Thus, the minimum loss that will occur with 5% probability is $7627 - 2424 = 5203$ USD, which is 68.22% of the current value of 7627 USD. The VaR expressed as a percentage is thus 68.33 and is marked by the point of contact of the black line with the green curve in Fig. 1.13.

If we wanted to calculate the VaR for $t = 18$, 19, 20, and 21 trading days as well, then we would of course have to take into account in the above computations that the strategy's remaining time horizon is then no longer 5 trading days but ends earlier when the payoffs are paid. However, that is a very simple exercise and will not be set out explicitly here.

Of course, for this example to actually contribute to a reliable risk estimate, it must be calculated with a volatility that varies over the strategy's life. Especially when the SPX falls sharply into the critical range, a strong increase in volatility is to be expected. This increasing volatility makes the two relevant options more expensive and increases the extent of the drop in the SPX for a given probability based on the model used.

So if we assume a deterministic dependence of the volatility on the price of the SPX of the form $\sigma_t = \sigma_0 \cdot \left(\frac{S(0)}{S(t)}\right)^a$, like we did several times in Section V with, as we have seen, quite a realistic value $a = 4$, then it is obvious how we need to adapt the above approach to arrive at the analogous result for dependent volatility:

With our parameter choice, we get $\sigma_t = 0.2 \cdot \left(\frac{2900}{S(t)}\right)^4$ of course. Whenever volatility is used in our calculations above with the SPX at $S(t)$, we will now use $\sigma_t = 0.2 \cdot \left(\frac{2900}{S(t)}\right)^4$ instead of $\sigma = 0.2$. If $X(t)$ is assumed, then we use $\sigma_t = 0.2 \cdot \left(\frac{2900}{X(t)}\right)^4$ instead of $\sigma = 0.2$.

In particular, we now have:

The value $S\left(t + \frac{5}{250}\right)$ will fall below the value of

$$X(t) = S(t) \cdot e^{-\left(\frac{\sigma_t^2}{2}\right) \cdot \frac{5}{250} - \sigma_t \cdot \sqrt{\frac{5}{250}} \cdot 1.64}$$

with a probability of approximately 5%, where $\sigma_t = 0.2 \cdot \left(\frac{2900}{S(t)}\right)^4$!

Figure 1.14 is the counterpart of Fig. 1.13, this time with volatility as a function of the underlying asset in the above form.

To better see the difference in results for constant volatility and for volatility as a function of $S(t)$, we superimposed the two bottom-right graphs from Figs. 1.13

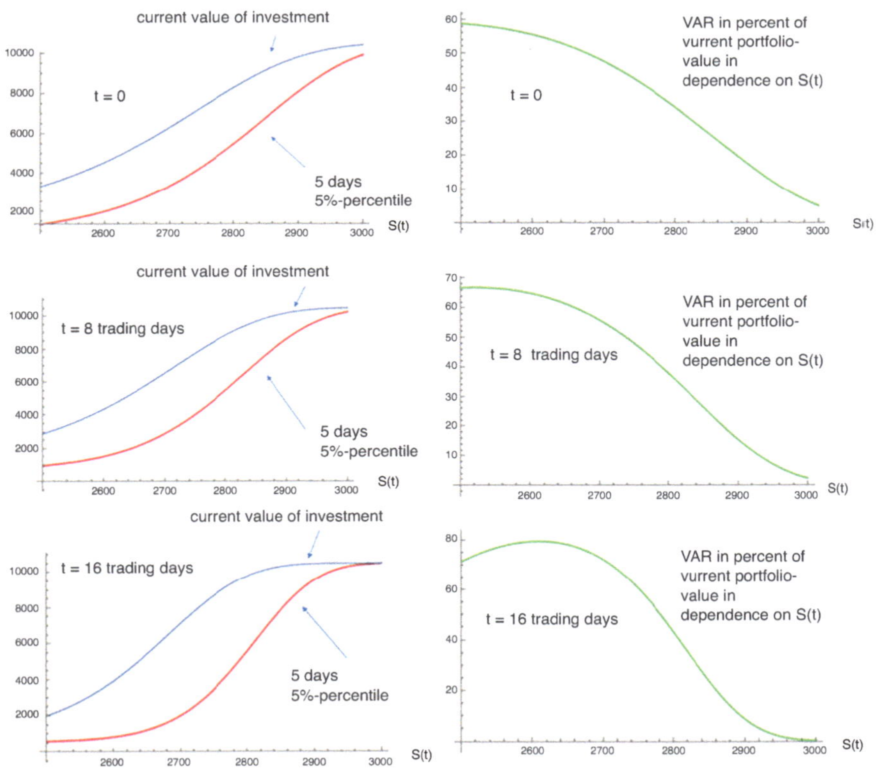

Fig. 1.14 Total value of the put option strategy (blue), the associated 5-day 5% percentile (red), and the 5-day VaR at 95% confidence level, as a percentage of the instantaneous portfolio value (green) for different time points t during the strategy's life, assuming volatility as a function of $S(t)$

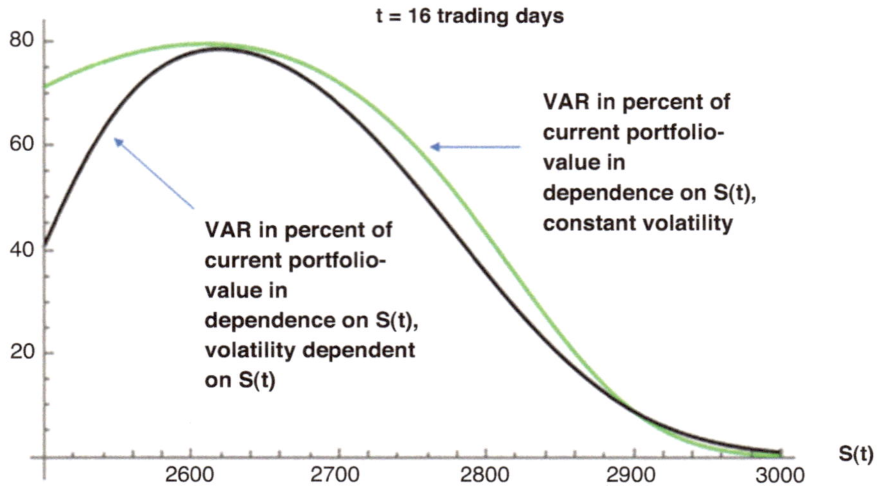

Fig. 1.15 Comparison of the VaR of the put strategy assuming constant volatility (black) and assuming volatility as a function of $S(t)$ (green)

and 1.14 in Fig. 1.15. The black curve shows the VaR as a percentage assuming constant volatility, and the green curve shows the VaR assuming volatility as a function of $S(t)$.

It clearly shows almost consistently a significantly higher value for the VaR with volatility as a function of $S(t)$!

1.8 Conditional VaR

The conditional value at risk (CVaR) has the same parameters as the ordinary VaR, that is, it relates to a certain future period (e.g. 10 days) and to a certain confidence level (e.g. 99%).

The CVaR specifies the average loss to be expected if the VaR event occurs.

Thus, to calculate the CVaR, it is necessary to know the VaR and the distribution of losses in the VaR case. So the CVaR contains substantially more information than just the regular VaR.

We want to illustrate the concept of CVaR using the profit/loss distribution of a strategy:

In the example shown in Fig. 1.16, the 10-day VaR at a 95% confidence level is −6%. Losses greater than 6% are basically possible, although the probability of higher losses decreases sharply the greater they become. The average loss weighted

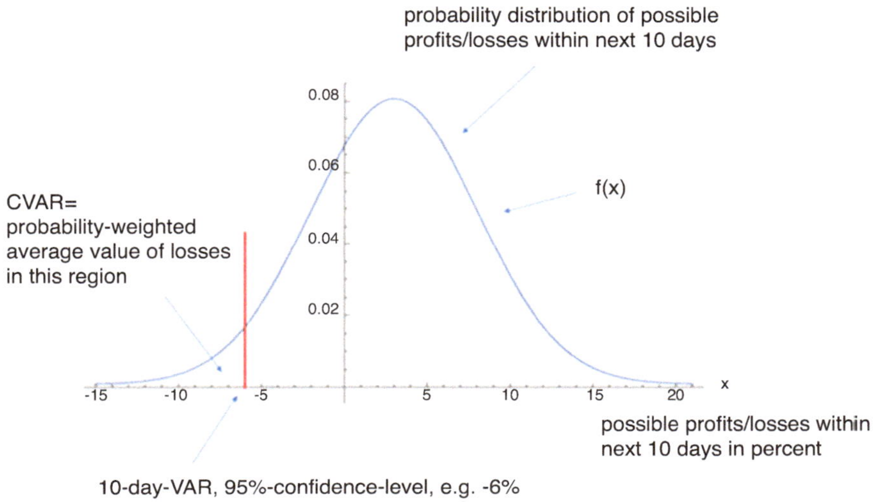

Fig. 1.16 Illustration of CVaR

with the respective probability is approximately somewhere around -9%. The CVaR would therefore be approximately -9%.

If the probability density $f(x)$ is known, then the CVaR at the confidence level of a% is calculated in the following way:

$$\text{CVAR} = \frac{100}{100-a} \cdot \int_{-\infty}^{\text{VAR}} x \cdot f(x)dx.$$

The integral part in the above formula should be clear: We calculate the average loss x weighted with the probability of its occurrence (the probabilistic density f(x)).

The factor $\frac{100}{100-a}$ may require some explanation: In our illustrative example, this factor would be $\frac{100}{100-95} = 20$. This factor represents precisely the reciprocal of the area under the density function f from $-\infty$ to the VaR. When multiplied by this factor, the function f is normalized from $-\infty$ to the VaR to a density function, and the integral then gives the expected value for losses greater than VaR, assuming that a loss greater than VaR occurs.

The informative value and relevance of the CVaR compared to the VaR are clearly seen from the following illustrative example (see Fig. 1.17):

The two loss distributions shown in Fig. 1.17 have the same VaR. The area under the density curve from $-\infty$ to VaR, marked by the red line at -8%, is equal to 5% of the total area in both cases.

However, the average loss in the area between $-\infty$ and VaR is (as per the graphs) around -14% in the upper picture and around -22% in the lower picture. Meaning: In both graphs, a loss of 8% or more occurs with approximately 5% probability. If

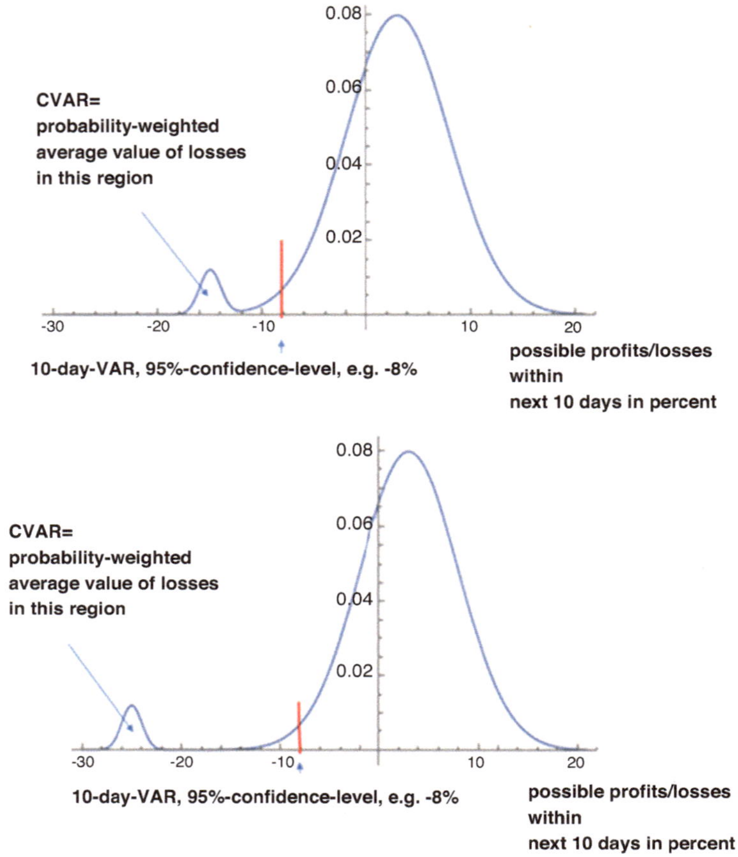

Fig. 1.17 Example: same VaR, different CVaR

a loss of 8% or more occurs, then this loss averages approximately -14% in the upper graph and approximately -22% in the lower graph.

The CVaR can again be estimated by historical simulation or else by modelling the portfolio and evaluating the above integral, either explicitly if possible or else using Monte Carlo simulation. In the following, we will estimate the CVaR for some of the VaR calculations we performed earlier.

1.9 CVaR Estimation: Some Examples

Example: Investment in the DAX (Example 1 from Above) We start with the DAX investment from Example 1 in Sect. 1.3 and use the data and notations from there. For illustration, we limit our discussion to the case where we perform the estimation based on the historical DAX data from 13 March 2006 to 13 March 2019, estimating the 10-day CVaR at 99% confidence level and the 1-day CVaR at 99% confidence level.

In that example we arranged all the 10-day returns of the past 13 years in ascending order $x_1 < x_2 < x_3 < \ldots < x_{m-1} < x_m$.

The VaR then resulted as x_i with index $i := \left[\frac{m}{100}\right]$.

The CVaR is now simply calculated on the basis of the average value of the 10-day returns that were less than or equal to x_i, i.e. based on

$$Y := \frac{1}{\left[\frac{m}{100}\right]} \cdot \sum_{k=1}^{\left[\frac{m}{100}\right]} x_k.$$

If we run this for our DAX data, we get a value for Y of -16.13% and thus a CVaR of 16,130 USD. The VaR we got was 12,823 USD.

To determine the 1-day CVaR at the 99% confidence level, we follow a completely analogous procedure, but using the historical 1-day returns. The resulting values are $Y = -5.25\%$, and we thus get a CVaR of 5250 USD. The VaR in this case was 4004 USD.

Example: Sum of Two Correlated Stocks (Example 3 from Above) To calculate the 10-day CVaR at the 99% confidence level, we use Monte Carlo simulation again and initially proceed in the exact same way as we did to determine the VaR. The 10-day returns are again modelled as follows:

$$S\left(\frac{10}{250}\right) = S\left(\frac{1}{25}\right) = 50 \cdot \left(e^{\sqrt{\frac{1}{25}} \cdot w^{(1)}} + e^{\sqrt{\frac{1}{25}} \cdot \left(\rho \cdot w^{(1)} + \sqrt{1-\rho^2} \cdot w^{(2)}\right)}\right)$$

with independent standard normally distributed $w^{(1)}$ and $w^{(2)}$. To determine the 1% percentile of this process using Monte Carlo, we generate 10,000 scenarios of 10-day price movements for each of various values of ρ, arrange them in ascending order, and select the one-hundredth scenario in this order as the VaR estimate.

The CVaR estimate, on the other hand, is based on the mean of the 100 smallest values of the 10-day paths of these 10,000 scenarios.

We perform this calculation again for different values for the correlation ρ, expanding the Fig. 1.10 from Sect. 1.6 to the following Fig. 1.18:

It is clear to see here that in all cases, the CVaR differs only little from the VaR. This fact also becomes obvious when looking at the example paths in Fig. 1.9. There are hardly any outliers with extremely high losses. The higher the negative correlation between the two stock returns, the more VaR and CVaR converge.

Example: Simple Put Option Strategy (Example 4 from Above) For purposes of calculating the 5-day CVaR at the 95% confidence level for the put option strategy

ρ	99%-percentile	1%-percentile	VAR	average of 100 smallest	CVAR
-1	110.9	97.6	**2.4**	97.5	**2.5**
-0.8	117.7	85.7	**14.3**	83.7	**16.3**
-0.6	123.9	80.5	**19.5**	78.7	**21.3**
-0.4	128.5	77	**23**	73.5	**26.5**
-0.2	133	73.8	**26.2**	70.8	**29.2**
0	137.5	71.1	**28.9**	67.3	**32.7**
0.2	140.8	68.4	**31.6**	65.1	**34.9**
0.4	146.2	66.3	**33.7**	63.7	**36.3**
0.6	149.4	64.8	**35.2**	61.5	**38.5**
0.8	153.1	63.2	**36.8**	59.4	**40.6**
1	155.8	61.3	**38.7**	57.7	**42.3**

Fig. 1.18 99% and 1% percentile of $S(t)$, VAR, and CVAR

example in Sect. 1.7, we pick a special choice for t, namely, $t = 16$. In addition, we will limit the discussion to the case with assumed constant volatility. (All other cases are again quite simple and can be performed analogously.) The best way to determine the CVaR here is by Monte Carlo simulation.

First, we generate, for example, 10,000 standard normally distributed random numbers, arrange them in ascending order, and choose the 500 (= 5%) smallest values $w_1, w_2, \ldots, w_{500}$. With each of these w_i, we calculate a possible value Y_i for $S\left(t + \frac{5}{250}\right)$ using $Y_i = S(t) \cdot e^{-0.02 \cdot \frac{5}{250} + 0.2 \cdot \sqrt{\frac{5}{250}} \cdot w_i}$. With this value we now calculate a possible value Z_i of $V(t+$ 5 days), thus

$$Z_i = 10,000 + 520 - 100\,(2700\mathcal{N}(-d_2(2700)) + Y_i \cdot \mathcal{N}(-d_1(2700))) + \\ +100\,(2600\mathcal{N}(-d_2(2600)) - Y_i \cdot \mathcal{N}(-d_1(2600)))$$

with $d_1(L) := \frac{\log\left(\frac{Y_i}{L}\right)+0.02\cdot\left(\frac{17}{250}-t\right)}{0.2\cdot\sqrt{\frac{17}{250}-t}}$ and $d_2(L) := \frac{\log\left(\frac{Y_i}{L}\right)-0.02\cdot\left(\frac{17}{250}-t\right)}{0.2\cdot\sqrt{\frac{17}{250}-t}}$.

We calculate the mean over these Z_i for $i = 1, 2, \ldots, 500$ to obtain an estimate for the mean of the 5% strongest price declines, and this mean then provides the basis for determining the CVaR.

The corresponding curve as a function of $S(t)$ is plotted as a green curve in Fig. 1.19, where it complements the graph at the bottom left of Fig. 1.13 (in that figure, the respective current value of the strategy (blue) and the 5-day 5% percentile (red) were plotted as a function of $S(t)$).

In the middle range, for values of the $S(t)$ between 2700 and 2900 points, one can see a clear difference between VaR and CVaR for this strategy. Stronger price

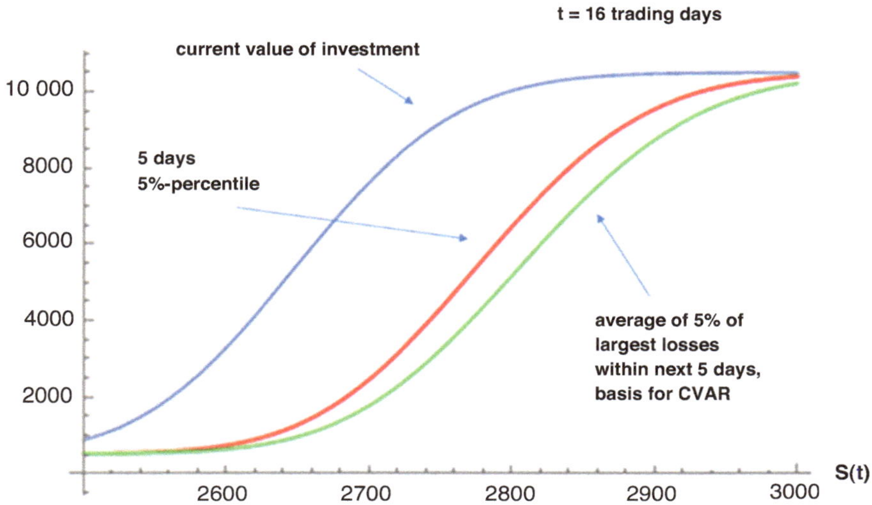

Fig. 1.19 Total value of the put option strategy (blue), the corresponding 5-day 5% percentile (red), and the mean of the 5% strongest price declines (green) assuming constant volatility

drops in the underlying asset can lead to much higher price losses than the VaR alone would indicate.

In the following sections, we are going to look more closely at a more specific risk management topic, namely, credit risk management. In particular, we are going to introduce two of the most commonly used credit risk management systems, J.P. Morgan's Credit Metrics system and Credit Suisse First Boston's Credit Risk+ system. These systems, which were originally developed for credit risk management, can also be used in adapted form for a wide variety of other risk management applications and are undoubtedly among the basic tools in the field of quantitative finance and financial engineering.

1.10 Credit Risk Management: Introduction

Ever since the Basel Committee on Banking Supervision pointed out in its 1998 consultative paper "Basel II" that the credit risk management of banks needed to be significantly improved, interest in this topic –including from the perspective of quantitative finance– has increased immensely, and various approaches to efficient credit risk management have been developed over the last two decades.

The main question in these approaches (especially in the *Credit Metrics* and *Credit Risk+* approaches presented below) is how much capital a bank should be required to hold to ensure that its exposure from lending activities will not endanger its solvency even in the face of severe turmoil on the credit markets.

A bank grants loans in a wide variety of forms for a wide variety of purposes to a wide variety of borrowers:

- Borrowers can be individuals; small, medium, or large companies; municipalities; states; countries; or even other banks.
- Loans can be collateralized or uncollateralized, and repayments (and possibly interest payments) can be made at the end of the loan period or continuously throughout the loan period.
- Banks can grant loans in the traditional form as individual loans or by purchasing bonds from the issuer of those bonds.

The funds that a bank provides borrowers with in the form of loans are essentially the funds that investors deposit with that bank in savings accounts or invest in that bank by purchasing bonds issued by that bank.

The cash flows generated by this process of investors investing (and withdrawing) funds and of borrowers simultaneously borrowing funds (and repaying them) lead to strong dynamics and fluctuation in terms of the **bank's own funds**, i.e. the bank's current cash positions.

These dynamics also entail risks of various kinds:

- A sudden surge in cash withdrawals by investors can lead to bottlenecks if the bank's own funds are too low.
- A spate of borrower defaults (i.e. inability of borrowers to repay their loans) can make it difficult for the bank to meet its regular payment obligations to investors if the bank's own funds are too low.

An adequate capital ratio is therefore essential for a bank to be able to absorb the majority of short-term risks arising from its lending business. On the other hand, that retained capital then lies idle in many cases and cannot be put to profitable use. It is therefore in the bank's interest to retain as little capital as possible.

This balancing act between the bank's interest in generating profits and the need for safeguards needs to be handled in the best way possible. The question then is:

What is the minimum amount of capital a bank must hold as own funds so that it is highly likely to survive the vast majority of critical situations in its lending business unscathed?

Until the end of the last century, the relevant capital adequacy rules imposed by most national bank supervisors were very simple. In highly simplified terms, they essentially said:

The minimum capital ratio that banks must maintain is 8% of the loan amounts granted. In the event that

- *A loan is granted to an OECD country, no additional own funds are required.*
- *A loan is granted to an OECD bank, own funds of only 1.6% of the loan amount are required.*

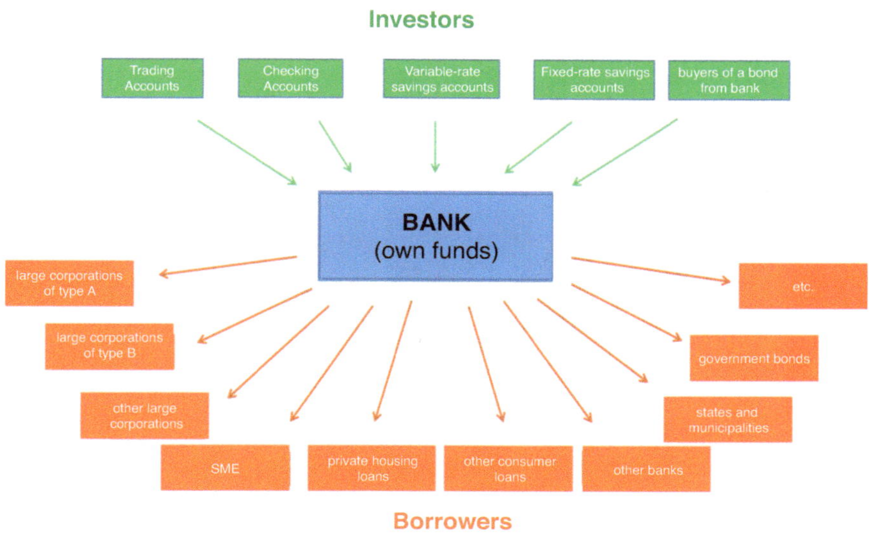

Fig. 1.20 Schematic representation of a bank's investor/borrower structure, diversified

- *A loan is secured by mortgage, own funds of only 4% of the loan amount are required.*

The Basel Committee on Banking Supervision (which has no legislative power but only an advisory function) strongly objected to this very simple regulation in its "Basel II" consultation paper in the early 2000s.

Among other things, it called on banks worldwide to implement much more sensitive credit risk management systems that, among others, take much greater account of the **"quality" of each loan** (and borrower) in their credit portfolio as well as of **correlation effects** between those loans in the portfolio.

Figure 1.20 schematically illustrates the "ideal type" of a bank's investor/borrower structure.

In the situation depicted above, the bank extends loans to many different types of borrowers. This represents diversification and largely prevents negative consequences for a bank's credit portfolio, which may definitely occur in situations such as the one shown in Fig. 1.21.

In this situation the following scenario might occur: Suppose that the funds granted as loans by the bank to private individuals and also residential construction companies were used primarily to build apartments. The borrowers plan to recover most of their borrowing costs through income generated by renting out these apartments. The rents are set accordingly. At one point, however, certain macroeconomic circumstances could cause prices in the real estate markets to suddenly drop significantly. The tenants of the apartments may no longer be willing to pay rent in the same amount as before, or they may seize the opportunity to buy

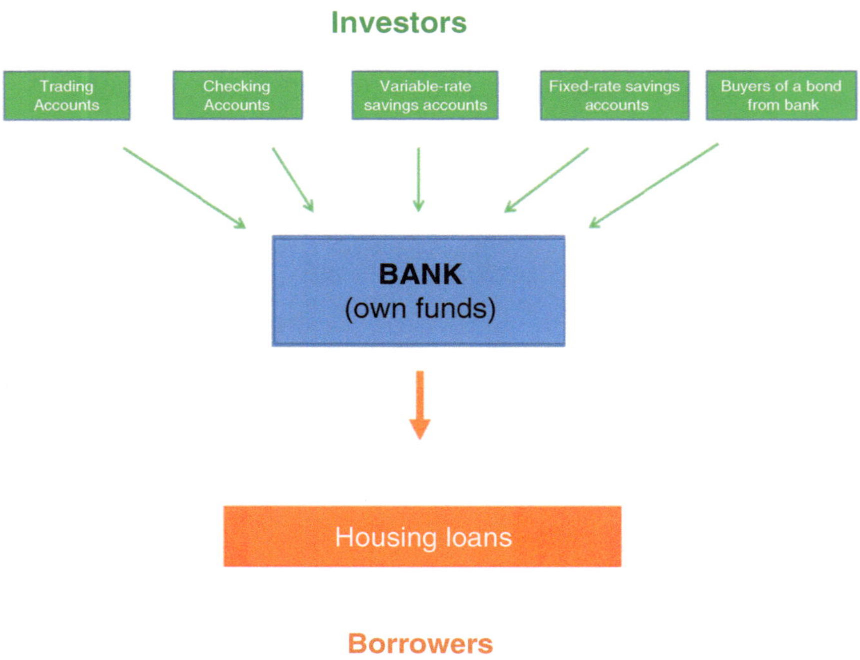

Fig. 1.21 Schematic representation of a bank's investor/borrower structure, not diversified

other apartments, now that they have become cheaper. The borrowers' rental income may therefore fall sharply, and they may no longer be able to pay the loan interest as agreed. A possible solution would now be for the borrowers to sell the apartments they built and use the proceeds from such sales to repay the loan. However, due to the plunge in real estate prices, the borrowers recover only a portion of the construction costs and are therefore only able to repay a portion of the original loan. In this situation, the bank will be faced with massive losses due to loans that cannot be repaid. Of course, word of such a situation would spread rapidly, and a considerable number of investors may then rush to withdraw their deposits at the bank. A too low amount of own funds would then be used up very quickly, and the bank may run into difficulties in satisfying investor claims for payouts. This may now prompt further investors, already panic-stricken, to push for payouts of their savings. A possible way out of the dilemma could now be for the bank to sell (pass on) these loans to other banks.

In principle, such a sale could look like the following Fig. 1.22:
The new bank (let's call it Z) transfers to the original bank (B) a sum in the amount of the loan and in return receives the borrower's promise of repayment of that sum at a certain time in the future and payment of corresponding interest on the loan. There is just one problem:

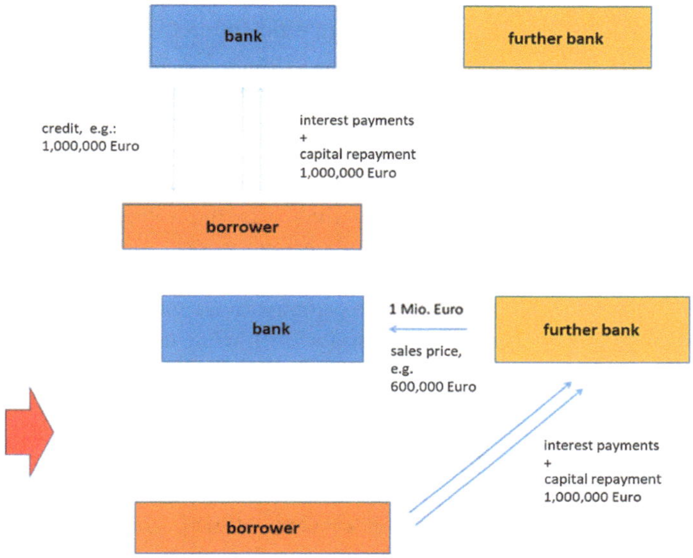

Fig. 1.22 Sale of a loan, schematic representation

Bank *Z*, just like bank *B*, knows that at least some of the current housing loan borrowers have, and will continue to have, substantial difficulties in repaying the loans. Therefore, Bank *Z* will likely not be willing to pay the price of 1,000,000 EUR to Bank *B* for taking over the loan, since it seems highly uncertain that it will actually receive 1,000,000 EUR from the borrower in the future. Bank *Z* may estimate the value of the loan (the repayment promise) to be worth only 600,000 EUR and thus offers Bank *B* to pay only this amount as purchase price. Bank *B*, in urgent need of this money to satisfy adamant investors, will probably have to accept this offer.

What happens then is easily calculated: The bank eventually sells its entire credit portfolio in this way at a price of 60% of the total loan amounts it extended. Add to that a capital ratio of 8% and the bank can satisfy at most 68% of its investors' claims. It may not be able to survive for long.

The massive problem in this example was triggered by the fact that a very large number of same-type borrowers were simultaneously confronted with payment difficulties due to **one single** market event and that the bank was exclusively dependent on the solvency of this group of borrowers.

Such correlations among a bank's borrowers (whether in this massive form or in a less concentrated form) should –according to Basel II– be an essential consideration and be adequately incorporated into the calculation of a bank's capital adequacy requirement.

There are various opinions and approaches as to how this should be done, and a number of models for implementing efficient credit risk management have been developed since the beginning of this century. We will present two of these approaches in more detail below, namely, the **"Credit Metrics"** system developed by J.P. Morgan and the **"Credit Risk+"** system developed by Credit Suisse First Boston.

Since these are **models (!)**, we need to bear in mind what we said earlier on the subject of "mathematical modelling". In the following, various model **assumptions** will be made which simulate reality more or less adequately, but which cannot be proven mathematically. Exact mathematics (including occasional approximation methods instead of exact computations) can then be performed only within a specifically defined and fixed model. Still, while the model assumptions are debatable, the calculations performed in the model are not.

1.11 Credit Metrics, Part I: Basic Approach

In our presentation, we follow the main contents of the [1] technical document that are necessary to understand the principle of Credit Metrics.

Somewhat simplified, the **principal approach of Credit Metrics** is as follows:

- We determine the value A_0 of a bank's credit portfolio at the current time 0.
- By A_1 we denote the random variable for the credit portfolio's value 1 year into the future.
- For information and control purposes, we determine the expected value and standard deviation of A_1.
- The actual proposal for determining own funds is then based on a VaR calculation for A_1: We determine the 1-year VaR at the 99% confidence level for A_1. This means: We determine a value X such that with a probability of 99%, the value of A_1 (the value of the credit portfolio in 1 year's time) will not fall below X.
- Thus, in 99% of all cases, the credit portfolio loss would be smaller than $A_0 - X$ (i.e. than the portfolio's present value minus 1% percentile). This potential loss should be covered by own funds.

In the following, we will execute this program step by step. To ensure a clearer representation, we will apply some simplifications to our problem:

Our credit portfolio will be assumed to consist exclusively of bonds, all with annual coupon payments. We also assume that at time 0, a coupon payment has just been made on each of these bonds. The next coupon payment for each of these bonds will occur exactly 1 year from now (just before A_1 is determined).

Further, we will assume that every issuer of the bonds in our credit portfolio has a Standard & Poor's rating.

1.12 Credit Metrics, Part II: Ratings and Fair Value of Bonds at Risk of Default

The (simplified) system used by the credit rating agency Standard & Poor's to classify the credit quality of companies, states, banks, etc. is

AAA
AA
A
BBB
BB
B
CCC
D

An entity (we collectively refer to all types of borrowers by "entity" in the following) with the proverbial triple-A rating AAA is an entity of the highest quality according to S&P's research and valuation approach. The quality of any such entity then gradually decreases to AA, A, BBB, BB, B, and CCC. A "D" rating denotes "Default" and is assigned to an entity that is already completely or even only partially (possibly temporarily) insolvent.

It is not our intention here to analyse the specific valuation approach used by S&P in more detail or to explore the exact meaning of different ratings. We will only address the key metrics that depend on such ratings and refer to them as needed.

In the following, as per our assumption, we will be dealing with bonds issued by rated entities. This means that all of the bonds discussed below have a rating. A bond's rating does not always correspond exactly with that of the issuing entity. Nevertheless, in the following we will refer to the rating of an entity and the rating of a bond (issued by that entity) as if they were the same.

A bond with a D rating does not necessarily have to be a bond from which no further payment can be expected. It may be a case of temporary insolvency, or that partial or late repayment has been agreed between the holder and the issuer of the bond, or that the issuer has been forced into bankruptcy and the bond holders can expect partial repayments from the bankruptcy estate.

For our further discussions and to execute the program developed in the previous section, we need to make certain assumptions regarding the approximate repayments that can be expected in the event of a bond's default (by "default" we mean that the bond is downgraded to a "D" rating).

This repayment amount, which we will refer to as the "recovery rate" in the following, is a random variable, of course. It will, among others, depend on how "safe" the bond is (meaning: is it collateralized and in what form, how does it rank in seniority when it comes to payments from a potential bankruptcy estate, etc.). In the Credit Metrics system, bonds are therefore categorized by seniority classes to account for this. These classes are senior secured, senior unsecured, senior

Table 1.1 Recovery rates by seniority class (according to Moody's)

Seniority Class	Mean (%)	Standard Deviation (%)
Senior Secured	53.80	26.86
Senior Unsecured	51.13	25.45
Senior Subordinated	38.52	23.81
Subordinated	32.74	20.18
Junior Subordinated	17.09	10.90

subordinated, subordinated, and junior subordinated. Senior secured refers to the highest credit quality class and junior subordinated to the lowest quality class. Credit Metrics uses statistics supplied by another major international rating firm, Moody's, which state the mean recovery rates for bonds in such classes as well the standard deviations for these recovery rates. One such table by Moody's (taken from the Credit Metrics technical document) is shown in Table 1.1.

We see a significant decrease in the mean recovery rates the lower the seniority is and at the same time a slight decrease in standard deviation. We will work with this table in the following.

So, we are at time 0 and have a portfolio **KP** consisting of bonds $K_1, K_2, \ldots, K_n$. We recall the relevant determinants and assumptions regarding these bonds:

Each of these loans K_i is a bond and has

- A certain face value N_i.
- A certain (constant) annual coupon C_i (as a percentage of the bond's face value).
- A certain remaining time horizon to maturity T_i. Here, T_i is an integer greater than or equal to 1, since it was assumed that a coupon payment had just taken place immediately before time 0.
- Each bond belongs to a certain seniority class, and we denote the mean of its recovery rate by R_i and the standard deviation of its recovery rate by σR_i.
- Each bond has a rating of X_i at time 0 (this rating can change over time).

By A_0 we denote the value of the credit portfolio KP at time 0. If by $A_0(K_i)$ we denote the value of the loan K_i at time 0, then

$$A_0 = A_0(K_1) + A_0(K_2) + \ldots + A_0(K_i) + \ldots + A_0(K_{n-1}) + A_0(K_n)$$

The value $A_0(K_i)$ is calculated as $N_i \cdot a_0(K_i)\backslash 100$, with $a_0(K_i)$ denoting the fair price of the loan/bond. (Due to the previously made assumption that coupon payments were made just immediately before, the fair price is equal to the fair value.)

We know the process for computing the fair price of a bond (of the above form) –under a certain material condition– from Volume I Section 1.12 (expressed as a percentage):

$$a_0(K_i) = \frac{C_i}{\left(1+f_{0,1}\right)^1} + \frac{C_i}{\left(1+f_{0,2}\right)^2} + \frac{C_i}{\left(1+f_{0,3}\right)^3} + \ldots + \\ + \frac{C_i}{\left(1+f_{0,T_i-1}\right)^{T_i-1}} + \frac{C_i+100}{\left(1+f_{0,T_i}\right)^{T_i}}$$

$f_{0,k}$ again denotes the ideal risk-free interest rate for investments from time 0 to time k.

However, this formula for the fair price is only applicable under a certain material condition, namely, that the loan in question is in no way at risk of default, thus under the assumption that all coupon payments and repayment of the face value will definitely be made. For AAA-rated bonds, this assumption can essentially be assumed to be satisfied (as we will make plausible later). Yet lower-rated bonds do not offer that certainty.

The fair price of a bond was derived based on the valuation of future payments, or more precisely, on the valuation of the **promise** of a future payment. In the case of a default-proof bond, this promise is equivalent to actual payment. In the case of a bond where default is possible, this promise is worth somewhat less than a payment that will definitely be made. The lower the bond's rating, the higher the likelihood that the payment promise will not be kept (that the bond will default), so, the lower the bond's rating is, the less the promise is worth.

In the case of the highest rating, that is, when payments will definitely be made, a payment of 1 EUR at time T has the value of $\frac{1}{\left(1+f_{0,T}\right)^T}$ EUR.

Here, $f_{0,T}$ is the risk-free interest rate for loans/for borrowers with the highest credit rating.

In the case of a lower rating X, the promise of such a payment has a lower value. We could denote this value by $\frac{1}{\left(1+f_{0,T}{}^X\right)^T}$, where $f_{0,T}{}^X$ would be a value that increases the worse that rating X is.

How did we determine the value $f_{0,T}$? From statistics on currently traded zero-coupon bonds issued by first-class issuers!

And this is exactly how we are going to determine the values $f_{0,T}{}^X$ for the different rating classes $X = AAA, AA, A, BBB, BB, B, CCC$, that is, from statistics on currently traded zero-coupon bonds issued by entities rated X!

Such statistics are again provided by various firms, and we will assume in the following that (based on such statistics) we have all these values $f_{0,T}{}^X$.

Thus, if we assume that we have the values ${f_{0,T}}^X$, then we can give the bond K_i with the rating X_i the fair price

$$a_0(K_i) = \frac{C_i}{\left(1+{f_{0,1}}^{X_i}\right)^1} + \frac{C_i}{\left(1+{f_{0,2}}^{X_i}\right)^2} + \frac{C_i}{\left(1+{f_{0,3}}^{X_i}\right)^3} + \ldots + \\ + \frac{C_i}{\left(1+{f_{0,T_i-1}}^{X_i}\right)^{T_i-1}} + \frac{C_i+100}{\left(1+{f_{0,T_i}}^{X_i}\right)^{T_i}}$$

and thus get

$$A_0(K_i) = \frac{N_i}{100} \cdot \left(\frac{C_i}{\left(1+{f_{0,1}}^{X_i}\right)^1} + \frac{C_i}{\left(1+{f_{0,2}}^{X_i}\right)^2} + \frac{C_i}{\left(1+{f_{0,3}}^{X_i}\right)^3} + \ldots + \right. \\ \left. + \frac{C_i}{\left(1+{f_{0,T_i-1}}^{X_i}\right)^{T_i-1}} + \frac{C_i+100}{\left(1+{f_{0,T_i}}^{X_i}\right)^{T_i}} \right).$$

Comment 1.3 It is important to clarify that in the case of payments that are not at risk of default, the value of a payment of one euro at time T does indeed represent the fair value $\frac{1}{(1+f_{0,T})^T}$ at time 0 in the strict financial mathematical sense. Any other value would –in a frictionless market– allow arbitrage. The prerequisite for this fact (for the construction of arbitrage opportunities) was that, in a frictionless market, a perfect market participant can both invest and borrow money at the interest rate $f_{0,T}$.

In the case of payments at risk of default, however, the value $\frac{1}{(1+{f_{0,T}}^X)^T}$ of such a promise made by a market participant rated X is only a **plausible value** and not a fair value in the strict financial mathematical sense!

(After all, the interest rate ${f_{0,T}}^X$ is set so that an X-rated borrower can borrow at that (higher) rate. It would be absurd to assume that the same borrower could also invest money (risk-free) at this higher (!) interest rate.)

1.13 Credit Metrics, Part III: Rating Transition Probabilities and Expected Value of the Loan Portfolio in 1 Year

We will now turn our attention to the value A_1 of the credit portfolio **KP** in 1 year (from the point of view of time 0). Of course, we again have

$$A_1 = A_1(K_1) + A_1(K_2) + \ldots + A_1(K_i) + \ldots + A_1(K_{n-1}) + A_1(K_n),$$

where $A_1(K_i)$ denotes the value of the loan K_i in 1 year. The value of the loan K_i depends on two parameters that are not yet known: One is the rating that K_i will have in 1 year (we denote it by Y_i), and the second parameter is the interest rates that will exist at time 1 until time t for different points in time $t > 1$.

The question as to the suitable interest rates can be solved relatively easily (plausibly (!)), by proceeding in the same way as in the case of payments that are not at risk of default: At time 0 we cannot yet know the actual interest rates for time 1. What we can do at time 0, however (as we saw in Volume I Section 1.10), is to generate for time 1 an interest rate up to time t of the form $f_{1,t}$ with a value of $f_{1,t} = \left(\frac{(1+f_{0,t})^t}{(1+f_{0,1})}\right)^{\frac{1}{t-1}} - 1$. This is in fact the best possible interest rate that can be secured at time 0 for the period $[1, t]$.

Thus we can make arrangements at time 0 (without investing any of our own money) so that, by investing $\frac{1}{(1+f_{1,t})^{t-1}}$ EUR at time 1, we can secure a payout of 1 EUR at time t. The value of a 1 EUR payout at time t therefore has a value of $\frac{1}{(1+f_{1,t})^{t-1}}$ EUR at time 1 from today's perspective.

It is plausible (!) therefore to proceed analogously in the case of payments at risk of default, i.e. payments with a lower rating.

We define $f_{1,t}{}^{(Y)} := \left(\frac{(1+f_{0,t}{}^{(Y)})^t}{(1+f_{0,1}{}^{(Y)})}\right)^{\frac{1}{t-1}} - 1$. With V we denote a promise that is given to us at time 1 by a Y-rated market participant for payment of 1 EUR at a future time t. At time 0, we set the value of V as $\frac{1}{(1+f_{1,t}{}^{(Y)})^{t-1}}$.

With this, we get for the value $A_1(K_i, Y_i)$ of the loan K_i at time 1, from the perspective of time 0 (in our model!), assuming that K_i has the rating Y_i at time 1, as follows (this holds for $Y_i \neq D$):

$$A_1(K_i, Y_i) = \frac{N_i}{100} \cdot \left(\frac{C_i}{\left(1 + f_{1,2}{}^{Y_i}\right)^1} + \frac{C_i}{\left(1 + f_{1,3}{}^{Y_i}\right)^2} + \ldots + \right.$$
$$\left. + \frac{C_i}{\left(1 + f_{1,T_i-1}{}^{Y_i}\right)^{T_i-2}} + \frac{C_i + 100}{\left(1 + f_{1,T_i}{}^{Y_i}\right)^{T_i-1}} \right).$$

So the actual value of this variable depends on the loan's actual rating Y_i in 1 year. This makes the value $A_1(K_i)$ a random variable.

Now, our first objective was to determine the expected value of A_1. We will have succeeded in doing so as soon as we can calculate the expected value $E(A_1(K_i))$ of $A_1(K_i)$.

If we knew the probabilities p_{X_i,Y_i} of the rating transition within 1 year from X_i to Y_i for all rating classes, then the expected value would be easy to calculate using

$$E(A_1(K_i)) = \sum_{Y \in Rat\setminus D} p_{X_i,Y} \cdot A_1(K_i, Y) + p_{X_i,D} \cdot R_i =$$

$$= \sum_{Y \in Rat\setminus D} p_{X_i,Y} \cdot \frac{N_i}{100} \cdot \left(\frac{C_i}{\left(1 + f_{1,2}{}^{Y}\right)^1} + \frac{C_i}{\left(1 + f_{1,3}{}^{Y}\right)^2} + \ldots + \right.$$

$$\left. + \frac{C_i}{\left(1 + f_{1,T_i-1}{}^{Y}\right)^{T_i-2}} + \frac{C_i + 100}{\left(1 + f_{1,T_i}{}^{Y}\right)^{T_i-1}} \right) + p_{X_i,D} \cdot R_i.$$

Here we denote by "Rat" the set of all rating classes, that is, $Rat = \{AAA, AA, A, BBB, BB, B, CCC, D\}$ and $Rat\setminus D$ denote the set of all rating classes except D.

Credit Metrics now proposes to choose the probabilities p_{X_i,Y_i} on the basis of statistics on such rating transition probabilities.

Such statistics are again available from various data providers. The Credit Metrics technical document gives the following table of Standard & Poor's 1-year rating transition probabilities.

By this statistic, the probability that a BBB-rated loan will be downgraded to B within 1 year is 1.17%, or 0.0117.

The probability that a BB-rated loan will default within 1 year is 1.06%, or 0.0106.

With this we can now determine the credit portfolio's expected value $E(A_1)$ in 1 year:

$$E(A_1) = \sum_{i=1}^{n} \left(\sum_{Y \in Rat\setminus D} p_{X_i,Y} \cdot A_1(K_i, Y) + p_{X_i,D} \cdot N_i \cdot R_i \right)$$

(Here, R_i is given as absolute values, not percentages, and N_i again denotes the face value of K_i.)

1.14 Credit Metrics, Part IV: Variance of a Credit Portfolio's Value in 1 Year

In the next planned step, we want to determine the variance of the random variable A_1 (= value of the credit portfolio in 1 year). We will approach this task step by step. In a first step, we want to determine the variance $V(A_1(K_i))$ of the value $A_1(K_i)$ of a loan K_i in 1 year. At first, this might seem a task that can be done in no time by using either the standard approach or the definition of the variance directly:

$$V(A_1(K_i)) = \sum_{Y \in Rat \backslash D} p_{X_i,Y} \cdot (A_1(K_i, Y) - E(A_1(K_i)))^2 + p_{X_i,D} \cdot (N_i \cdot \sigma R_i)^2$$

(Here, σR_i is given as absolute values, not percentages, and N_i again denotes the face value of K_i.) Unfortunately, however, this variance formula is not correct here. The error is in the last summand. It is not possible to simply set the variance of the recovery rate as a factor here. The correct approach is examined in the following excursus.

The **question** we want to find an answer to can be generally stated as follows: *We consider a random variable Z which can attain the possible values* $Z_1, Z_2, \ldots, Z_{s-1}$ *with probabilities* $p_1, p_2, \ldots, p_{s-1}$ *or, with probability* p_s, *the value of another random variable U. Thus we have* $p_1 + p_2 + \ldots + p_{s-1} + p_s = 1$. *Furthermore, we know the expected value* $E(U)$ *and the variance* $V(U)$ *of* U. *The expected value of Z is given by* $E(Z) = p_1 \cdot Z_1 + p_2 \cdot Z_2 + \ldots + p_{s-1} \cdot Z_{s-1} + p_s \cdot E(U)$ *of course. We now want to find the variance* $V(Z)$ *of the random variable Z.*
To determine $V(Z)$, we assume that the random variable U has density φ. This gives us

$$V(Z) = \sum_{i=1}^{s-1} p_i \cdot (Z_i - E(Z))^2 + p_s \cdot \int_{-\infty}^{\infty} \varphi(u) \cdot (u - E(Z))^2 du.$$

We evaluate the integral that occurs here:
$\int_{-\infty}^{\infty} \varphi(u) \cdot (u - E(Z))^2 du =$
$\int_{-\infty}^{\infty} \varphi(u) \cdot u^2 \, du - 2 \cdot E(Z) \cdot \int_{-\infty}^{\infty} \varphi(u) \cdot u \, du + (E(Z))^2 \cdot \int_{-\infty}^{\infty} \varphi(u) \, du =$
$E(U^2) - 2 \cdot E(Z) \cdot E(U) + (E(Z))^2 = V(U) + (E(U))^2 - 2 \cdot E(Z) \cdot E(U) + (E(Z))^2 = V(U) + (E(U) - E(Z))^2$.
From this it follows that

$$V(Z) = \sum_{i=1}^{s-1} p_i \cdot (Z_i - E(Z))^2 + p_s \cdot V(U) + p_s \cdot (E(U) - E(Z))^2.$$

We see that an additional summand $p_s \cdot (E(U) - E(Z))^2$ occurs here. If we denote the value $E(U)$ by Z_s, then we can summarize the variance $V(Z)$ as follows:

$$\mathbf{V(Z) = \sum_{i=1}^{s} p_i \cdot (Z_i - E(Z))^2 + p_s \cdot V(U).}$$

As noted above: The summation now runs from 1 to s and Z_s stands for $E(U)$.

We now use this finding to compute the variance $V(A_1(K_i))$. The correct formula is now:

$$V(A_1(K_i)) =$$

$$= \sum_{Y \in Rat} p_{X_i,Y} \cdot (A_1(K_i, Y) - E(A_1(K_i)))^2 + p_{X_i,D} \cdot (N_i \cdot \sigma R_i)^2,$$

where $A_1(K_i, D) := N_i \cdot R_i$.

1.15 Credit Metrics, Part V: Variance of a Credit Portfolio's Value in 1 Year

If the values $A_1(K_1), A_1(K_2), \ldots, A_1(K_n)$ of the loans in 1 year were independent of one another, then determining the variance of the credit portfolio would obviously be very easy. We would then have:

$$V(A_1) = V(A_1(K_1)) + V(A_1(K_2)) + \ldots + V(A_1(K_n))$$

But actually, and this is the reason and the point of the whole effort here, these values are, of course, not independent of each other. Instead:

$$V(A_1) = \sum_{i=1}^{n} V(A_1(K_i)) + \sum_{\substack{i,j=1 \\ i \neq j}}^{n} cov(A_1(K_i), A_1(K_j))$$

And let us also take into account that
$V(A_1(K_i) + A_1(K_j)) = V(A_1(K_i)) + V(A_1(K_j)) + 2 \cdot cov(A_1(K_i), A_1(K_j))$, hence
$cov(A_1(K_i), A_1(K_j)) = \frac{1}{2} \cdot (V(A_1(K_i) + A_1(K_j)) - (V(A_1(K_i)) + V(A_1(K_j))))$.
And so by substituting, we get

$$V(A_1) = \sum_{\substack{i,j=1 \\ i < j}}^{n} V(A_1(K_i) + A_1(K_j)) - (n-2) \cdot \sum_{i=1}^{n} V(A_1(K_i)).$$

Thus, to calculate $V(A_1)$, it is sufficient to determine the variance $V(A_1(K_i) + A_1(K_j))$ pairwise for each of two loans K_i and K_j.

The expected value of $A_1(K_i) + A_1(K_j)$ is, of course, simply the sum of the two expected values of $A_1(K_i)$ and $A_1(K_j)$.

Let us now assume that we know the joint rating transition probabilities of the two loans K_i and K_j, that is, the values $p_{X_i \to Y_i, X_j \to Y_j}$ which indicate the probability that the loan K_i will be rated Y_i in 1 year and that the loan K_j will be rated Y_j in 1 year.

Before going any further, we note: These values $p_{X_i \to Y_i, X_j \to Y_j}$ **cannot** be values that depend only on the rating classes X_i, X_j, Y_i, Y_j, because they most certainly also depend on the correlation between the two **specific** loans K_i and K_j!

So these values cannot be read from general statistics on rating transition probabilities, and moreover, a general relationship of the form $p_{X_i \to Y_i, X_j \to Y_j} = p_{X_i \to Y_i} \cdot p_{X_j \to Y_j}$ will also not hold. Determining or, more precisely, estimating these probabilities will therefore be a topic that we will deal with separately. But suppose we had these values $p_{X_i \to Y_i, X_j \to Y_j}$ for the two loans K_i and K_j. Then we could determine the variance $V(A_1(K_i) + A_1(K_j))$ as follows:

$$V(A_1(K_i) + A_1(K_j)) =$$

$$= \sum_{Y_i, Y_j \in Rat} p_{X_i \to Y_i, X_j \to Y_j} \cdot (A_1(K_i, Y_i) + A_1(K_j, Y_j) - E(A_1(K_i) + A_1(K_j)))^2 +$$

$$+ p_{X_i, D} \cdot (N_i \cdot \sigma R_i)^2 + p_{X_j, D} \cdot (N_j \cdot \sigma R_j)^2$$

again with $A_1(K_i, D) := N_i \cdot R_i$ or $A_1(K_j, D) := N_j \cdot R_j$.

The approach in principle is clear; the necessary adaptation for cases in which Y_i or Y_j is equal to D can be derived in the same way as above in the case of **one** credit.

Determining the variance of the credit portfolio's value in 1 year is now only a matter of estimating reasonably convincing values $p_{X_i \to Y_i, X_j \to Y_j}$ for the two loans K_i and K_j. This will be done in the next section based on J.P. Morgan's asset value approach, whereupon the determination of the variance of the credit portfolio's value in 1 year will then have been completed.

1.16 Credit Metrics, Part VI: The Asset Value Model for Determining Joint Credit Rating Transition Probabilities

To determine the joint rating transition probabilities of two loans, J.P. Morgan proposes in its Credit Metrics system to use the asset values of the borrowers (the bond issuers). Again, we are not going to discuss in further detail the exact meaning of an issuer's (varying (!)) asset value. However, to be able to work with a concrete value in the following, let us make the following assumption:

We assume that the issuers of the bonds K_i and K_j (the borrowers) are listed public companies, and we will use their stock price as their asset value (or as a suitable equivalent to their asset value) in the following:

J.P. Morgan then goes on to make the following **3 asset value model assumptions**:

Asset value model, assumption 1:
The returns on the issuers' asset values (stock prices) are normally distributed.

Asset value model, assumption 2:
changes correspond one to one to returns on the asset values.

Asset value model, assumption 3:
Return pairs of the asset values (the stock prices) of two issuers are bi-normally distributed.

The **first of these assumptions** was already discussed in detail in Volume I Section 4.1, and we noted there that, although systematic deviations of the return distributions from the normal distribution do occur ("fat tails"), we can accept this assumption as a working hypothesis in some cases (including in our case here).

The second and third of these assumptions need to be explained in more detail:

The **second assumption** of the asset value model is to be understood as follows:

The returns on the stock prices of company (borrower) K follow a normal distribution according to assumption 1, that is, approximately the distribution $N(\mu, \sigma^2)$. If we assume this, then for each $z \in \mathbb{R}$, we can determine the probability that the stock price return from now to a year from now is less than z. This probability is $\frac{1}{\sqrt{2\pi} \cdot \sigma} \cdot \int_{-\infty}^{z} e^{-\frac{(x-\mu)^2}{2 \cdot \sigma^2}} dx$.

Let the current rating of K be X.
From Standard & Poor's table of 1-year rating transitions in Fig. 1.23, we can read the probability that K will be rated D in 1 year, that is, the probability $p_{X,D}$.

Initial rating	Rating at year-end (%)							
	AAA	AA	A	BBB	BB	B	CCC	Default
AAA	90.81	8.33	0.68	0.06	0.12	0	0	0
AA	0.70	90.65	7.79	0.64	0.06	0.14	0.02	0
A	0.09	2.27	91.05	5.52	0.74	0.26	0.01	0.06
BBB	0.02	0.33	5.95	86.93	5.30	1.17	0.12	0.18
BB	0.03	0.14	0.67	7.73	80.53	8.84	1.00	1.06
B	0	0.11	0.24	0.43	6.48	83.46	4.07	5.20
CCC	0.22	0	0.22	1.30	2.38	11.24	64.86	19.79

Fig. 1.23 Standard & Poor's 1-year rating transition probabilities

Assumption 2 of the asset value model now states that this downgrade of K from X to a D rating occurs if and only if the stock price return of K by the end of 1 year is smaller than a certain value z (which we will denote by Z_D^X). Accepting this assumption, we can also calculate this value Z_D^X. Of course, we then have

$$p_{X,D} = W(\text{return} < Z_D^X) = \frac{1}{\sqrt{2\pi} \cdot \sigma} \cdot \int_{-\infty}^{Z_D^X} e^{-\frac{(x-\mu)^2}{2 \cdot \sigma^2}} dx.$$

In this equation (as per our assumption), all values except Z_D^X are known, and we can now (e.g. using the Mathematica software) calculate Z_D from this equation.

Yet Assumption 2 of the asset value model also states that K will have a rating of CCC in 1 year if the stock price return from today to 1 year from now falls below a certain bound Z_{CCC}^X, but not if it falls below Z_D^X, since that would cause a D rating. The probability of a CCC rating by the end of 1 year can again be read from the Fig. 1.23, namely, the value tabulated there as $p_{X,CCC}$. This gives us the equation

$$p_{X,CCC} = W(Z_D^X < \text{return} < Z_{CCC}^X) = \frac{1}{\sqrt{2\pi} \cdot \sigma} \cdot \int_{Z_D^X}^{Z_{CCC}^X} e^{-\frac{(x-\mu)^2}{2 \cdot \sigma^2}} dx$$

In this equation (as per our assumption), all values except Z_{CCC}^X are known (Z_D^X was already determined in the previous step), and we can calculate Z_{CCC}^X from this equation.

Proceeding in this way, we determine the values Z_B^X, Z_{BB}^X, Z_{BBB}^X, Z_A^X, and Z_{AA}^X. Z_{AA}^X, for example, is determined such that the borrower K is rated AA in 1 year if and only if the next 1-year return on the issuer's stock price is exactly between Z_A^X and Z_{AA}^X.

It then follows of course that the borrower K will be AAA rated in 1 year if and only if the next 1-year return on the issuer's stock price is above Z_{AA}^X.

In reality, we are going to determine the values Z_D^X, Z_{CCC}^X, Z_B^X, Z_{BB}^X, Z_{BBB}^X, Z_A^X, and Z_{AA}^X in a slightly different way, like this:

We start again as above but then perform a substitution $y := \frac{x-\mu}{\sigma}$, so that $dx = \sigma \cdot dy$, and then get

$$p_{X,D} = \frac{1}{\sqrt{2\pi} \cdot \sigma} \cdot \int_{-\infty}^{Z_D^X} e^{-\frac{(x-\mu)^2}{2 \cdot \sigma^2}} dx = \frac{1}{\sqrt{2\pi}} \cdot \int_{-\infty}^{\overline{Z}_D^X} e^{-\frac{y^2}{2}} dy,$$

where

$$\overline{Z}_D^X = \frac{Z_D^X - \mu}{\sigma}$$

From the equation $p_{X,D} = \frac{1}{\sqrt{2\pi}} \cdot \int_{-\infty}^{\overline{Z}_D^X} e^{-\frac{y^2}{2}} dy$, we calculate the value $\overline{Z}_D^X$ which is completely independent of μ and σ and depends only on the value $p_{X,D}$.

The value Z_D^X is then simply determined by $Z_D^X = \sigma \cdot \overline{Z}_D^X + \mu$. However –and this is some good news upfront– we will see that just knowing $\overline{Z}_D^X$ is already completely sufficient, in fact, the values μ and σ will never be needed in the following. So only the value $\overline{Z}_D^X$ and the similarly calculated values $\overline{Z}_{CCC}^X, \ldots, \overline{Z}_{AA}^X$ will be relevant in what follows!

Thus, having calculated $\overline{Z}_D^X$ in this way, the equation $p_{X,CCC} = \frac{1}{\sqrt{2\pi}\cdot\sigma} \cdot \int_{Z_D^X}^{Z_{CCC}^X} e^{-\frac{(x-\mu)^2}{2\cdot\sigma^2}} dx$ will, again by substitution $y := \frac{x-\mu}{\sigma}$, become the equation

$p_{X,CCC} = \frac{1}{\sqrt{2\pi}} \cdot \int_{\overline{Z}_D^X}^{\overline{Z}_{CCC}^X} e^{-\frac{y^2}{2}} dy$, from which the value $\overline{Z}_{CCC}^X$, which depends only on $p_{X,CCC}$ (and implicitly on $p_{X,D}$), can then be determined.

In this way we obtain the values $\overline{Z}_Y^X$, which can be calculated purely from the Fig. 1.23.

Recalculating these values is therefore only ever necessary when an adapted version of the rating transition probabilities table is issued.

We have calculated the $\overline{Z}_Y^X$ for $X = AAA, AA, A, BBB, BB, B, CCC$ and $Y = D, CCC, B, BB, BBB, A, AA$ (we will need the relevant values when running a concrete example later anyway) and listed them in the following Fig. 1.23.
This sufficiently clarifies the stated Assumption 2 of the asset value model.

Now let's examine **Assumption 3 of the asset value model**.

The bi-normal distribution was already discussed in Volume I Section 4.2. Assuming that we have accepted Assumption 1, that is, the normal distribution of stock price returns, then the assumption of the bi-normal distribution of pairs of stock price returns of the two companies K_i and K_j is also acceptable.

So the premises are: If the returns x_l of the stock price of K_i and the returns y_l of the stock price of K_j are distributed as $N(\mu_1, \sigma_1^2)$ resp. $N(\mu_2, \sigma_2^2)$ and if there is a correlation ρ between the return series, then the return pairs (x_l, y_l) follow a bi-normal distribution with density

$$f(x, y) = \frac{1}{2\pi\sigma_1\sigma_2\sqrt{1-\rho^2}} \cdot \exp\left(-\frac{1}{2\cdot(1-\rho^2)} \cdot \left(\frac{(x-\mu_1)^2}{\sigma_1^2} - 2\rho\frac{x-\mu_1}{\sigma_1} \cdot \frac{y-\mu_2}{\sigma_2} + \frac{(y-\mu_2)^2}{\sigma_2^2}\right)\right)$$

(Here we write $\exp(z)$ for e^z.)

This is the content of Assumption 3.

Note that we can already compute the probabilities $p_{X_i \to Y_i, X_j \to Y_j}$ for the two loans K_i and K_j. We have

$$p_{X_i \to Y_i, X_j \to Y_j} = W\left(x_l \in \left[Z_{\widetilde{Y}_i}^{X_i}, Z_{Y_i}^{X_i}\right] \text{ and } y_l \in \left[Z_{\widetilde{Y}_j}^{X_j}, Z_{Y_j}^{X_j}\right]\right) =$$

$$= \int_{Z_{\widetilde{Y}_i}^{X_i}}^{Z_{Y_i}^{X_i}} \int_{Z_{\widetilde{Y}_j}^{X_j}}^{Z_{Y_j}^{X_j}} \frac{1}{2\pi\sigma_1\sigma_2\sqrt{1-\rho^2}} \cdot$$

$$\cdot \exp\left(-\frac{1}{2\cdot(1-\rho^2)} \cdot \left(\frac{(x-\mu_1)^2}{\sigma_1^2} - 2\rho\frac{x-\mu_1}{\sigma_1} \cdot \frac{y-\mu_2}{\sigma_2} + \frac{(y-\mu_2)^2}{\sigma_2^2}\right)\right) dx\, dy.$$

The notations used here are as follows:

- x_l and y_l are the 1-year stock price returns of K_i and of K_j.
- $\widetilde{Y}_i$ and $\widetilde{Y}_j$ are the rating classes that are exactly one grade below Y_i and Y_j.
- The Z values are the values determined above at which rating changes occur. In addition, we set $Z_{AAA}^X := \infty$ and $Z_{\widetilde{D}}^X := -\infty$.

Substituting $s := \frac{x-\mu_1}{\sigma_1}$ and $t := \frac{y-\mu_2}{\sigma_2}$ in the last integral, we get the following simplified version

$$p_{X_i \to Y_i, X_j \to Y_j} = \int_{\overline{Z}_{\widetilde{Y}_i}^{X_i}}^{\overline{Z}_{Y_i}^{X_i}} \int_{\overline{Z}_{\widetilde{Y}_j}^{X_j}}^{\overline{Z}_{Y_j}^{X_j}} \frac{1}{2\pi\sqrt{1-\rho^2}} \cdot$$

$$\cdot \exp\left(-\frac{1}{2\cdot(1-\rho^2)} \cdot (s^2 - 2\rho \cdot s \cdot t + t^2)\right) ds\, dt$$

So here we do not need to know the trends or the volatilities μ_1, μ_2, σ_1, and σ_2 for any intents or purposes—not even for computing the values $\overline{Z}$, since they have already been computed (see, e.g. the Fig. 1.24)! We only need the correlation ρ between the two stock price return series!

To determine the correlation ρ, we can actually resort to empirical stock price estimates. However, the assumption that there actually is a stock price for the

Rating X	Rating Y						
	AA	A	BBB	BB	B	CCC	Default
AAA	-1.329	-2.382	-2.911	-3.036	$-\infty$	$-\infty$	$-\infty$
AA	2.454	-1.963	-2.382	-2.848	-2.948	-3.540	$-\infty$
A	3.130	1.985	-1.507	-2.300	-2.716	-3.195	-3.239
BBB	3.542	2.697	1.530	-1.493	-2.178	-2.748	-2.911
BB	3.473	2.937	2.393	1.368	-1.232	-2.041	-2.304
B	∞	3.037	2.688	2.414	1.456	-1.324	-1.626
CCC	2.875	2.875	2.634	2.115	1.739	1.022	-0.849

Fig. 1.24 Table of the $\overline{Z}_Y^X$ values based on the rating transition probabilities Fig. 1.23

issuer in question is not necessarily always fulfilled. J.P. Morgan therefore also recommends using empirical data and statistics when choosing ρ. Every borrower K_i and every borrower K_j should be allocated to sectors subdivided by region and industry. Such sectors could be, for example, "automotive industry/Europe" or "airlines/Asia", etc. Then, one should use statistics on typical average correlation values between stock prices from different sectors.

Joint rating transition probabilities are then calculated based on J.P. Morgan's asset value model, and the variance $V(A_1(\mathbf{KP}))$ of the credit portfolio's 1-year value $A_1(\mathbf{KP})$ can then be computed based on that.

1.17 Credit Metrics, Part VII: Determining the Percentile of A_1 Using Monte Carlo Simulation

Now, finally, we get to the real goal of our analysis. After all, the actual task was to determine the 1-year VaR at a confidence level of 99% for A_1. We will use Monte Carlo simulation for this purpose. As we will see in a moment, we already have all the necessary parameters to do this.

However, we start off with a little preparatory work, namely, normalizing the values that we will be using in the following:

Our starting position is the value A_0, the current value of the credit portfolio. Based on a very rough estimate, we assume, of course, that this value is roughly close to the total face value of the entire credit portfolio (under normal circumstances, a loan's value is approximately equivalent to its principal amount, its face value). We now normalize the value to a percentage of the total face value. This means: Instead of A_0, we are going to work with a value of the form $\frac{100}{N_1+N_2+\ldots+N_n} \cdot A_0$ in the following.

Similarly, instead of A_1, we are going to work with the value $\frac{100}{N_1+N_2+\ldots+N_n} \cdot A_1$, that is, normalized to a percentage of the total face value. Having stated this here, we will not explicitly emphasize this any more in the following.

Based on our modelling (!), the value A_1 is now only dependent on how the issuers' (the borrowers') stock price returns change from today to 1 year from now. The stock return movements uniquely determine what ratings borrowers will have in 1 year, and these 1-year borrower ratings in turn uniquely determine the value A_1.

The normalized value A_1 can never be much above 100 (just as the price of a bond will never be much above 100). But the normalized value A_1 may well be considerably less than 100.

If the stock price returns of borrowers exhibit a very positive trend over 1 year, then their ratings will be upgraded, and then the A_1 value can be expected to be significantly, but not much above 100 (in the range of about 110).

If the stock price returns of borrowers exhibit a very negative trend over 1 year, then there will be rating downgrades and also issuer defaults, and we will have to expect A_1 values well below 100 (roughly in the 50 to 80 range).

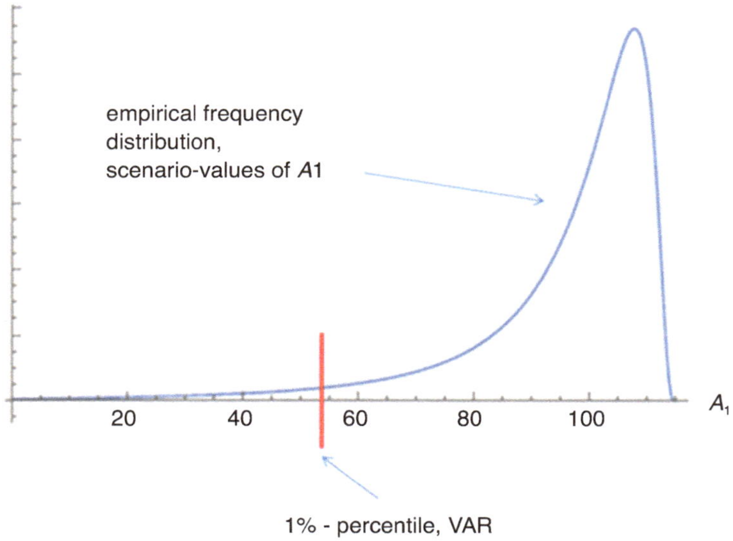

Fig. 1.25 Approximate empirical frequency distribution of the (normalized) A_1 of a credit portfolio's value in 1 year

If the stock price returns of borrowers exhibit a 1-year trend in line with average expectations, then A_1 will be roughly in the range around 100. This is the most likely case.

We now use Monte Carlo simulation to simulate in a representative manner various possible scenarios for A_1. From the resulting empirical distribution, which we expect (based on the above considerations) to have approximately the following shape (illustrated in Fig. 1.25), from this empirical distribution, we then estimate the 1% percentile, that is, the VaR that we are looking for.

More specifically, we need to proceed in the following way (we first outline the algorithm in general terms and then go into the details of each step):

1. Random construction of M (= number of scenarios in the simulation, e.g. $M = 10{,}000$) vectors $(x_1, x_2, \ldots, x_n)$ of possible stock price returns for borrowers $K_1, K_2, \ldots, K_n$ over the next year. The x_i here are to be chosen as normally distributed. Since we noted, in deriving the procedure, that trend and volatility of individual stock prices are irrelevant, **we can work with standard normally distributed x_i .** The construction of the vectors must correspond to the correlation structure between the borrowers' stock prices.
2. For each of the M random vectors $(x_1, x_2, \ldots, x_n)$, we now compute the new borrower ratings as determined by the returns $x_1, x_2, \ldots, x_n$.
3. Based on the borrowers' new rating structure caused by the scenario $(x_1, x_2, \ldots, x_n)$, we now compute the A_1 value of the credit portfolio in 1

year based on this return scenario. This gives us M possible credit portfolio values $A_1^{(1)}, A_1^{(2)}, \ldots, A_1^{(M)}$.

4. An estimate of the 1% percentile is then made from these values by arranging them in ascending order $A_1^{(u_1)} < A_1^{(u_2)} < \ldots < A_1^{(u_M)}$ and using the value $A_1^{(u_\theta)}$ with index $\theta = \left[\frac{M}{100}\right]$ as the estimated value.

Some more detailed remarks regarding the above steps are now called for:

Ad (1)

We already know from Volume I Section 4.7 how to proceed in order to obtain the appropriate correlation between the values $x_1, x_2, \ldots, x_n$: We first have to do a Cholesky decomposition $COR = C \cdot C^T$ with a lower triangular matrix C for the correlation matrix COR. Then we generate a vector $(y_1, y_2, \ldots, y_n)$ of independent standard normally distributed random numbers. We obtain the vector $(x_1, x_2, \ldots, x_n)$ by multiplying the Cholesky matrix C by the vector $(y_1, y_2, \ldots, y_n)$.

Ad (3)

In principle, it is clear how, based on the new ratings for each of the borrowers, the resulting A_1 value is to be calculated for this scenario. There is only one imponderable here: If one of the return values x_i is so small that it results in a D rating for borrower K_i, then it is not clear what $A_1(K_i)$ value we should expect in this scenario. In terms of the recovery rate, we only know its mean and its standard deviation, based on the seniority class to which K_i belongs, but we do not know its explicit value. So in this case, we have no choice but to determine this value again, using a random variable. The generated (normalized (!)) random value for $A_1(K_i)$ should be somewhere between 0 and 100 and have a mean and a standard deviation that correspond to the mean and the standard deviation of the respective recovery rate. J. P. Morgan proposes generating a random value according to a **beta distribution** for this purpose. How to generate random numbers with respect to arbitrary given distributions was already set out in Volume II Section 2.23. Therefore, we only need to explain here which form, which parameters, and which density the beta distribution has.

The beta distribution:

Beta distributions are very practical whenever random numbers are needed that lie in a certain range $[a, b]$ and should have a given mean M and a given standard deviation S.

We will show below how, using beta distribution, random numbers z can be generated on the interval $[0, 1]$ with mean μ and standard deviation σ. Once we master that, the general task is easily done in the following way:

We generate beta-distributed random numbers x on $[0, 1]$ with expected value $\mu = \frac{M-a}{b-a}$ and with standard deviation $\sigma = \frac{S}{b-a}$. Then we set $y :=$

(continued)

$a + (b - a) \cdot x$ and y lies in $[a, b]$ and has expected value M and standard deviation S, which is easily verified.

The density of a beta distribution depends on two parameters $p > 0$ and $q > 0$, is denoted by $B_{p,q}(x)$, and is of the form $B_{p,q}(x) = \frac{x^{p-1} \cdot (1-x)^{q-1}}{\int_0^1 u^{p-1} \cdot (1-u)^{q-1} du}$ for $x \in [0, 1]$ and $B_{p,q}(x) = 0$ for x outside the unit interval.

The expected value μ of a beta distribution is $\mu = \frac{p}{p+q}$, and the standard deviation σ is given by $\sigma^2 = \frac{pq}{(p+q+1) \cdot (p+q)^2}$.

For given (desired) parameters μ and σ^2 as expected value and variance of the beta distribution, we need to choose suitable p and q. And we get those from the equations $\mu = \frac{p}{p+q}$ and $\sigma^2 = \frac{pq}{(p+q+1) \cdot (p+q)^2}$, which give us the equations $p = \frac{\mu^2 - \mu^3 - \mu \cdot \sigma^2}{\sigma^2}$ and $q = (1 - \mu) \cdot \frac{\mu - \mu^2 - \sigma^2}{\sigma^2}$.

In Fig. 1.26 we present the density function $B_{p,q}(x)$ for different values of p and q. We have:

Blue curve:	$p = 2/q = 2$,	$\mu = 0.5/\sigma = 0.2236$
Red curve:	$p = 3/q = 3$,	$\mu = 0.5/\sigma = 0.18898$
Green curve:	$p = 2/q = 3$,	$\mu = 0.4/\sigma = 0.2$
Turquoise curve:	$p = 3/q = 2$,	$\mu = 0.6/\sigma = 0.2$
Magenta curve:	$p = 0.5/q = 2$,	$\mu = 0.2/\sigma = 0.2138$
Yellow curve:	$p = 2/q = 0.5$,	$\mu = 0.8/\sigma = 0.2138$
Black curve:	$p = 0.5/q = 0.5$,	$\mu = 0.5/\sigma = 0.3536$
Olive green curve:	$p = 0.5/q = 0.1$,	$\mu = 0.8333/\sigma = 0.2946$

Since $\mu = \frac{p}{p+q}$ and $p > 0$ and $q > 0$, we see immediately that μ can attain arbitrary values (strictly) between 0 and 1. A very small q gives a mean close to 1 and a very large q gives a mean close to 0.

For a given μ strictly between 0 and 1, we recognize from the formulas $p = \frac{\mu^2 - \mu^3 - \mu\sigma^2}{\sigma^2}$ and $q = (1 - \mu) \cdot \frac{\mu - \mu^2 - \sigma^2}{\sigma^2}$ that we can find suitable parameters $p > 0$ and $q > 0$ if and only if σ^2 is chosen such that σ^2 is greater than 0 and such that $\mu - \mu^2 - \sigma^2 > 0$ holds, i.e. that $\sigma^2 < \mu - \mu^2$ is satisfied.

So, to summarize:

For every μ with $0 < \mu < 1$ and every σ^2 with $\sigma^2 < \mu - \mu^2$, there are positive parameters p and q, such that the associated beta distribution has mean μ and variance σ^2.

Example 1.4 If we want to simulate a recovery rate for a loan in the "senior subordinated" class, for instance, then we read the mean of the recovery rate 38.52% and the standard deviation of the recovery rate 23.81% from the Table 1.1.

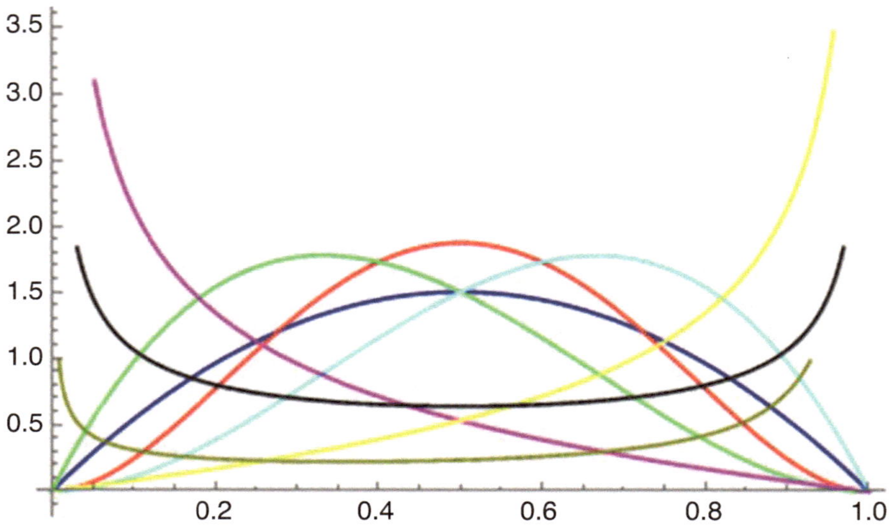

Fig. 1.26 Density function $B_{p,q}(x)$ of the beta function for different values p and q

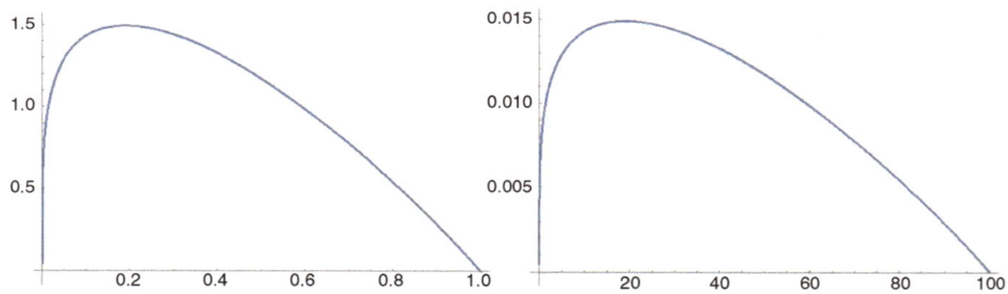

Fig. 1.27 Underlying density function of the beta distribution for simulation of the recovery rates of a "senior subordinated" loan on the left and actual density function on the right

The simulated values should be in the range [0, 100]. We will therefore generate beta-distributed random numbers in the range [0, 1] with $\mu = 0.3852$ and with $\sigma = 0.2381$. This is possible because $0.2381^2 < 0.3852 - 0.3852^2$. For the parameters p and q, we need to choose the following values:

$$p = \frac{\mu^2 - \mu^3 - \mu\sigma^2}{\sigma^2} = 1.2239 \text{ und } q = (1-\mu) \cdot \frac{\mu - \mu^2 - \sigma^2}{\sigma^2} = 1.9534$$

The associated density function $B_{p,q}(x)$ at [0, 1] has the shape shown in Fig. 1.27 (left). The density function for the random numbers in the range [0, 100], which is what we were actually interested in, is then obtained by
$\widetilde{B}_{p,q}(x) := \frac{B_{p,q}\left(\frac{x}{100}\right)}{100}$ on [0, 100] (Fig. 1.27, right).

Incidentally, in [2], a combination of Monte Carlo and quasi-Monte Carlo methods is used to approximate the percentile, which is definitely advantageous in some situations and certainly fares no worse than pure MC or pure QMC approaches.

1.18 Credit Metrics, Part VIII: Detailed Example with Variants

In this section, we will analyse a practical example of a credit portfolio and (based on the chosen data) analyse the portfolio's risk using the J. P. Morgan method. We will do this in three variants, for three different dependence structures between the loans, i.e. for three different correlation matrices, with otherwise identical parameters.

We start by providing the relatively high number of parameters that we need for our example. First, we provide the required interest rates ${f_{0,k}}^Y$. We use the following Table 1.2 for this:

Interest Rates ${f_{0,k}}^{(Y)}$ in Percent

We calculate the forward rates right away here, i.e. ${f_{1,t}}^{(Y)} := \left(\frac{(1+{f_{0,t}}^{(Y)})^t}{(1+{f_{0,1}}^{(Y)})}\right)^{\frac{1}{t-1}} - 1$ for $Y = AAA, AA, A, BBB, BB, B, CCC$ and $t = 2, 3, \ldots, 10$ and obtain the values (rounded to three decimals).

The **forward interest rates ${f_{1,k}}^{(Y)}$ in percent** are given in Table 1.3:

We use recovery rates and rating transition probabilities, and hence also the threshold values $\overline{Z}_Y^X$, as given in the Tables 1.4, 1.5, and 1.6. For completeness, we list the tables again below:

Recovery rates:

In the following, we denote the seniority classes in the above order by the numbers 1, 2, 3, 4, and 5.

Rating transition probabilities $p_{X,Y}$:

Threshold values $\overline{Z}_Y^X$:

To illustrate the method and the procedure in principle, we will examine a portfolio of only 20 loans $K_1, K_2, \ldots, K_{20}$. Let all loans be denominated in

Table 1.2 Interest rates $f_{0,k}^{(Y)}$ in percent

Y/k	1	2	3	4	5	6	7	8	9	10
AAA	− 0.4	− 0.2	0	0.1	0.2	0.4	0.7	1.0	1.2	1.4
AA	− 0.2	− 0.1	0.2	0.4	0.5	0.7	0.9	1.2	1.4	1.6
A	0	0	0.3	0.6	0.7	0.9	1.2	1.5	1.7	1.8
BBB	0.2	0.4	0.8	1.0	1.3	1.5	1.7	1.8	2.0	2.3
BB	0.5	0.9	1.2	1.4	1.8	2.1	2.4	2.7	2.9	3.1
B	0.8	1.1	1.4	1.7	2.0	2.3	2.7	2.9	3.2	3.5
CCC	1.5	1.7	1.8	2.2	2.5	2.8	3.0	3.5	3.8	4.2

Table 1.3 Interest rates $f_{1,k}^{(Y)}$ in percent

Y/k	2	3	4	5	6	7	8	9	10
AAA	0.	0.2	0.267	0.35	0.56	0.884	1.201	1.401	1.601
AA	0.	0.4	0.6	0.675	0.88	1.084	1.401	1.601	1.801
A	0.	0.45	0.8	0.875	1.08	1.401	1.716	1.914	2.001
BBB	0.6	1.101	1.268	1.576	1.762	1.952	2.03	2.227	2.536
BB	1.301	1.551	1.701	2.127	2.423	2.72	3.018	3.204	3.393
B	1.4	1.701	2.001	2.302	2.602	3.02	3.203	3.503	3.804
CCC	1.9	1.95	2.434	2.751	3.061	3.252	3.788	4.091	4.504

Table 1.4 Recovery rates by seniority class (according to Moody's)

Seniority class	Mean (%)	Standard deviation (%)
Senior secured	53.80	26.86
Senior unsecured	51.13	25.45
Senior subordinated	38.52	23.81
Subordinated	32.74	20.18
Junior subordinated	17.09	10.90

Table 1.5 Standard & Poor's 1-year rating transition probabilities

Initial	Rating at year-end (%)							
rating	AAA	AA	A	BBB	BB	B	CCC	Default
AAA	90.81	8.33	0.68	0.06	0.12	0	0	0
AA	0.70	90.65	7.79	0.64	0.06	0.14	0.02	0
A	0.09	2.27	91.05	5.52	0.74	0.26	0.01	0.06
BBB	0.02	0.33	5.95	86.93	5.30	1.17	0.12	0.18
BB	0.03	0.14	0.67	7.73	80.53	8.84	1.00	1.06
B	0	0.11	0.24	0.43	6.48	83.46	4.07	5.20
CCC	0.22	0	0.22	1.30	2.38	11.24	64.86	19.79

Table 1.6 Table of the $\overline{Z}_Y^X$ values based on the rating transition probabilities Fig. 1.23

Rating	Rating Y						
X	AA	A	BBB	BB	B	CCC	Default
AAA	−1.329	−2.382	−2.911	−3.036	−1000	$-\infty$	$-\infty$
AA	2.454	−1.963	−2.382	−2.848	−2.948	−3.540	$-\infty$
A	3.130	1.985	−1.507	−2.300	−2.716	−3.195	−3.239
BBB	3.542	2.697	1.530	−1.493	−2.178	−2.748	−2.911
BB	3.473	2.937	2.393	1.368	−1.232	−2.041	−2.304
B	∞	3.037	2.688	2.414	1.456	−1.324	−1.626
CCC	2.875	2.875	2.634	2.115	1.739	1.022	−0.849

Table 1.7 Credit parameters for the respective loans K_i

	Face value N_j	Rating X_i	Maturity T_i	Coupon $C_i\%$	Seniority class	$A_0(K_i)$	$E(A_1(K_i))$
K_1	300,000	BB	4	1	1	295,435	292,582
K_2	500,000	A	6	3	1	562,042	545,721
K_3	300,000	CCC	2	0.5	1	292,982	269,573
K_4	300,000	AA	7	0.5	1	292,046	289,416
K_5	700,000	BB	3	0	2	675,392	675,735
K_6	500,000	AA	3	4	2	556,972	535,757
K_7	500,000	AAA	8	0.2	2	469,587	465,987
K_8	700,000	B	6	1	2	650,220	633,337
K_9	900,000	A	5	1	3	913,549	902,601
K_{10}	500,000	AAA	5	0.5	3	507,515	502,326
K_{11}	700,000	B	5	3	3	734,183	695,670
K_{12}	900,000	CCC	2	5	3	958,006	813,426
K_{13}	300,000	BB	6	5	4	350,111	334,945
K_{14}	900,000	AA	3	3	4	975,568	946,437
K_{15}	300,000	A	10	2	4	307,287	300,243
K_{16}	900,000	BBB	8	4	4	1,051,561	1,013,874
K_{17}	300,000	AAA	4	0.5	5	304,809	301,789
K_{18}	900,000	A	5	2	5	957,947	937,827
K_{19}	700,000	B	3	5	5	773,939	712,621
K_{20}	300,000	AAA	3	0.25	5	302,256	300,173

EUR. As parameters for the respective loans, we select the following data in the Table 1.7 (the red values in the columns $A_0(K_i)$ and $E(A_1(K_i))$ will be calculated later):

Credit parameters:

Finally, we need a correlation matrix for the borrowers' asset value returns. That is, we need a positive definite 20×20 matrix. Generating such a positive definite 20×20 matrix a priori is not all that simple, however. So for those interested, we will describe a possible approach and a corresponding program in our software in more detail in an excursus in the following Section 1.19.

For our example, we will consider three different dependency structures. In **Case 1** we will consider independent loans, so the correlation matrix COR_1 will be the 20×20 unit matrix. Similarly, the associated Cholesky decomposition matrix C_1 will then also be the 20×20 unit matrix.

In **Case 2** we will use a correlation matrix COR_2 with mainly clearly positive dependencies. This will be the following matrix (with rounded values here for better representation):

1.	0.77	0.23	0.53	0.96	0.64	0.42	0.12	0.07	0.43	0.91	0.12	0.49	0.49	0.79	0.67	0.85	0.46	0.26	0.72
0.77	1.	0.38	0.67	0.79	0.7	0.64	0.36	0.62	0.39	0.84	0.57	0.53	0.89	0.67	0.93	0.91	0.78	0.46	0.67
0.23	0.38	1.	0.62	0.25	0.4	0.4	0.45	0.32	0.82	0.43	0.79	0.28	0.39	0.6	0.58	0.5	0.74	0.54	0.79
0.53	0.67	0.62	1.	0.55	0.63	0.87	0.69	0.62	0.68	0.75	0.75	0.64	0.75	0.64	0.76	0.77	0.77	0.54	0.83
0.96	0.79	0.25	0.55	1.	0.66	0.49	0.22	0.14	0.51	0.93	0.19	0.52	0.54	0.81	0.71	0.88	0.55	0.48	0.74
0.64	0.7	0.4	0.63	0.66	1.	0.75	0.68	0.44	0.49	0.72	0.5	0.86	0.64	0.8	0.7	0.82	0.7	0.46	0.67
0.42	0.64	0.4	0.87	0.49	0.75	1.	0.88	0.74	0.56	0.66	0.72	0.81	0.81	0.58	0.71	0.75	0.75	0.62	0.66
0.12	0.36	0.45	0.69	0.22	0.68	0.88	1.	0.64	0.59	0.41	0.73	0.83	0.6	0.52	0.5	0.53	0.7	0.67	0.53
0.07	0.62	0.32	0.62	0.14	0.44	0.74	0.64	1.	0.23	0.35	0.8	0.45	0.87	0.19	0.7	0.5	0.74	0.5	0.3
0.43	0.39	0.82	0.68	0.51	0.49	0.56	0.59	0.23	1.	0.65	0.65	0.5	0.39	0.79	0.59	0.65	0.75	0.74	0.9
0.91	0.84	0.43	0.75	0.93	0.72	0.66	0.41	0.35	0.65	1.	0.43	0.64	0.68	0.87	0.83	0.95	0.73	0.54	0.85
0.12	0.57	0.79	0.75	0.19	0.5	0.72	0.73	0.8	0.65	0.43	1.	0.48	0.76	0.47	0.74	0.58	0.88	0.66	0.64
0.49	0.53	0.28	0.64	0.52	0.86	0.81	0.83	0.45	0.5	0.64	0.48	1.	0.61	0.74	0.57	0.7	0.63	0.44	0.6
0.49	0.89	0.39	0.75	0.54	0.64	0.81	0.6	0.87	0.39	0.68	0.76	0.61	1.	0.52	0.89	0.81	0.83	0.53	0.58
0.79	0.67	0.6	0.64	0.81	0.8	0.58	0.52	0.19	0.79	0.87	0.47	0.74	0.52	1	0.73	0.88	0.74	0.56	0.9
0.67	0.93	0.58	0.76	0.71	0.7	0.71	0.5	0.7	0.59	0.83	0.74	0.57	0.89	0.73	1.	0.91	0.91	0.61	0.77
0.85	0.91	0.5	0.77	0.88	0.82	0.75	0.53	0.5	0.65	0.95	0.58	0.7	0.81	0.88	0.91	1.	0.83	0.61	0.85
0.46	0.78	0.74	0.77	0.55	0.7	0.75	0.7	0.74	0.75	0.73	0.88	0.63	0.83	0.74	0.91	0.83	1.	0.78	0.8
0.26	0.46	0.54	0.54	0.48	0.46	0.62	0.67	0.5	0.74	0.54	0.66	0.44	0.53	0.56	0.61	0.61	0.78	1.	0.62
0.72	0.67	0.79	0.83	0.74	0.67	0.66	0.53	0.3	0.9	0.85	0.64	0.6	0.58	0.9	0.77	0.85	0.8	0.62	1.

The corresponding Cholesky matrix C_2 is the following matrix (again with rounded values for better representation):

1.	0.	0.	0.	0.	0.	0.	0.	0.	0.	0.	0.	0.	0.	0.	0.	0.	0.	0.	0.
0.77	0.62	0.	0.	0.	0.	0.	0.	0.	0.	0.	0.	0.	0.	0.	0.	0.	0.	0.	0.
0.23	0.32	0.91	0.	0.	0.	0.	0.	0.	0.	0.	0.	0.	0.	0.	0.	0.	0.	0.	0.
0.53	0.41	0.39	0.61	0.	0.	0.	0.	0.	0.	0.	0.	0.	0.	0.	0.	0.	0.	0.	0.
0.96	0.07	0.	0.	0.25	0.	0.	0.	0.	0.	0.	0.	0.	0.	0.	0.	0.	0.	0.	0.
0.64	0.32	0.16	0.14	0.05	0.65	0.	0.	0.	0.	0.	0.	0.	0.	0.	0.	0.	0.	0.	0.
0.42	0.49	0.15	0.6	0.18	0.29	0.24	0.	0.	0.	0.	0.	0.	0.	0.	0.	0.	0.	0.	0.
0.12	0.42	0.31	0.52	0.29	0.49	0.19	0.22	0.	0.	0.	0.	0.	0.	0.	0.	0.	0.	0.	0.
0.07	0.9	0.	0.33	0.04	0.07	0.09	0.	0.19	0.	0.	0.	0.	0.	0.	0.	0.	0.	0.	0.
0.43	0.08	0.76	0.17	0.33	0.01	0.19	0.11	0.14	0.06	0.	0.	0.	0.	0.	0.	0.	0.	0.	0.
0.91	0.21	0.15	0.17	0.14	0.01	0.01	0.06	0.09	0.07	0.14	0.	0.	0.	0.	0.	0.	0.	0.	0.
0.12	0.75	0.56	0.23	0.07	0.07	0.12	0.05	0.	0.02	0.06	0.06	0.	0.	0.	0.	0.	0.	0.	0.
0.49	0.23	0.09	0.38	0.09	0.59	0.17	0.36	0.01	0.03	0.02	0.01	0.03	0.	0.	0.	0.	0.	0.	0.
0.49	0.08	0.01	0.22	0.01	0.03	0.19	0.05	0.	0.02	0.	0.	0.01	0.04	0.	0.	0.	0.	0.	0.
0.79	0.09	0.42	0.01	0.16	0.27	0.07	0.21	0.07	0.	0.04	0.05	0.1	0.	0.04	0.	0.	0.	0.	0.
0.67	0.65	0.23	0.06	0.07	0.01	0.02	0.01	0.14	0.05	0.01	0.07	0.02	0.02	0.09	0.11	0.	0.	0.	0.
0.85	0.4	0.18	0.12	0.15	0.13	0.1	0.01	0.04	0.04	0.03	0.02	0.06	0.	0.02	0.	0.	0.	0.	0.
0.46	0.66	0.45	0.1	0.22	0.12	0.07	0.11	0.18	0.04	0.03	0.	0.	0.	0.	0.	0.	0.01	0.	0.
0.26	0.41	0.37	0.13	0.76	0.05	0.05	0.03	0.08	0.	0.	0.	0.	0.	0.	0.	0.	0.	0.	0.
0.72	0.18	0.61	0.2	0.11	0.01	0.08	0.06	0.	0.	0.	0.	0.01	0.	0.	0.01	0.	0.	0.	0.

Finally, in **Case 3** we will use a correlation matrix COR_3 with relatively many negative dependencies. This is going to be the following matrix (with rounded values for better representation):

1	0.72	-0.93	0.78	0.76	-0.65	-0.32	0.17	0.07	0.36	-0.43	-0.73	0.89	-0.68	-0.01	0.33	-0.39	-0.82	0.5	-0.65
0.72	1.	−0.45	0.15	0.89	−0.32	−0.46	−0.19	0.68	0.42	−0.46	−0.3	0.89	0.02	−0.5	0.07	−0.07	−0.57	0.88	−0.84
−0.93	−0.45	1.	−0.9	−0.49	0.56	0.36	−0.17	0.29	−0.26	0.17	0.88	−0.73	0.84	−0.18	−0.39	0.27	0.67	−0.3	0.48
0.78	0.15	−0.9	1.	0.36	−0.61	0.01	0.44	−0.47	0.18	−0.33	−0.75	0.48	−0.97	0.4	0.49	0.15	−0.67	−0.07	−0.1
0.76	0.89	−0.49	0.36	1.	−0.57	−0.12	0.09	0.62	0.34	−0.74	−0.22	0.81	−0.18	−0.37	0.16	−0.36	−0.72	0.59	−0.66
−0.65	−0.32	0.56	−0.61	−0.57	1.	−0.3	−0.58	−0.08	−0.06	0.45	0.2	−0.44	0.63	−0.24	0.08	0.	0.85	0.05	0.35
−0.32	−0.46	0.36	0.01	−0.12	−0.3	1.	0.64	−0.02	−0.12	−0.42	0.54	−0.48	−0.07	0.41	−0.01	0.76	−0.19	−0.65	0.65
0.17	−0.19	−0.17	0.44	0.09	−0.58	0.64	1.	−0.23	0.22	−0.41	0.15	0.01	−0.47	0.48	0.03	0.45	−0.55	−0.39	0.37
0.07	0.68	0.29	−0.47	0.62	−0.08	−0.02	−0.23	1.	0.27	−0.5	0.42	0.35	0.6	−0.55	−0.2	−0.41	−0.19	0.6	−0.47
0.36	0.42	−0.26	0.18	0.34	−0.06	−0.12	0.22	0.27	1.	−0.45	−0.21	0.49	−0.05	−0.09	0.59	−0.33	−0.39	0.54	−0.11
−0.43	−0.46	0.17	−0.33	−0.74	0.45	−0.42	−0.41	−0.5	−0.45	1.	−0.07	−0.44	0.15	−0.08	−0.42	−0.22	0.57	−0.16	0.01
−0.73	−0.3	0.88	−0.75	−0.22	0.2	0.54	0.15	0.42	−0.21	−0.07	1.	−0.59	0.7	−0.1	−0.51	0.33	0.37	−0.31	0.4
0.89	0.89	−0.73	0.48	0.81	−0.44	−0.48	0.01	0.35	0.49	−0.44	−0.59	1.	−0.33	−0.21	0.22	−0.61	−0.69	0.77	−0.76
−0.68	0.02	0.84	−0.97	−0.18	0.63	−0.07	−0.47	0.6	−0.05	0.15	0.7	−0.33	1.	−0.53	−0.37	−0.22	0.59	0.23	0.03
−0.01	−0.5	−0.18	0.4	−0.37	−0.24	0.41	0.48	−0.55	−0.09	0.08	−0.1	−0.21	−0.53	1.	0.01	0.54	−0.15	−0.52	0.47
0.33	0.07	−0.39	0.49	0.16	0.08	−0.01	0.03	−0.02	0.59	−0.42	−0.51	0.22	−0.37	0.01	1.	0.11	−0.18	0.13	0.19
−0.39	−0.7	0.27	0.15	−0.36	0.	0.76	0.45	−0.41	−0.33	−0.22	0.33	−0.61	−0.22	0.54	0.11	1.	0.18	−0.85	0.85
−0.82	−0.57	0.67	−0.67	−0.72	0.85	−0.19	−0.55	−0.19	−0.39	0.57	0.37	−0.69	0.59	−0.15	−0.18	0.18	1.	−0.29	0.42
0.5	0.88	−0.3	−0.07	0.59	0.05	−0.65	−0.39	0.6	0.54	−0.16	−0.31	0.77	0.23	−0.52	0.13	−0.85	−0.29	1.	−0.76
−0.65	−0.84	0.48	−0.1	−0.66	0.35	0.65	0.37	−0.47	−0.11	0.01	0.4	−0.76	0.03	0.47	0.19	0.85	0.42	−0.76	1.

And the corresponding Cholesky matrix C_3 is the following matrix (again with rounded values for better representation):

1.	0.	0.	0.	0.	0.	0.	0.	0.	0.	0.	0.	0.	0.	0.	0.	0.	0.	0.	0.
0.72	0.69	0.	0.	0.	0.	0.	0.	0.	0.	0.	0.	0.	0.	0.	0.	0.	0.	0.	0.
−0.93	0.31	0.2	0.	0.	0.	0.	0.	0.	0.	0.	0.	0.	0.	0.	0.	0.	0.	0.	0.
0.78	−0.6	0.08	0.15	0.	0.	0.	0.	0.	0.	0.	0.	0.	0.	0.	0.	0.	0.	0.	0.
0.76	−0.48	0.34	0.08	0.19	0.	0.	0.	0.	0.	0.	0.	0.	0.	0.	0.	0.	0.	0.	0.
−0.65	0.21	−0.48	0.48	−0.27	0.08	0.	0.	0.	0.	0.	0.	0.	0.	0.	0.	0.	0.	0.	0.
−0.32	−0.33	0.86	−0.07	−0.05	0.07	0.16	0.	0.	0.	0.	0.	0.	0.	0.	0.	0.	0.	0.	0.
0.17	−0.45	0.66	−0.13	−0.19	−0.4	−0.09	0.33	0.	0.	0.	0.	0.	0.	0.	0.	0.	0.	0.	0.
0.07	0.91	0.33	−0.1	0.05	−0.03	0.05	−0.19	0.01	0.	0.	0.	0.	0.	0.	0.	0.	0.	0.	0.
0.36	0.23	0.06	0.19	−0.47	−0.61	0.33	−0.16	0.08	0.19	0.	0.	0.	0.	0.	0.	0.	0.	0.	0.
−0.43	−0.22	−0.74	−0.36	−0.12	0.19	−0.13	0.17	0.07	0.01	0.08	0.	0.	0.	0.	0.	0.	0.	0.	0.
−0.73	0.33	0.51	−0.14	0.06	−0.11	−0.12	0.1	0.14	−0.09	0.02	0.	0.	0.	0.	0.	0.	0.	0.	0.
0.89	0.35	−0.08	0.02	−0.11	−0.12	−0.08	0.01	0.07	−0.02	−0.16	−0.09	0.03	0.	0.	0.	0.	0.	0.	0.
−0.68	0.73	−0.05	0.01	−0.05	−0.02	0.01	0.01	−0.01	0.02	−0.02	−0.02	−0.01	0.01	0.	0.	0.	0.	0.	0.
−0.01	−0.71	0.2	−0.18	−0.37	0.	−0.18	−0.21	0.08	−0.25	−0.06	−0.07	0.35	−0.11	0.11	0.	0.	0.	0.	0.
0.33	−0.24	−0.03	0.64	−0.14	−0.14	0.56	−0.2	0.08	0.12	−0.02	−0.01	0.01	−0.11	0.11	0.	0.	0.	0.	0.
−0.39	−0.61	0.56	0.32	0.04	0.13	−0.17	−0.12	0.07	−0.05	−0.02	−0.01	−0.01	0.	0.	−0.01	0.	0.	0.	0.
−0.82	0.03	−0.45	0.28	0.09	0.06	−0.16	−0.09	0.12	−0.01	0.01	−0.01	0.	0.	0.	−0.01	0.	0.	0.	0.
0.5	0.74	−0.31	0.01	−0.26	−0.07	0.11	−0.01	0.05	0.01	−0.06	0.06	0.	−0.03	0.	−0.01	0.	−0.01	0.	0.
−0.65	−0.55	0.33	0.35	−0.22	−0.08	0.03	−0.01	0.01	0.	−0.01	−0.01	0.	0.	0.	−0.01	0.	0.	0.	0.

Note that the portfolio's total face value is 11,400,000 EUR.

We start our calculations with the simple exercise of determining the portfolio's current value A_0: For every loan K_i, the current value $A_0(K_i)$ is given by

$$A_0(K_i) = \frac{N_i}{100} \cdot \left(\frac{C_i}{\left(1 + f_{0,1}{}^{X_i}\right)^1} + \frac{C_i}{\left(1 + f_{0,2}{}^{X_i}\right)^2} + \frac{C_i}{\left(1 + f_{0,3}{}^{X_i}\right)^3} + \ldots + \right.$$
$$\left. + \frac{C_i}{\left(1 + f_{0,T_i-1}{}^{X_i}\right)^{T_i-1}} + \frac{C_i + 100}{\left(1 + f_{0,T_i}{}^{X_i}\right)^{T_i}} \right).$$

We thus obtain for $A_0(K_i)$ the values that we already presented in the credit parameters table above. The sum of these current credit values is the **current portfolio value A_0**, resulting in a total portfolio value of **11,931,407 EUR**. Normalized to a percentage of the face value, this means **104.661 %**.

In the next step, we determine the values $A_1(K_i, Y)$ for $i = 1, 2, \ldots, 20$ and $Y = AAA, AA, A, BBB, BB, B$, and CCC by

$$A_1(K_i, Y) = \frac{N_i}{100} \cdot \left(\frac{C_i}{\left(1 + f_{1,2}{}^{Y}\right)^1} + \frac{C_i}{\left(1 + f_{1,3}{}^{Y}\right)^2} + \ldots + \right.$$
$$\left. + \frac{C_i}{\left(1 + f_{1,T_i-1}{}^{Y}\right)^{T_i-2}} + \frac{C_i + 100}{\left(1 + f_{1,T_i}{}^{Y}\right)^{T_i-1}} \right)$$

Table 1.8 Values of $A(K_i, Y)$ for $i = 1, \dots 20$ and $Y = AAA, \dots CCC$

	AAA	AA	A	BBB	BB	B	CCC
K_1	306,571	303,579	301,808	297,676	293,913	291,367	287,736
K_2	560,411	552,117	547,042	530,050	514,297	510,035	499,473
K_3	301,498	301,499	301,500	299,700	297,626	297,6334	295,877
K_4	293,402	289,962	284,668	275,664	263,728	259,237	255,770
K_5	697,200	694,425	693,737	684,832	678,769	676,775	673,473
K_6	537,919	535,858	535,347	528,613	523,972	522,471	519,921
K_7	466,709	460,331	450,543	440,915	412,420	407,292	391,628
K_8	715,324	704,297	697,528	674,996	654,022	648,422	634,429
K_9	923,256	911,606	904,549	880,272	861,752	855,953	841,339
K_{10}	502,985	496,579	492,694	479,353	469,184	466,008	457,983
K_{11}	773,726	764,294	758,601	738,901	723,826	719,057	707,168
K_{12}	944,996	944,999	945,000	939,360	932,857	931,944	927,376
K_{13}	365,926	360,698	357,510	346,775	336,862	334,147	327,469
K_{14}	950,291	946,617	945,706	933,752	925,538	922,871	918,368
K_{15}	311,473	306,343	301,287	288,580	269,546	261,176	247,546
K_{16}	1,073,145	1,059,421	1,039,722	1,017,664	960,458	949,479	917,896
K_{17}	302,089	299,118	297,357	293,273	289,551	287,024	283,426
K_{18}	959,023	947,135	939,947	915,144	896,193	890,227	875278
K_{19}	767,059	764,146	763,424	753,864	747,258	745,130	741,494
K_{20}	300,296	299,104	298,809	294,978	292,368	291,511	290,089

The values of $\boldsymbol{A_1(K_i, Y)}$ **for** $\boldsymbol{i = 1, 2, \dots, 20}$ **and** $\boldsymbol{Y = AAA, AA, A, BBB, BB, B}$, **and** $\boldsymbol{CCC}$ are given in the following Table 1.8:
Using these values and

$$E\left(A_1(K_i)\right) = \sum_{Y \in Rat \setminus D} p_{X_i,Y} \cdot \frac{N_i}{100} \cdot \left(\frac{C_i}{\left(1 + f_{1,2}{}^Y\right)^1} + \frac{C_i}{\left(1 + f_{1,3}{}^Y\right)^2} + \dots + \right.$$
$$\left. + \frac{C_i}{\left(1 + f_{1,T_i-1}{}^Y\right)^{T_i-2}} + \frac{C_i + 100}{\left(1 + f_{1,T_i}{}^Y\right)^{T_i-1}} \right) + p_{X_i,D} \cdot R_i$$

we get the expected values for each K_i in 1 year. We already gave these values in the last column of the credit parameters Table 1.7 further above.

The sum of these values gives the **expected value** $\boldsymbol{E(A_1)}$ of the credit portfolio in 1 year. It amounts to **11,470,040 EUR**. Normalized to a percentage of the face value, this means **100,614%**.

We are not going to calculate the volatility $V(A_1)$ of the credit portfolio's value in 1 year explicitly, but we will include this volatility in the Monte Carlo simulation for determining the 1% percentile. In this Monte Carlo simulation, we will once again approximate the expected value $E(A_1)$. This exactly calculated expected value will then serve as a benchmark for the convergence quality of the Monte Carlo simulation.

Thus we now get to the Monte Carlo simulation of the variance of A_1 and of the 1% percentile of A_1, where we follow the algorithm described in the previous section: We generate $M = 10{,}000$ vectors $(y_1, y_2, \ldots, y_{20})$ of independent standard normally distributed random numbers. Then we multiply each of these vectors by the respective Cholesky matrix C_1 (no multiplication is necessary in Case 1, since C_1 is the 20×20 unit matrix), C_2 (in Case 2), or C_3 (in Case 3) and thus obtain $M = 10{,}000$ vectors $(x_1, x_2, \ldots, x_{20})$ of standard normally distributed random numbers with correct correlation.

Each of these vectors represents a possible asset value return scenario for the respective borrower, and so, based on each such vector, the borrower's new rating at the end of 1 year is then calculated assuming this scenario. For example, for the first loan K_1 and the corresponding asset value return x_1, we proceed as follows:

We find the current rating of K_1, which is BB. For the following we therefore have to use the threshold values $\overline{Z}_Y^{BB}$, thus

$$\overline{Z}_{AAA}^{BB} = \infty, \overline{Z}_{AA}^{BB} = 3.473, \overline{Z}_{A}^{BB} = 2.937, \overline{Z}_{BBB}^{BB} = 2.393, \overline{Z}_{BB}^{BB} = 1.368,$$

$$\overline{Z}_{B}^{BB} = -1.232, \overline{Z}_{CCC}^{BB} = -2.041, \overline{Z}_{D}^{BB} = -2.304, Z_{\tilde{D}}^{X} := -\infty$$

We then determine between which two of the above consecutive values the return x_1 lies and thus get the new rating. For example, if x_1 has a value of 1.5, then x_1 is between $\overline{Z}_{BB}^{BB}$ and $\overline{Z}_{BBB}^{BB}$. The new rating is therefore BBB.

This gives us a vector $(Y_1, Y_2, \ldots, Y_{20})$ of new ratings. Based on these ratings, we calculate the values $A_1(K_i, Y_i)$ that occur in this scenario, using the above-stated formula

$$A_1(K_i, Y_i) = \frac{N_i}{100} \cdot \left(\frac{C_i}{\left(1 + f_{1,2}{}^{Y_i}\right)^1} + \frac{C_i}{\left(1 + f_{1,3}{}^{Y_i}\right)^2} + \ldots + \right.$$
$$\left. + \frac{C_i}{\left(1 + f_{1,T_i-1}{}^{Y_i}\right)^{T_i-2}} + \frac{C_i + 100}{\left(1 + f_{1,T_i}{}^{Y_i}\right)^{T_i-1}} \right).$$

However, if $Y_i = D$, further simulation of the value $A_1(K_i, D)$ using a suitable beta-distributed random number is necessary. We use Mathematica to generate the required beta-distributed random numbers. Yet we still need the suitable parameters p and q. Depending on the seniority class of K_i, we have to use a different set of parameters p and q. In the following Table 1.9, we see the parameters p and q that are to be used for each seniority class. (How these p and q were calculated was explained at the end of the previous section.)

Table 1.9 Parameters p and q for various seniority classes

Seniority class	p	q
Senior secured	1.3155	1.1297
Senior unsecured	1.4612	1.3966
Senior subordinated	1.2239	1.9534
Subordinated	1.4430	2.9645
Junior subordinated	1.8673	9.0588

Parameters for the Beta Distribution by Seniority Class

We demonstrate the procedure using a specific scenario for Case 2 (i.e. with the nontrivial Cholesky matrix C_2):

Uncorrelated standard normally distributed random numbers $(y_1, y_2, \ldots, y_{20}) =$

$\{-1.584, -1.183, -1.423, -1.092, 2.382, 1.611, 2.315, 0.629, -0.810, 1.091, 0.846, 0.135, -0.091, -0.525, 0.3405, 0.392, 1.628, -0.406, -0.124, -1.68\}$

Correlated asset values $(x_1, x_2, \ldots, x_{20}) =$

$\{-1.584, -1.976, -2.062, -2.582, -1.002, -0.605, -0.672, 0.391, -1.284, -0.754, -1.542, -1.433, 0.243, -1.428, -0.879, -2.014, -1.331, -1.393, 0.418, -1.912\}$

New ratings $(Y_1, Y_2, \ldots, Y_{20})(1 = AAA, \ldots, 8 = D)$:
$\{6, 4, 8, 4, 5, 2, 1, 6, 3, 1, 7, 8, 5, 2, 3, 5, 2, 3, 6, 2\}$

Values $A_1(K_i, Y_i)$:
$\{291368, 530050, 198459, 275664, 678770, 535859, 466710, 648423, 904549, 502985, 707168, 241326, 336862, 946617, 301288, 960458, 299118, 939947, 745131, 299105\}$

Here, the 3rd and the 12th values were each created by generating a beta-distributed random number between 0 and 1 and multiplying this random number by the loan's face value. The parameters used to generate the beta-distributed random number depended on the seniority classes (1 and 3, respectively) of the two loans involved.

The sum of these 20 values is the total value of the credit portfolio A_1 in 1 year under this scenario. In our case, this total value would be 10,809,877 EUR. We then normalize each of these values to a percentage of the face value of 11,400,000. In our case, this results in a percentage value of 94.824%.

Proceeding in this way, using Mathematica, we generate 10,000 such normalized values and call them $A_1^{(1)}, A_1^{(2)}, \ldots, A_1^{(10{,}000)}$. We then take the mean over these values, which gives us an approximation for the expected value $E(A_1)$. The quality of the estimate can be verified against the exact value calculated earlier for

$E(A_1)$. From these values we calculate the standard deviation and thus obtain an approximation for the standard deviation $\sigma(A_1)$.

And finally, we compute the 1% percentile from these values: We arrange the values $A_1^{(1)}, A_1^{(2)}, \ldots, A_1^{(10,000)}$ in ascending order and choose the 100th value in this ordered sequence as the estimate for the 1% percentile.

We then do this for each of the three cases, one after the other. First, we simply state, in each case, the resulting simulation values for the expected value (which should essentially coincide approximately with the exact value of 100,614, regardless of the chosen correlation matrix), variance, and 1% percentile (as well as the resulting proposal for the required level of own funds).

Subsequently we will also illustrate the convergence of the simulation results: The respective graphs show that, when simulating a total of 10,000 scenarios in each case, the results converge visibly as the number of scenarios increases. It is particularly striking to see how the simulation results for the expected value converge very quickly towards the exact expected value.

Results

Case 1:
Expected value: 100.635
Standard deviation: 2.905
1% percentile: 91.021
Own funds as a percentage of face value: A_0 minus 1%−percentile = 104.661 − 91.021 = 13.64%
The convergence behaviour for expected value, standard deviation, and the 1% percentile of A_1 for this case is given in the following Fig. 1.28.

Case 2:
Expected value: 100.598
Standard deviation: 4.327
1% percentile: 84.080
Own funds as a percentage of face value: A_0 minus 1%−percentile = 104.661 − 84.080 = 20.58%
As for case 1, the subsequent plots Fig. 1.29 describe the most important properties of the second case.

Case 3:
Expected value: 100.612
Standard deviation: 3.079
1% percentile: 91.205
Own funds as a percentage of face value: A_0 minus 1%−percentile = 104.661 − 91.205 = 13.46%. The convergence behaviour for expected value, standard deviation, and the 1% percentile of A_1 for the last case is given in the Fig. 1.30.

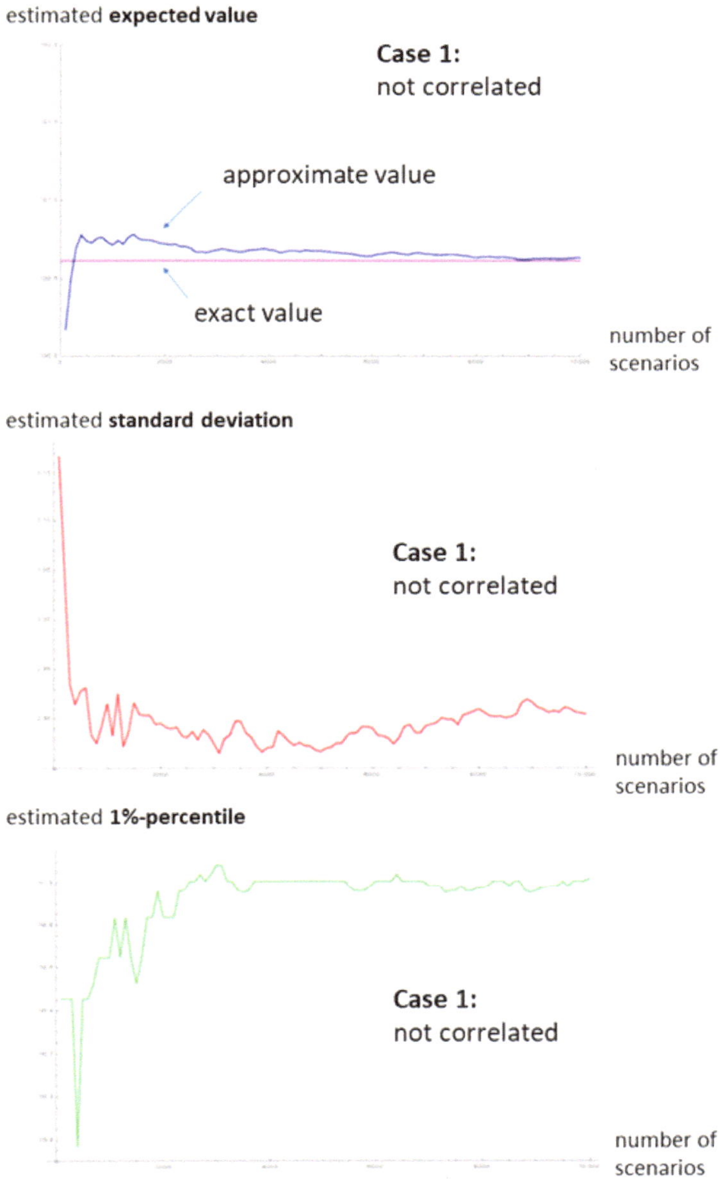

Fig. 1.28 Convergence behaviour for expected value, standard deviation, and 1% percentile of A_1 in example Case 1

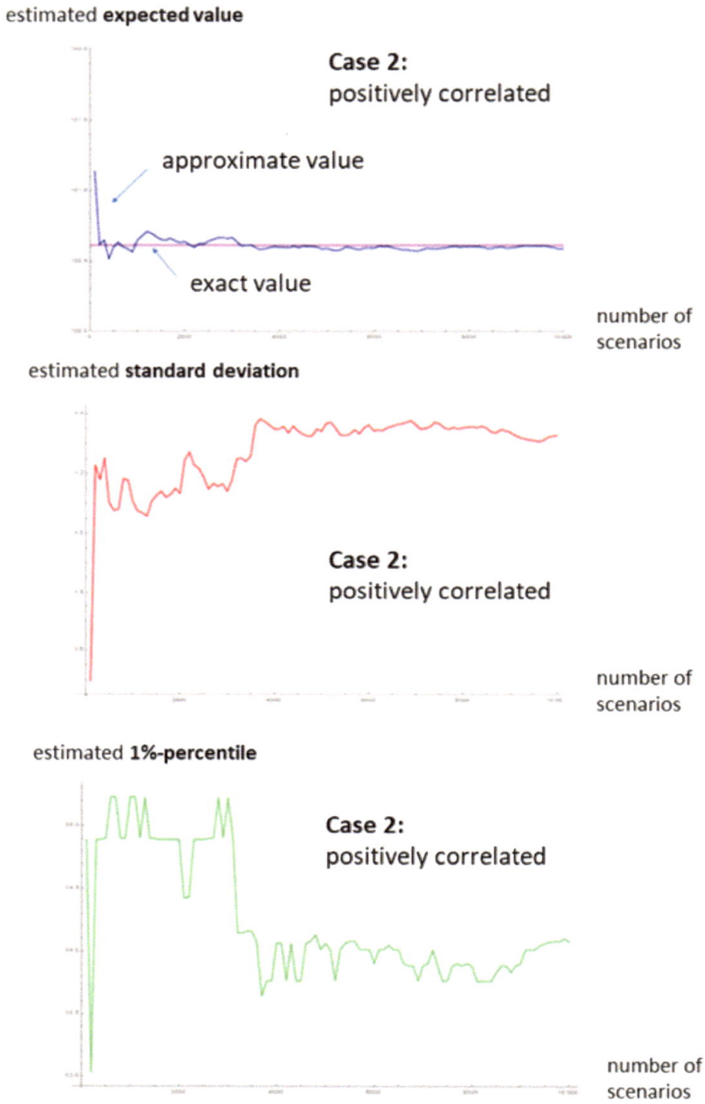

Fig. 1.29 Convergence behaviour for expected value, standard deviation, and 1% percentile of A_1 in example Case 2

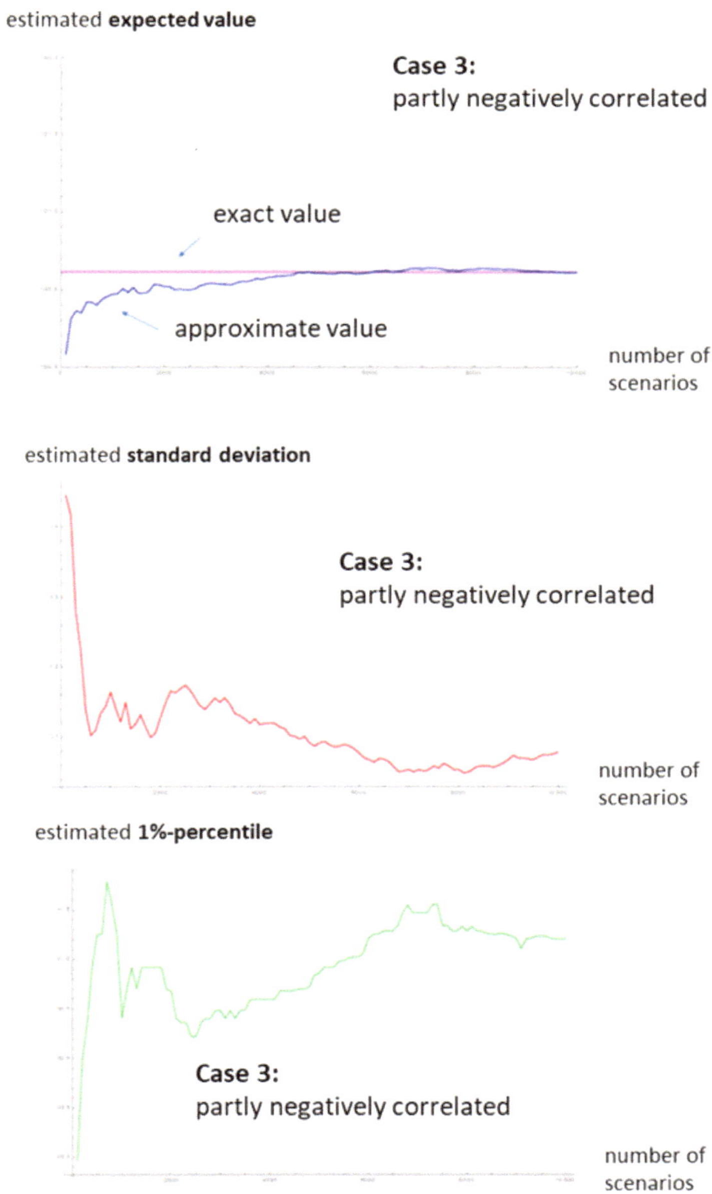

Fig. 1.30 Convergence behaviour for expected value, standard deviation, and 1% percentile of A_1 in example Case 3

Thus, while Case 1 and Case 3 yield fairly similar results for variance and percentile, we note that the variance in Case 2, where the loan asset values exhibit stronger positive correlation, is significantly higher than in the other two cases and the 1% percentile is significantly lower. This means there is a higher probability for higher losses to occur. The proposed capital requirement increases accordingly in Case 2.

1.19 Excursus: Generating Positive Definite Random Correlation Matrices

This section and the associated software were created by Lukas Wögerer.

We have worked with correlation matrices several times in this book so far, including in Volume I Chapter 4 and in the previous section. A matrix M can be used as a correlation matrix if it is symmetric and positive semidefinite, all entries are between -1 and 1, and the entries in the main diagonal are all equal to 1. In the following we will discuss how to generate such matrices.

Method 1

Positive semidefinite matrices can be constructed fairly quickly, since, for any real square matrix A in which there is at least one nonzero entry in each row, the product $B = A.A^{\mathsf{T}}$ is symmetric and positive semidefinite. In general, however, the result is of course not yet a correlation matrix. Normalization can remedy this.

Select $D = diag\,(B)$, that is, a matrix which contains the entries of B in the diagonal and otherwise only zeros. Then the matrix $M = D^{-1/2}BD^{-1/2}$ is symmetric, positive semidefinite, its diagonal elements are all equal to 1, and the remaining entries are not greater than 1. Thus the constructed matrix M is a (positive semidefinite) correlation matrix.

Example Suppose we want to generate a 4×4 correlation matrix. First we choose an arbitrary real 4×4 matrix. In this example, we randomly select the entries of our matrix from the interval $[-1, 1]$.

$$A = \begin{pmatrix} 0.84 & -0.47 & -0.2 & 0.38 \\ 0.31 & 0.39 & 0.57 & 0.84 \\ 0.42 & -0.29 & 0.18 & 0.91 \\ -0.82 & 0.2 & 0.56 & 1 \end{pmatrix}$$

We further calculate the matrices B and D and get:

$$B = A.A^{\mathsf{T}} = \begin{pmatrix} 1.1109 & 0.2823 & 0.7989 & -0.5148 \\ 0.2823 & 1.2787 & 0.8841 & 0.983 \\ 0.7989 & 0.8841 & 1.121 & 0.6084 \\ -0.5148 & 0.983 & 0.6084 & 2.026 \end{pmatrix}$$

$$D^{-1/2} = diag(B)^{-1/2} = \begin{pmatrix} \frac{1}{\sqrt{1.1109}} & 0 & 0 & 0 \\ 0 & \frac{1}{\sqrt{1.2787}} & 0 & 0 \\ 0 & 0 & \frac{1}{\sqrt{1.121}} & 0 \\ 0 & 0 & 0 & \frac{1}{\sqrt{2.026}} \end{pmatrix}$$

The 4×4 correlation matrix obtained in this way has the form

$$M = D^{-1/2} B D^{-1/2} = \begin{pmatrix} 1 & 0.237 & 0.716 & -0.343 \\ 0.237 & 1 & 0.738 & 0.611 \\ 0.716 & 0.738 & 1 & 0.404 \\ -0.343 & 0.611 & 0.404 & 1 \end{pmatrix}$$

Unfortunately, this approach has a major drawback: It is not well suited for very large matrices. The values outside the diagonals tend to become very small. So we construct a 15×15 correlation matrix using the same procedure:

$$M = \begin{pmatrix}
1. & -0.01 & 0.37 & 0.15 & 0.53 & 0.19 & -0.16 & -0.14 & -0.12 & 0.44 & 0.07 & 0.02 & -0.49 & -0.13 & -0.14 \\
-0.01 & 1. & -0.14 & 0.17 & -0.09 & -0.13 & 0.32 & 0.29 & -0.64 & -0.05 & -0.49 & 0.25 & 0.09 & 0.12 & -0.55 \\
0.37 & -0.14 & 1. & -0.05 & 0.04 & 0.09 & 0.18 & 0.06 & 0.61 & 0.21 & 0.02 & -0.04 & 0.15 & -0.27 & 0.16 \\
0.15 & 0.17 & -0.05 & 1. & -0.31 & 0.29 & 0.32 & -0.23 & -0.4 & 0.21 & 0.29 & -0.13 & 0.31 & -0.09 & 0.06 \\
0.53 & -0.09 & 0.04 & -0.31 & 1. & 0.43 & -0.03 & -0.53 & 0. & 0. & -0.13 & 0.1 & -0.42 & -0.1 & -0.2 \\
0.19 & -0.13 & 0.09 & 0.29 & 0.43 & 1. & 0.25 & -0.72 & -0.09 & 0.01 & -0.1 & 0.04 & 0.04 & -0.13 & 0.02 \\
-0.16 & 0.32 & 0.18 & 0.32 & -0.03 & 0.25 & 1. & -0.32 & -0.27 & 0.06 & -0.22 & 0.43 & 0.33 & -0.23 & -0.18 \\
-0.14 & 0.29 & 0.06 & -0.23 & -0.53 & -0.72 & -0.32 & 1. & -0.01 & 0.06 & -0.15 & -0.15 & 0.21 & 0.07 & -0.27 \\
-0.12 & -0.64 & 0.61 & -0.4 & 0. & -0.09 & -0.27 & -0.01 & 1. & -0.19 & 0.2 & -0.23 & 0.13 & -0.03 & 0.57 \\
0.44 & -0.05 & 0.21 & 0.21 & 0. & 0.01 & 0.06 & 0.06 & -0.19 & 1. & 0.44 & 0.11 & 0.05 & -0.28 & -0.13 \\
0.07 & -0.49 & 0.02 & 0.29 & -0.13 & -0.1 & -0.22 & -0.15 & 0.2 & 0.44 & 1. & -0.33 & -0.1 & -0.49 & 0.19 \\
0.02 & 0.25 & -0.04 & -0.13 & 0.1 & 0.04 & 0.43 & -0.15 & -0.23 & 0.11 & -0.33 & 1. & -0.12 & 0.27 & 0.03 \\
-0.49 & 0.09 & 0.15 & 0.31 & -0.42 & 0.04 & 0.33 & 0.21 & 0.13 & 0.05 & -0.1 & -0.12 & 1. & 0.12 & 0.15 \\
-0.13 & 0.12 & -0.27 & -0.09 & -0.1 & -0.13 & -0.23 & 0.07 & -0.03 & -0.28 & -0.49 & 0.27 & 0.12 & 1. & 0.43 \\
-0.14 & -0.55 & 0.16 & 0.06 & -0.2 & 0.02 & -0.18 & -0.27 & 0.57 & -0.13 & 0.19 & 0.03 & 0.15 & 0.43 & 1.
\end{pmatrix}$$

We see that many of the entries are close to zero. The most significant correlation is only 0.72, and in total, there are only 14 values that are significantly larger than 0.5.

This behaviour can be seen even more clearly in the distribution of matrix entries. For this purpose, we generated a 20×20, a 100×100, and a 500×500 correlation matrix and visualized the distributions of the entries using histograms in Fig. 1.31.

The correlation matrix entries are all around zero, and the larger the desired matrix is to be, the more they concentrate there. Of course, this does not apply to the diagonal elements, which are represented by the bar on the right edge of the histogram.

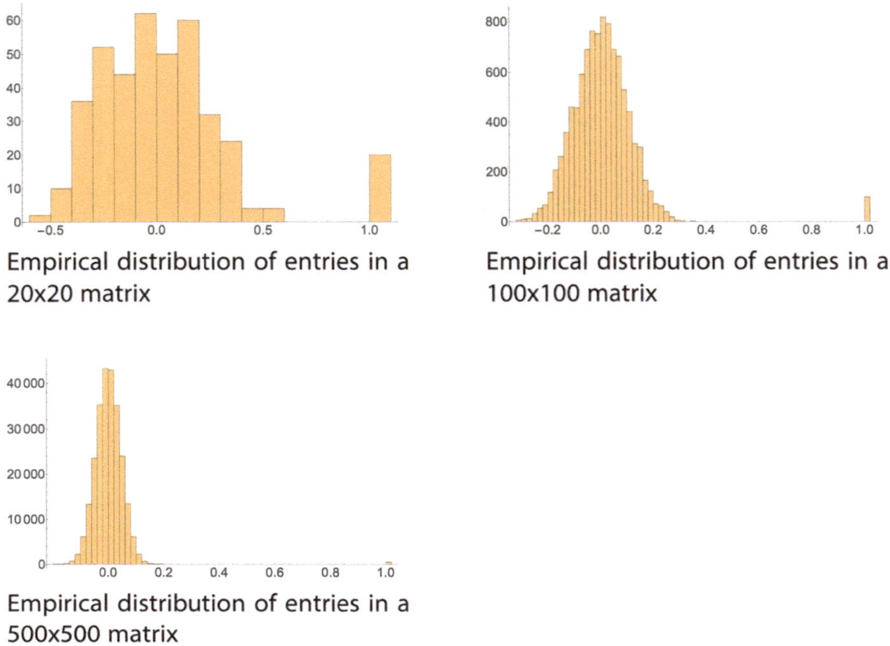

Fig. 1.31 Empirical distributions of randomly generated 20×20, 100×100, 500×500 correlation matrices

So this method is less suitable for large matrices. On the other hand, it is well suited for generating small correlation matrices, since it can be implemented quickly and easily.

Method 2

Another way to create correlation matrices is to increase the size of an existing correlation matrix step by step. We motivate the idea with an example:

We have a correlation matrix M of dimension 3×3 and its Cholesky decomposition:

$$M = \begin{pmatrix} 1 & 0.5 & -0.7 \\ 0.5 & 1 & 0.2 \\ -0.7 & 0.2 & 1 \end{pmatrix}$$

$$= \begin{pmatrix} 1 & 0 & 0 \\ 0.5 & 0.866025 & 0 \\ -0.7 & 0.635085 & 0.326599 \end{pmatrix} \begin{pmatrix} 1 & 0.5 & -0.7 \\ 0 & 0.866025 & 0.635085 \\ 0 & 0 & 0.326599 \end{pmatrix}$$

We note here that the Cholesky matrix C (for which $C^T \cdot C = M$) always matches M in the first row (i.e. the first column of C^T matches M).

We want to construct a correlation matrix of size 4×4 from this. We choose the first correlation $\rho_{1,4}$ arbitrarily from the open interval $(-1, 1)$, for example, 0.8, and extend our matrix by one dimension.

$$M = \begin{pmatrix} 1 & 0.5 & -0.7 & 0.8 \\ 0.5 & 1 & 0.2 & \rho_{2,4} \\ -0.7 & 0.2 & 1 & \rho_{3,4} \\ 0.8 & \rho_{2,4} & \rho_{3,4} & 1 \end{pmatrix}$$

$$= \begin{pmatrix} 1 & 0 & 0 & 0 \\ 0.5 & 0.866025 & 0 & 0 \\ -0.7 & 0.635085 & 0.326599 & 0 \\ 0.8 & x & y & z \end{pmatrix} \begin{pmatrix} 1 & 0.5 & -0.7 & 0.8 \\ 0 & 0.866025 & 0.635085 & x \\ 0 & 0 & 0.326599 & y \\ 0 & 0 & 0 & z \end{pmatrix}$$

The task now is to choose the two missing correlations. Note that not every arbitrary choice yields a correlation matrix, of course! Multiplying out the right side, we get:

$$M = \begin{pmatrix} 1 & 0.5 & -0.7 & 0.8 \\ 0.5 & 1. & 0.2 & 0.866025x + 0.4 \\ -0.7 & 0.2 & 1. & 0.635085x + 0.326599y - 0.56 \\ 0.8 & 0.866025x + 0.4 & 0.635085x + 0.326599y - 0.56 & x^2 + y^2 + z^2 + 0.64 \end{pmatrix}$$

Comparing the entries, we can immediately read the restriction $x^2 + y^2 + z^2 = 0.36$. In particular, x must lie in the interval $[-0.6, 0.6]$. For $y, z \neq 0$ to hold, x must even lie in the open interval $(-0.6, 0.6)$.

Together with the equation

$$\rho_{2,4} = 0.4 + 0.8665025 \cdot x$$

we obtain for the choice of the next correlation:

$$\rho_{2,4} \in (0.4 - 0.8665025 \cdot 0.6,\ 0.4 + 0.8665025 \cdot 0.6) = (-0.119615,\ 0.919615)$$

As an example, we choose $\rho_{2,4} = -0.1$.

In exactly the same way, we then conclude:

$$\rho_{3,4} \in (-0.98,\ -0.873)$$

Choosing $\rho_{3,4} = -0.95$, we get the correlation matrix

$$M = \begin{pmatrix} 1 & 0.5 & -0.7 & 0.8 \\ 0.5 & 1 & 0.2 & -0.1 \\ -0.7 & 0.2 & 1 & -0.95 \\ 0.8 & -0.1 & -0.95 & 1 \end{pmatrix}$$

$$= \begin{pmatrix} 1 & 0 & 0 & 0 \\ 0.5 & 0.866025 & 0 & 0 \\ -0.7 & 0.635085 & 0.326599 & 0 \\ 0.8 & -0.57735 & -0.07144 & 0.146842 \end{pmatrix} \cdot$$

$$\cdot \begin{pmatrix} 1 & 0.5 & -0.7 & 0.8 \\ 0 & 0.866025 & 0.635085 & -0.57735 \\ 0 & 0 & 0.326599 & -0.07144 \\ 0 & 0 & 0 & 0.146842 \end{pmatrix}$$

With this approach, in contrast to the first method, we have a more direct influence on the correlations.

In general, an $n \times n$ correlation matrix can be expanded as follows:

We have: $M = \left(\rho_{ij}\right) \in \mathbb{R}^{n \times n}$ positive definite correlation matrix

Step 1: Calculate Cholesky decomposition $C = \left(c_{ij}\right)$ from M and set $\widetilde{C} = \left(\widetilde{c}_{ij}\right) = \mathbf{0} \in \mathbb{R}^{(n+1)\times(n+1)}$

Step 2: Set $\widetilde{c}_{ij} := c_{ij}$ for all $i, j = 1, \ldots, n$

Step 3: Choose correlation $\rho_{1,n+1} \in (-1, 1)$ and set $\widetilde{c}_{1,n+1} = \rho_{1,n+1}$

Step 4: Calculate stepwise for $k = 2, \ldots, n$:

$$\alpha_k = \sum_{i=1}^{k-1} \widetilde{c}_{ik}\widetilde{c}_{i,n+1}$$

$$\beta_k = 1 - \sum_{i=1}^{k-1} \widetilde{c}_{i,n+1}^2$$

$$I_k = \left(\alpha_k - \widetilde{c}_{kk}\sqrt{\beta_k}, \alpha_k + \widetilde{c}_{kk}\sqrt{\beta_k}\right)$$

Choose correlation $\rho_{k,n+1} \in I_k$ and set $\widetilde{c}_{k,n+1} = \frac{\rho_{k,n+1} - \alpha_k}{\widetilde{c}_{kk}}$.

Step 5: Calculate β_{n+1} in the same way as β_k in step 4, set $\widetilde{c}_{n+1,n+1} = \sqrt{\beta_{n+1}}$, and return $\widetilde{M} = \widetilde{C}^{\mathsf{T}}\widetilde{C}$.

The (random) correlation matrices can be generated using our software.

1.20 Credit Risk+, Part I: Basic Concept and Expected Number of Defaults Assuming Independent Loans

The advantage of the Credit Risk+ system developed by Credit Suisse First Boston is that it makes do with much less basic data in terms of loans in the portfolio and interest rate structures of bonds at risk of default. The results are obtained by means of explicit mathematics and not by simulation and can therefore be obtained much faster in many cases.

However, the *Credit Metrics* analysis goes substantially deeper and into more detail than the *Credit Risk+* analysis. While *Credit Metrics* attempts to estimate the value of the entire credit portfolio in 1 year and derive a verifiable capital requirement from the VaR of this credit portfolio value, *Credit Risk+* addresses only the portfolio's expected default value in the coming year. So the capital requirement proposed by *Credit Risk+* is based only on statements about the expected loss amount that will result from defaulting loans. We will explain exactly what the own funds requirement is in *Credit Risk+* at the end of our discussion of *Credit Risk+*. Until then, our goal will be to accurately analyse the random variable: "Expected loss amount of a credit portfolio over the next year".

The data and information that we will need regarding the credit portfolio are the following:

We are again looking at a **portfolio of n loans** $K_1, K_2, K_3, \ldots, K_n$.

Their **face values** are again denoted by $N_1, N_2, N_3, \ldots, N_n$.

Each of the loans is assigned a specific **recovery rate** R_i. For this purpose, we select (from the Table 1.1) for each loan the mean recovery rate of the seniority class it belongs to (the standard deviation given in the Table 1.1 is not relevant for what follows).

Each loan K_i is assigned a **default probability** p_i in 1 year. If we again assume that the borrowers are rated according to Standard & Poor's, then the rating transition value $p_{X,D}$ (where X denotes the respective borrower's rating) can be used for this purpose.

We further make a choice of regional, industry-defined (and also "artificial") sectors S_k, $k = 1, 2, 3, \ldots, m$. Each borrower is allocated to either one single sector or to several sectors on a percentage basis. Further, the average annual default probabilities associated with a sector S_k are assumed to have a certain mean μ^{S_k} and a certain standard deviation σ^{S_k}. We will give more details later.

We now approach the desired goal in several steps, making the assumptions more realistic as we go along.

In the first step, we **assume that the loans in the portfolio are independent** of one another. Under this (unrealistic) assumption, we want to calculate the **distribution of the random variable Y "number of defaulting loans in the portfolio in the next year"**.

The random variable Y can be represented in the form $Y = Y_1 + Y_2 + Y_3 + \ldots + Y_i + \ldots + Y_n$, where Y_i is a random variable that attains the value 1 if the loan K_i

defaults in the coming year and attains the value 0 if the loan K_i does not default in the coming year.

Expected value and variance of Y are very easily calculated (assuming independent loans, that is, independent Y_i).

The expected value of a single Y_i is given by $E(Y_i) = p_i \cdot 1 + (1-p_i) \cdot 0 = p_i$, and the variance of a single Y_i is given by $V(Y_i) = p_i \cdot (1-p_i)^2 + (1-p_i) \cdot (0-p_i)^2 = p_i \cdot (1-p_i)$. So, for the expected value $E(y)$ and the variance $V(Y)$ of Y, we have:

$$E(Y) = \mu := \sum_{i=1}^{n} p_i \text{ und } V(Y) = \sum_{i=1}^{n} p_i \cdot (1-p_i) = \sum_{i=1}^{n} p_i - \sum_{i=1}^{n} p_i^2$$

In a reasonably "sensible" credit portfolio, most of the p_i will be very small, and only very few of the p_i will attain somewhat larger values. Recall once again the default probabilities (stated in absolute numbers, not as percentages) given in the Fig. 1.23:

$$\begin{aligned}
&p_{AAA,D} = p_{AA,D} = 0 &&\rightarrow\ p_{AAA,D}{}^2 = p_{AA,D}{}^2 = 0\\
&p_{A,D} = 0.0006 &&\rightarrow\ p_{A,D}{}^2 = 0.00000036\\
&p_{BBB,D} = 0.0018 &&\rightarrow\ p_{BBB,D}{}^2 = 0.00000324\\
&p_{BB,D} = 0.0106 &&\rightarrow\ p_{BB,D}{}^2 = 0.00011236\\
&p_{B,D} = 0.052 &&\rightarrow\ p_{B,D}{}^2 = 0.002704\\
&p_{CCC,D} = 0.1979 &&\rightarrow\ p_{CCC,D}{}^2 = 0.0391644
\end{aligned}$$

Thus, in most cases, p_i^2 is significantly smaller than p_i. Therefore, in a reasonably "sensible" credit portfolio, $\sum_{i=1}^{n} p_i^2$ is significantly smaller than $\mu = \sum_{i=1}^{n} p_i$.

So normally, $V(Y) = \sum_{i=1}^{n} p_i - \sum_{i=1}^{n} p_i^2 \approx \sum_{i=1}^{n} p_i = \mu$.

Thus, the variance is approximately as large as the expected value of the random variable Y.

Here is an example for illustration: In a credit portfolio consisting of approximately

1000 AAA rated
2000 AA rated
5000 A rated
3000 BBB rated
2000 BB rated
1000 B rated
300 CCC rated

loans, we have $\mu = \sum_{i=1}^{n} p_i = 140.97$ and $V(Y) = \sum_{i=1}^{n} p_i - \sum_{i=1}^{n} p_i^2 = 126.28$.

What we need now is the distribution of the random variable Y. To determine the distribution, we use the method of generating functions. We denote by a_k the probability that Y attains the value k, i.e. that exactly k loans will default in the coming year. The generating function $G_Y(x)$ of the random variable Y is then defined as a power series

$$G_Y(x) := \sum_{k=0}^{\infty} a_k \cdot x^k.$$

We now use the following **result** from probability theory on generating functions:

If a random variable Z is the sum of a finite number of independent random variables Z_i, *hence* $Z = Z_1 + Z_2 + \ldots + Z_n$, *then the generating function* $G_Z(x)$ *is the product of the generating functions* G_{Z_i}, *hence* $G_Z(x) = \prod_{i=1}^{n} G_{Z_i}(x)$.

A possible **proof** of this theorem could be as follows:
The function $G_Z(x) = \sum_{k=0}^{\infty} W(Z = k) \cdot x^k$ is, after all, on closer inspection, precisely the expected value of the random variable x^Z. Therefore

$$G_Z(x) = E(x^Z) = E\left(x^{Z_1+Z_2+\ldots+Z_n}\right) = E\left(x^{Z_1} \cdot x^{Z_2} \cdot \ldots \cdot x^{Z_n}\right) =$$

$$= E(x^{Z_1}) \cdot E(x^{Z_2}) \cdot \ldots \cdot E(x^{Z_n}) = G_{Z_1}(x) \cdot G_{Z_2}(x) \cdot \ldots \cdot G_{Z_n}(x)$$

and thus the theorem has been proven.
Here, we simply made use of the fact that the expected value of the product of independent random variables is equal to the product of the expected values.

Thus, for our random variable Y, we get $G_Y(x) = \prod_{i=1}^{n} G_{Y_i}(x)$.

The generating function $G_{Y_i}(x)$ of the random variable Y_i can be easily calculated directly:

$G_{Y_i}(x) = (1 - p_i) \cdot x^0 + p_i \cdot x^1 = 1 + p_i(x - 1)$ and therefore $G_Y(x) = \prod_{i=1}^{n}(1 + p_i \cdot (x - 1))$.

Taking the logarithm, we get

$$\log(G_X(x)) = \sum_{i=1}^{n} \log(1 + p_i \cdot (x - 1)).$$

Now we use another approximation, namely, $\log(1 + z) \approx z$ for small values of x. So, in the above product, we replace the value $\log(1 + p_i \cdot (x - 1))$ with the value $p_i \cdot (x - 1)$. (Again we can assume that, in the vast majority of cases, the probability p_i is normally indeed a small value and for the variable x we can assume that – for our requirements here– it is a value between 0 and 1.) Thus, $p_i \cdot (x - 1)$ will usually be so small that the inaccuracy caused by the approximation is negligible. The power series expansion $z - \frac{z^2}{2} + \frac{z^3}{3} - \frac{z^4}{4} + \ldots$ for the function $\log(1+z)$ shows that for a small z, an error of the order of about z^2 is committed here if we replace

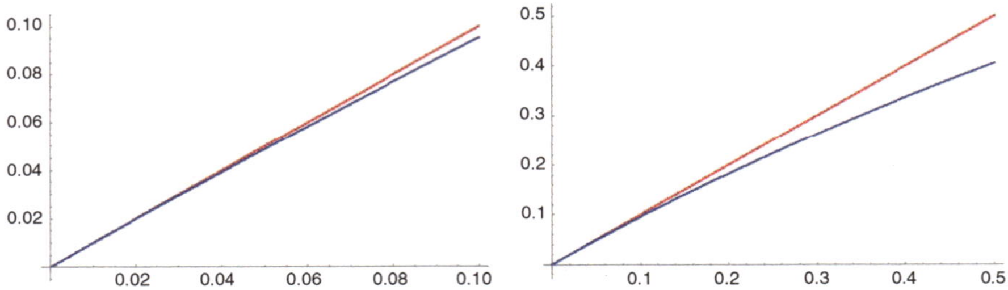

Fig. 1.32 The functions $\log(1+z)$ in blue and z in red for $z \in [0, 0.1]$ (left) and for $z \in [0, 0.5]$ (right)

$\log(1+z)$ by z. See also Fig. 1.32, where we plotted the functions $\log(1+z)$ in blue and z in red in the left image for $z \in [0, 0.1]$ and in the right image for $z \in [0, 0.5]$.

By this approximation we obtain
$\log(G_Y(x)) \approx \sum_{i=1}^{n} p_i \cdot (x-1) = \mu \cdot (x-1)$ and thus $G_Y(x) \approx e^{\mu \cdot (x-1)}$.

Now we use the power series representation of the function $e^z = \sum_{k=0}^{\infty} \frac{z^k}{k!}$. This gives us:

$$G_Y(x) \approx e^{\mu \cdot (x-1)} = e^{-\mu} \cdot e^{\mu \cdot x} = e^{-\mu} \cdot \sum_{k=0}^{\infty} \frac{(\mu \cdot x)^k}{k!} = \sum_{k=0}^{\infty} e^{-\mu} \cdot \frac{\mu^k}{k!} \cdot x^k.$$

Comparing this result with the original definition of $G_Y(x)$, we get:

$$G_Y(x) := \sum_{k=0}^{\infty} a_k \cdot x^k \approx \sum_{k=0}^{\infty} e^{-\mu} \cdot \frac{\mu^k}{k!} \cdot x^k$$

and by coefficient comparison $a_k = W(Y=k) \approx e^{-\mu} \cdot \frac{\mu^k}{k!}$ for all $k = 0, 1, 2, \ldots$.

A random variable Z with non-negative integer values k, for which $W(Y = k) = e^{-\mu} \cdot \frac{\mu^k}{k!}$ for all k (with an arbitrary positive constant μ), is called **Poisson-distributed**.

Thus, the random variable Y approximately follows a Poisson distribution. Y cannot have an exact Poisson distribution, of course, since the value $e^{-\mu} \cdot \frac{\mu^k}{k!}$ is positive for all k and the random variable Y cannot attain a value above n (it is not possible for more than n loans to default, since the portfolio only consists of n loans). For large k, however, the value quickly becomes vanishingly small.

To illustrate this, let us take a look at the above example portfolio, which consists of

1000 AAA rated
2000 AA rated
5000 A rated
3000 BBB rated

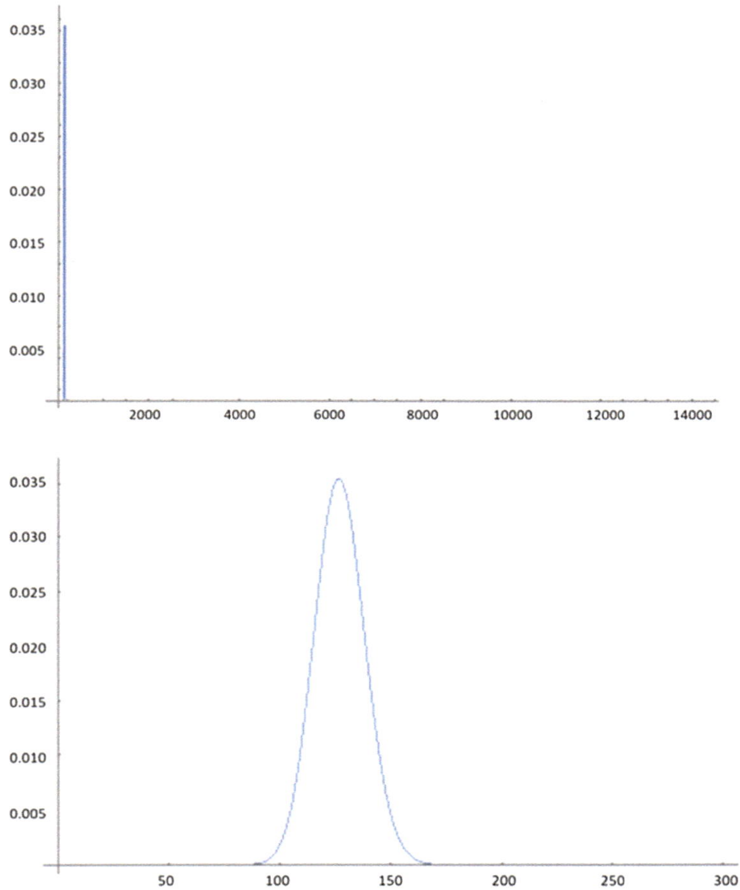

Fig. 1.33 Approximate probability distribution of the number of defaults in the example portfolio (lower chart showing zoomed area from upper chart)

2000 BB rated
1000 B rated
300 CCC rated

loans and whose expected value μ for the number of defaults in the coming year is 126.28. The portfolio consists of 14,300 loans. In Fig. 1.33 (upper chart), we see on the x-axis the possible values k that Y can attain (i.e. the number of defaults that can occur in principle) and on the y-axis the values $e^{-\mu} \cdot \frac{\mu^k}{k!}$, i.e. the approximate probabilities for these numbers to occur. We see that the default numbers will most likely be in a narrow range around the expected value $\mu = 126.28$. We have therefore zoomed out this area in more detail in Fig. 1.33 (lower chart).

For values k greater than 14,300, the occurring probabilities are so infinitesimally small that they can practically be treated as 0.

By the way, the expected value of a Poisson distribution is actually equal to μ. The variance of a Poisson distribution is also equal to μ. Recall that we noted earlier that the actual variance of the random variable Y is equal to the value $\sum_{i=1}^{n} p_i - \sum_{i=1}^{n} p_1^2$, which is normally very close to μ.

1.21 Credit Risk+, Part II: Expected Loss Amount Assuming Independent Loans

Of much greater interest than the **number** of credit defaults expected in the coming year is in fact the total **credit loss amount** to be expected in the coming year! So, in the following, we will focus on the random variable **Z = loss amount due to defaults in the credit portfolio over the coming year**. For the time being, we are going to stick to our case of independent loans.

In a first step, we round and divide the face amounts of the loans into "bands". To this end, we choose a suitable unit norm L (e.g. $L = 100{,}000$ EUR), and all "credit risk amounts" are then rounded to the nearest multiple of that unit norm L.

The amount of the credit product K_i that may potentially default, i.e. the "credit risk amount" of K_i is deemed to be the face amount N_i less the recovery rate R_i of K_i.

We denote this credit risk amount of K_i, rounded to L, by $v_i \cdot L$.

The **"credit band"** V_u denotes the set of all K_i for which $v_i = u$. We denote the credit bands occurring in the credit portfolio by $V_{u_1}, V_{u_2}, \ldots, V_{u_m}$. By Z_i we denote the loss amount in the coming year in the band V_{u_i}. Then $Z \approx Z_1 + Z_2 + \ldots + Z_m$. The "$\approx$" symbol is used instead of the equal sign because of rounding errors when normalizing to multiples of L.

We proceed with the generating function $G_Z(x)$, which we can now define as follows:

$G_Z(x) := \sum_{k=0}^{\infty} W(Z = k \cdot L) \cdot x^k$, since the (rounded) loss amounts can only be multiples of L. Because of the assumption of independent loans, we get $G_Z(x) = \prod_{i=1}^{m} G_{Z_i}(x)$. So let us consider $G_{Z_i}(x)$ and bear in mind that Z_i can only be a multiple of $u_i \cdot L$ and further that $Z_i = k \cdot u_i \cdot L$ occurs only if the number Y of defaults in the band V_{u_i} is exactly equal to k! It follows that:

$$G_{Z_i}(x) = \sum_{k=0}^{\infty} W(Z_i = k \cdot L) \cdot x^k = \sum_{k=0}^{\infty} W(Z_i = k \cdot u_i \cdot L) \cdot x^{k \cdot u_i} =$$

$$= \sum_{k=0}^{\infty} W(Y = k) \cdot x^{k \cdot u_i} = \sum_{k=0}^{\infty} W(Y = k) \cdot (x^{u_i})^k \approx e^{\mu_i \cdot (x^{u_i} - 1)}.$$

In the last equation, we used the fact that the penultimate expression represents the generating function of Y in the band V_{u_i} applied to the variable x^{u_i} (instead of x) and that we already know an approximation to the generating function of Y. Observe, however, that instead of μ, we now have to use the sum μ_i of the default probabilities p_l of only the loans in the band V_{u_i}. This now finally gives us

$$G_Z(x) = \prod_{i=1}^{m} G_{Z_i}(x) = \prod_{i=1}^{m} e^{\mu_i \cdot (x^{u_i} - 1)} = e^{\sum_{i=1}^{m} \mu_i \cdot (x^{u_i} - 1)} = e^{\sum_{i=1}^{m} \mu_i \cdot x^{u_i} - \mu}$$
$$=: e^{Q(x) - \mu}$$

In the following we set $Q(x) := \sum_{i=1}^{m} \mu_i \cdot x^{u_i}$.

For later purposes, we note the following: Recall that the generating function $G_Y(x)$ of the number Y of defaults was of the form $e^{\mu \cdot (x-1)}$. The generating function $G_Z(x)$ of the loss amounts is

$$G_Z(x) = e^{Q(x) - \mu} = e^{\mu \cdot (\widetilde{Q}(x) - 1)} = G_Y\left(\widetilde{Q}(x)\right) \tag{1.1}$$

where $\widetilde{Q}(x) := \sum_{i=1}^{m} \frac{\mu_i}{\mu} \cdot x^{u_i}$.

In order to run a coefficient comparison again and thus determine $W(Z = k \cdot L)$ (at least approximately), we would need to know the power series representation $\sum_{k=0}^{\infty} A_k \cdot x^k$ of the function $e^{\sum_{i=1}^{m} \mu_i \cdot (x^{u_i} - 1)}$. Then $W(Z = k \cdot L)$ would be $\approx A_k$.

For the coefficients of the power series representation $\sum_{k=0}^{\infty} b_k \cdot x^k$ of a function $f(x)$, we have:

$$b_k = \frac{1}{k!} \cdot \left. \frac{d^k f}{dx^k} \right|_{x=0}$$

Using this fact, we will attempt to determine the coefficients A_k of the power series of $G_z(x)$ via the derivatives of the function $e^{Q(x)-\mu}$.

Instead of $\left. \frac{d^k f}{dx^k} \right|_{x=0}$, we are going to write $f(x)^{(k)}(0)$ for short in the following.

And we will use the Leibniz product rule, namely:

$(f(x) \cdot g(x))^{(s)} = \sum_{i=0}^{s} \binom{s}{i} \cdot (f(x))^{(i)} \cdot (g(x))^{(s-i)}$, and we will use it in the following computation for $s = k - 1$, $f(x) = Q'(x)$ and $g(x) = e^{Q(x)-\mu}$.

With that we get

$$A_k = \frac{1}{k!} \cdot \frac{d^k \left(e^{Q(x)-\mu}\right)}{dx^k}\bigg|_{x=0} = \frac{1}{k!} \cdot \left(e^{Q(x)-\mu}\right)^{(k)}(0) =$$
$$= \frac{1}{k!}\left(Q'(x) \cdot e^{Q(x)-\mu}\right)^{(k-1)}(0) =$$
$$= \frac{1}{k!} \cdot \sum_{i=0}^{k-1} \binom{k-1}{i} \cdot (Q'(x))^{(i)} \cdot \left(e^{Q(x)-\mu}\right)^{(k-1-i)}(0) =$$
$$= \frac{1}{k!} \cdot \sum_{i=0}^{k-1} \binom{k-1}{i} \cdot (Q(x))^{(i+1)}(0) \cdot (k-1-i)! \cdot A_{k-1-i} =$$
$$= \frac{1}{k!} \cdot \sum_{j=1}^{k} \binom{k-1}{j-1} \cdot (Q(x))^{(j)}(0) \cdot (k-j)! \cdot A_{k-j} =$$
$$= \frac{1}{k} \cdot \sum_{j=1}^{k} \frac{1}{(j-1)!} \cdot (Q(x))^{(j)}(0) \cdot A_{k-j}$$

We see that in this way, we obtain a recursion for the values A_k. A_k is expressed here by the values $A_0, A_1, \ldots, A_{k-1}$. To obtain a really useful recursion for the successive computation of A_k, we only need the correct initial value A_0 and an explicit representation for $(Q(x))^{(j)}(0)$.

A_0 is very easily determined, since $A_0 = G(0) = e^{Q(0)-\mu} = e^{-\mu}$. Now let's get to $(Q(x))^{(j)}(0) = \left(\sum_{i=1}^{m} \mu_i \cdot x^{u_i}\right)^{(j)}(0)$. This is the constant term of the polynomial F, which is obtained by j-times differentiation of the polynomial $\sum_{i=1}^{m} \mu_i \cdot x^{u_i}$. However, this constant term of F is different from 0 only if j is equal to one of the exponents u_i. If $j = u_i$, then the constant term is given by $\mu_i \cdot (u_i)!$

With this we get

$$\boldsymbol{W(Z = k \cdot L)} \approx \boldsymbol{A_k} = \frac{1}{k} \cdot \sum_{j=1}^{k} \frac{1}{(j-1)!} \cdot (Q(x))^{(j)}(0) \cdot A_{k-j} =$$
$$= \sum_{\substack{i \\ u_i \le k}} \boldsymbol{\frac{u_i}{k} \cdot \mu_i \cdot A_{k-u_i}}$$

So, although this procedure did not give us an explicit representation for the coefficients A_k, we found a recursion that can be used to successively compute A_k in a fast and efficient manner.

To summarize, we have found the (approximate) probability distribution of the random variable Z for the total credit loss amount in a year (as a multiple of the norm value L). However, all of this is assuming independent loans!

For this reason, we next want to introduce a first level of dependencies between loans.

1.22 Credit Risk+, Part III: Allocating Loans to Mutually Independent Sectors

Dependencies between loans are modelled in the Credit Risk+ system by introducing a number of sectors $S_1, S_2, \ldots, S_w$. The number of sectors shouldn't be too large, however (e.g. 10 to 20 sectors). They can represent major regions or broad industries. Usually, an artificial sector is also assumed, which in a way is meant to represent the influence of the global financial market. Developments in different sectors are considered to be independent of each other. In a first step, we are going to assume that each loan K_i is allocated in its entirety to exactly one of the sectors. In the last step, each loan is then allocated to several sectors on a percentage basis, thereby achieving variability and a finer dependency structure between the loans. We will give more details later.

So, in this first step, where each loan is allocated in its entirety to exactly one sector, loans from different sectors are independent of each other. Loans within the same sector become implicitly interdependent in the following way:

We consider one of the sectors S_i and analyse historical data for this sector. In particular, we consider the random variable H_i = estimated average percentage of defaulting loans in the S_i sector within 1 year. By μ^{S_i} we denote the estimated expected value of H_i and by σ^{S_i} the estimated standard deviation of H_i. (For later purposes: H_i does *not* denote the actual number of defaults in the respective sector but the estimated (!) average percentage of defaults (estimated from year to year on the basis of various technical and fundamental parameters). Recall, by comparison, the beginning of our Credit Risk+ discussion where we essentially assumed a constant value μ as the average number of defaults... !)

An implicit interdependence between the loans is thus intuitively given as follows: Difficult market phases specific to a sector S_i will, if the loans within the sector are highly correlated, lead to an above-average number of defaults during these phases. Conversely, defaults are unlikely to occur in very good market phases. Thus, for such a sector, the standard deviation σ^{S_i} will attain a relatively large value. However, in the case of a rather broadly and loosely defined sector, that is, if there are only minor correlations between loans in that sector, then the number of defaults will tend to fluctuate little from year to year and will tend to be close to the long-term average μ^{S_i}. Difficult market phases will generally have less of an impact on such a sector as a whole. The standard deviation σ^{S_i} of such a sector will thus be rather small. Conversely, small values of σ^{S_i} will implicitly (!) indicate weak interdependencies among loans in the sector S_i while large σ^{S_i} will imply stronger interdependencies.

For the random variable H_i, for which we already have the mean μ^{S_i} and standard deviation σ^{S_i}, we now also assume a distribution, namely, the gamma distribution. The distribution is defined for positive real values. The density f of the gamma distribution depends on two parameters α and β and is given by

$$f(x) := \frac{1}{\beta^{\alpha} \cdot \Gamma(\alpha)} \cdot e^{-\frac{x}{\beta}} \cdot x^{\alpha-1} \text{ for } x \geq 0.$$

Here Γ denotes the gamma function; hence, $\Gamma(z) := \int_0^{\infty} e^{-x} \cdot x^{z-1} dx$.

The expected value μ of this distribution is given by $\alpha \cdot \beta$ and the standard deviation σ by $\alpha \cdot \beta^2$.

Thus, the density of the random variable H_i in the sector S_i is a function f_i of the above form, in which the parameters α_i and β_i are to be chosen such that $\alpha_i \cdot \beta_i = \mu^{S_i}$ and $\alpha_i \cdot \beta_i^2 = \sigma^{S_i}$ are satisfied; hence, $\beta_i = \frac{\sigma^{S_i}}{\mu^{S_i}}$ and $\alpha_i = \frac{(\mu^{S_i})^2}{\sigma^{S_i}}$.

We illustrate in Fig. 1.34 the density of the gamma function for realistic values of μ^{S_i} and σ^{S_i}. Remember: These are the expected value and standard deviation of the percentage of credit defaults in the S_i sector within 1 year. A pessimistic assumption would be $\mu^{S_i} = 3\%$ and $\sigma^{S_i} = 5\%$, for example. This gives us $\alpha_i = 1.8$ and $\beta_i = 1.66\ldots$.

We see in Fig. 1.34 (on the left, where the density was plotted for x from 0 to 120) that the density is tightly concentrated around the expected value $\mu^{S_i} = 3\%$. We have therefore additionally plotted the relevant range for x from 0 to 10 in the right part of the figure. The values for $x > 100$ (which should not actually occur) are negligibly small.

In this setting, we are again interested in finding the probability distribution of the number of defaults in the entire credit portfolio. We denote the relevant generating function by $F(z)$, which is given by $F(z) = \prod_{i=1}^{w} F_i(z)$, with $F_i(z)$ being the generating function for the number of defaults in the sector S_i.

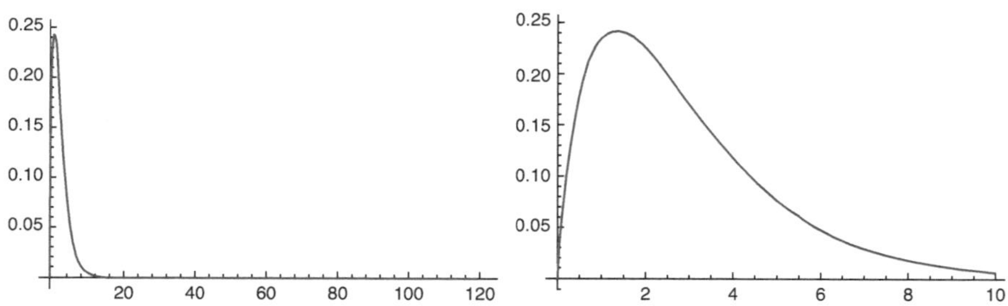

Fig. 1.34 Density of the gamma distribution with expected value 3 and standard deviation 5

To determine $F_i(z)$, we now proceed again with the support of heuristic considerations:

If the expected value of the number of defaults were assumed to be essentially constant as x (like we did at the beginning of our Credit Risk+ discussion, where $x = \mu$), then we would already know the generating function from our earlier considerations: It would have the form $F_i(z) = e^{x\cdot(z-1)}$. Yet x is a gamma-distributed random variable with density $f_i(x)$. We can therefore assume (heuristics !) that the actual generating function is a density-weighted average of the functions $e^{x\cdot(z-1)}$; thus (for clarity, we will use α and β instead of α_i and β_i in the following):

$$F_i(z) = \int_0^\infty e^{x\cdot(z-1)} \cdot f_i(x)dx =$$

$$= \frac{1}{\beta^\alpha \Gamma(\alpha)} \cdot \int_0^\infty e^{x(z-1)} \cdot e^{-\frac{x}{\beta}} \cdot x^{\alpha-1} dx =$$

$$= \frac{1}{\beta^\alpha \Gamma(\alpha)} \int_0^\infty e^{x\left(z-1-\frac{1}{\beta}\right)} \cdot x^{\alpha-1} dx =$$

$$\left(\text{we substitute } y = -x\left(z-1-\tfrac{1}{\beta}\right) \text{ and get } dy = -dx \cdot \left(z-1-\tfrac{1}{\beta}\right)\right)$$

$$= \frac{1}{\beta^\alpha \Gamma(\alpha)} \int_0^\infty e^{-y} \cdot y^{\alpha-1} \cdot \frac{1}{\left(1+\frac{1}{\beta}-z\right)^{\alpha-1}} \cdot \frac{1}{1+\frac{1}{\beta}-z} dy =$$

$$= \frac{1}{\beta^\alpha \Gamma(\alpha)} \cdot \frac{1}{\left(1+\frac{1}{\beta}-z\right)^{\alpha}} \cdot \int_0^\infty e^{-y} \cdot y^{\alpha-1} dy =$$

$$= \frac{1}{\left(1+\frac{1}{\beta}-z\right)^{\alpha} \cdot \beta^\alpha}$$

Thus, for the generating function $F(z)$ of the total number of defaults, we get:

$$F(z) = \prod_{i=1}^{w} \frac{1}{\left(1+\frac{1}{\beta_i}-z\right)^{\alpha_i} \cdot \beta_i^{\alpha_i}}$$

With this, the main mathematical ideas that form the basis of this concept have been set out. Further elaborations regarding ways to achieving the final result are analogous to the above, and we are therefore not going to derive the final result in detail here (we refer interested readers to the Credit Risk+ technical document [3]).

The next step would be to determine the generating function $G(x)$ of the total loss amount in the credit portfolio within 1 year. This function $G(x)$ is again defined by $G(x) := \sum_{k=0}^{\infty} W(Z = k \cdot L) \cdot x^k$ and is obtained (in the same way as above, after dividing the loans in the sectors into bands, etc.) by:

$$G(x) = \prod_{i=1}^{w} \left(\frac{1}{\beta_i^{\alpha_i}} \cdot \left(1 + \frac{1}{\beta_i} - \widetilde{Q}_i(x) \right)^{-\alpha_i} \right)$$

where $\widetilde{Q}_i(x) := \sum_{i=1}^{m(i)} \frac{\mu_j^{(i)}}{\mu^{(i)}} \cdot x^{u_j^{(i)}}$.

Some of the parameters remain to be defined:

In each of the sectors $S_1, S_2, \ldots, S_w$, the credit risk amounts are again rounded to multiples of a norm value L and divided into bands. The sector S_i contains credit bands with credit risk amounts of the form $u_1^{(i)} \cdot L, u_2^{(i)} \cdot L, u_3^{(i)} \cdot L, \ldots, u_{m(i)}^{(i)} \cdot L$.

$\mu_j^{(i)}$ is the sum of default probabilities for loans in the sector S_i in the band with credit risk amount $u_j^{(i)} \cdot L$.

And finally, we have $\mu^{(i)} = \sum_{j=1}^{m(i)} \mu_j^{(i)}$.

Now, the representation for $G(x)$ is not at all surprising: Here too, just as in the case of independent loans, the generating function $G(x)$ of the loss amount is equal to the generating function $F(x)$ of the number of defaults applied to the function $\widetilde{Q}(x)$ (which is a counterpart to the earlier function $\widetilde{Q}(x)$). So what we have here is an analog of the Formula (1.1).

1.23 Credit Risk+, Part IV: Allocation of Loans to Mutually Independent Sectors and Determination of Required Capital Ratio

As announced, we will now proceed with the final step: Here, loans are no longer allocated to one single sector each; rather, each loan is distributed among several sectors on a percentage basis. The respective percentage is calculated according to how strongly the fact that the loan belongs to the respective sector is likely to influence its performance.

For Toyota as a borrower, for example, 30% could be allocated to the Asia sector, 50% to the automotive manufacturing sector, and 20% to an "artificial sector" such as World Economic Development.

So, a sequence of w non-negative weights $\theta_{l,i}$ for $i = 1, 2, \ldots, w$ is allocated to each loan K_l. $\theta_{l,i}$ gives the percentage (in absolute numbers, not as a percentage) at which K_l is allocated to the sector S_i. For each l, we have $\sum_{i=1}^{w} \theta_{l,i} = 1$.

The result, i.e. the generating function for the portfolio's loss amount, changes only in obvious ways. The generating function $G(x)$ is again given by

$$G(x) = \prod_{i=1}^{w} \left(\frac{1}{\beta_i^{\alpha_i}} \cdot \left(1 + \frac{1}{\beta_i} - \widetilde{Q}_i(x)\right)^{-\alpha_i} \right),$$

where $\widetilde{Q}_i(x) := \sum_{j=1}^{m(i)} \frac{\mu_j^{(i)}}{\mu^{(i)}} \cdot x^{u_j^{(i)}}$.

However, as stated above, the parameters that occur here need to be redefined in obvious ways: In each of the sectors $S_1, S_2, \ldots, S_w$, the credit risk amounts are again rounded to multiples of a norm value L and divided into bands. The sector S_i contains bands with credit risk amounts of the form $u_1^{(i)} \cdot L, u_2^{(i)} \cdot L, u_3^{(i)} \cdot L, \ldots, u_{m(i)}^{(i)} \cdot L$.

$\mu_j^{(i)}$ is the sum of the default probabilities p_l of the loan products K_l lying in the band with credit risk amount $u_j^{(i)} \cdot L$; however, these default probabilities p_l have to be weighted by the factor $\theta_{l,i}$.

And finally again, we have $\mu^{(i)} = \sum_{j=1}^{m(i)} \mu_j^{(i)}$.

If we now expand the function $G(x)$ into a power series $G(x) = \sum_{k=0}^{\infty} A_k \cdot x^k$, then we are rewarded with approximate values A_k for the probability W that we were looking for *(loss amount in the coming year* $= k \cdot L$*)*. Expanding $G(x)$ into a power series can be done with an approach similar to the one used in the case of independent loans, that is, by deriving a recursion for the coefficients A_k (this approach is described in Section A_{10} of the Credit Risk+ technical document). However, we would suggest here to entrust this task to one of the extremely powerful mathematical software packages such as Mathematica.

Due to the minor approximations made in the course of deriving the relevant formulas, it is advantageous for what follows to slightly adapt the results $A_1, A_2, A_3, \ldots$.

Let $K \cdot L$ be the maximum possible loss amount (which would occur if all loans defaulted). Then let us look only at the results for $A_1, A_2, A_3, \ldots, A_K$. The sum A of these approximate probabilities usually will deviate from 1, even if only slightly so. Instead of using the values $A_1, A_2, A_3, \ldots, A_K$, it is therefore computationally more efficient to work with normalized values $\overline{A_k} := \frac{A_k}{A}$ and to use these values as approximations for W *(loss amount in the coming year* $= k \cdot L$*)*.

Credit Risk+ now proposes the following credit provisioning (i.e. required ratio of own funds): Calculate the 1% percentile of the credit portfolio's loss level in the coming year. This can be done by approximation as follows:

We add $\overline{A_1} + \overline{A_2} + \overline{A_3} + \ldots + \overline{A_T}$ up until this value first exceeds 0.99. This then means that there is an approximately 99% probability that the portfolio's credit loss amount will not exceed the value of $T \cdot L$ in the coming year. Thus, by retaining the sum of $T \cdot L$ in own funds, the losses from defaulting loans in the portfolio can be compensated in 99% of all cases. The proposal is thus to retain $T \cdot L$ EUR in own funds.

In the following section, we will demonstrate the algorithm presented here in more detail, by working out an example.

1.24 Credit Risk+, Part V: An Illustrative Example

In this section we want to illustrate the procedure described above by means of a practical –but, of course, for the sake of clarity, again not realistic– numerical example. We consider 3 sectors S_1, S_2, and S_3 with the parameters given in the subsequent Table 1.10 and 20 loans with the following parameters from Table 1.11.

The bands have the multipliers $u_1^{(i)} = 3, u_2^{(i)} = 5, u_3^{(i)} = 7$ and $u_4^{(i)} = 9$ for all $i = 1, 2, 3$.

The following Table 1.12 sort the given data by sectors and bands:

Table 1.10 Sectors S_i and their parameters

	μ^{S_i}	σ^{S_i}
S_1	0.03	0.01
S_2	0.02	0.02
S_3	0.06	0.05

Table 1.11 Considered loans and their corresponding parameters

	Credit risk amount	Rounded multiple of 100,000	Default probability	Share in S_1	Share in S_2	Share in S_3
K_1	340,000	3	0.05	1	0	0
K_2	480,000	5	0.03	1	0	0
K_3	300,000	3	0.1	0.5	0.5	0
K_4	280,000	3	0.02	0.5	0.5	0
K_5	690,000	7	0.05	0.5	0	0.5
K_6	500,000	5	0.02	0.5	0	0.5
K_7	530,000	5	0.01	0	0.5	0.5
K_8	720,000	7	0.08	0	1	0
K_9	940,000	9	0.03	0	1	0
K_{10}	480,000	5	0.01	0	0	1
K_{11}	740,000	7	0.06	0	0	1
K_{12}	890,000	9	0.15	0.25	0.25	0.5
K_{13}	280,000	3	0.05	0.25	0.25	0.5
K_{14}	930,000	9	0.02	0.25	0.25	0.5
K_{15}	270,000	3	0.03	0.25	0.5	0.25
K_{16}	855,000	9	0.04	0.25	0.5	0.25
K_{17}	310,000	3	0.01	0.25	0.5	0.25
K_{18}	900,000	9	0.03	0.5	0.25	0.25
K_{19}	710,000	7	0.07	0.5	0.25	0.25
K_{20}	260,000	3	0.01	0.5	0.25	0.25

Table 1.12 Loans K_i sorted by sectors and bands

Sector 1	Band	Default probability	Share
K_1	3	0.05	1
K_3	3	0.1	0.5
K_4	3	0.02	0.5
K_{13}	3	0.05	0.25
K_{15}	3	0.03	0.25
K_{17}	3	0.01	0.25
K_{20}	3	0.01	0.5
Sector 1	Band	Default probability	Share
K_2	5	0.03	1
K_6	5	0.02	0.5
Sector 1	Band	Default probability	Share
K_5	7	0.05	0.5
K_{19}	7	0.07	0.5
Sector 1	Band	Default probability	Share
K_{12}	9	0.15	0.25
K_{14}	9	0.02	0.25
K_{16}	9	0.04	0.25
K_{18}	9	0.03	0.5
Sector 2	Band	Default probability	Share
K_3	3	0.1	0.5
K_4	3	0.02	0.5
K_{13}	3	0.05	0.25
K_{15}	3	0.03	0.5
K_{17}	3	0.01	0.5
K_{20}	3	0.01	0.25
Sector 2	Band	Default probability	Share
K_7	5	0.01	0.5
Sector 2	Band	Default probability	Share
K_8	7	0.08	1
K_{19}	7	0.07	0.25
Sector 2	Band	Default probability	Share
K_9	9	0.03	1
K_{12}	9	0.15	0.25
K_{14}	9	0.02	0.25
K_{16}	9	0.04	0.5
K_{18}	9	0.03	0.25
Sector 3	Band	Default probability	Share
K_{13}	3	0.05	0.5
K_{15}	3	0.03	0.25
K_{17}	3	0.01	0.25
K_{20}	3	0.01	0.25

(continued)

Table 1.12 (continued)

Sector 3	Band	Default probability	Share
K_6	5	0.02	0.5
K_7	5	0.01	0.5
K_{10}	5	0.01	1
Sector 3	Band	Default probability	Share
K_5	7	0.05	0.5
K_{11}	7	0.06	1
K_{19}	7	0.07	0.25
Sector 3	Band	Default probability	Share
K_{12}	9	0.15	0.5
K_{14}	9	0.02	0.5
K_{16}	9	0.04	0.25
K_{18}	9	0.03	0.25

Next, we compute the values $\mu_j^{(i)}$ for sectors $i = 1, 2, 3$ and multiplier indices $j = 1, 2, 3, 4$. Recall that $u_1^{(i)} = 3$, $u_2^{(i)} = 5$, $u_3^{(i)} = 7$ and $u_4^{(i)} = 9$ for all $i = 1, 2, 3$. It is therefore (see the first ordered table above for sector 1 and the multiplier $u_1^{(i)} = 3$):
$\mu_1^{(1)} = 1 \times 0.05 + 0.5 \times 0.1 + 0.5 \times 0.02 + 0.25 \times 0.05 + 0.25 \times 0.03 + 0.25 \times 0.01 + 0.5 \times 0.01 = 0.1375$

The other values are calculated in the same way:

$\mu_2^{(1)} = 0.04$
$\mu_3^{(1)} = 0.06$
$\mu_4^{(1)} = 0.0675$
$\mu_1^{(2)} = 0.095$
$\mu_2^{(2)} = 0.005$
$\mu_3^{(2)} = 0.0975$
$\mu_4^{(2)} = 0.1$
$\mu_1^{(3)} = 0.0375$
$\mu_2^{(3)} = 0.025$
$\mu_3^{(3)} = 0.1025$
$\mu_4^{(3)} = 0.1025$

Furthermore, $\mu_j^{(i)} = \sum_{j=1}^{4} \mu_j^{(i)}$ for $i = 1, 2, 3$, thus

$\mu^{(1)} = 0.305$
$\mu^{(2)} = 0.2975$
$\mu^{(3)} = 0.2675$

Table 1.13 Parameters α_i, β_i, μ^{S_i}, and σ^{S_i} for the different sectors

	μ^{S_i}	σ^{S_i}	α_i	β_i
S_1	0.06	0.03	0.12	0.5
S_2	0.07	0.02	0.245	0.286...
S_3	0.08	0.05	0.128	0.625

Table 1.14 First twenty members of the obtained series

A_0	A_1	A_2	A_3	A_4	A_5	A_6	A_7	A_8	A_9
0.8417	0	0	0.0356	0	0.0091	0.0026	0.0375	0.0013	0.039
A_{10}	A_{11}	A_{12}	A_{13}	A_{14}	A_{15}	A_{16}	A_{17}	A_{18}	A_{19}
0.0047	0.0002	0.0061	0.0006	0.0042	0.0008	0.0057	0.0009	0.003	0.0013

To determine the generating function $G(x) = \prod_{i=1}^{3}\left(\frac{1}{\beta_i^{\alpha_i}} \cdot \left(1 + \frac{1}{\beta_i} - \widetilde{Q}_i(x)\right)^{-\alpha_i}\right)$ with $\widetilde{Q}_i(x) := \sum_{j=1}^{4} \frac{\mu_j^{(i)}}{\mu^{(i)}} \cdot x^{u_j^{(i)}}$, we now need the parameters α_i and β_i for $i = 1, 2, 3$. Remember: $\alpha_i = \frac{(\mu^{S_i})^2}{\sigma^{S_i}}$ and $\beta_i = \frac{\sigma^{S_i}}{\mu^{S_i}}$. So we get the following Table 1.13:

Therefore

$$\widetilde{Q}_1(x) = 0.45082x^3 + 0.131148x^5 + 0.196721x^7 + 0.221311x^9$$
$$\widetilde{Q}_2(x) = 0.319328x^3 + 0.0168067x^5 + 0.327731x^7 + 0.336134x^9$$
$$\widetilde{Q}_3(x) = 0.140187x^3 + 0.0934579x^5 + 0.383178x^7 + 0.383178x^9$$

and

$G(x) = 0.9525 \cdot \left(1 - 0.3333\, \widetilde{Q}_1(x)\right)^{-0.12} \cdot 0.9403 \cdot \left(1 - 0.2222\, \widetilde{Q}_2(x)\right)^{-0.245} \cdot 0.9397 \cdot \left(1 - 0.3846\, \widetilde{Q}_2(x)\right)^{-0.128}$

Using Mathematica, we expand this function into a power series, and using the first 20 members of this series we get the subsequent Table 1.14.

As is easily calculated, the highest possible loss, if all the loans were to default, is $114 \times 100,000$ EUR (rounded). Therefore, all values A_0 to A_{114} are relevant.

Adding these values calculated by Mathematica, we get exactly –up to the eighth decimal– the value 1. Thus we can use these values directly in our further computations, without prior normalization.

We get $A_0 + A_1 + \ldots + A_{17} = 0.9898\ldots$ and $A_0 + A_1 + \ldots + A_{17} + A_{18} = 0.9929\ldots$

The proposed amount of own funds would thus be $18 \times 100,000$. That is 15.79% of the rounded total face value of $114 \times 100,000$ EUR.

For comparison, we are now going to use the same loans and the same credit parameters to determine the distribution of loss amounts under the assumption that the loans are independent, i.e. without sector allocations. We recall the formula for the relevant generating function $G_Z(x)$:

$G_Z(x) = e^{Q(x)-\mu}$ where $Q(x) = \sum_{i=1}^{4} \mu_i \cdot x^{u_i}$.

For the u_i, we have to consider again the values 3, 5, 7, and 9, μ is the sum of all default probabilities, and μ_i is the sum of default probabilities of the loans in band i. We get

$\mu_1 = 0.27$
$\mu_2 = 0.07$
$\mu_1 = 0.26$
$\mu_1 = 0.27$

and thus
$\mu = 0.87$

Hence $Q(x) = 0.27x^3 + 0.07x^5 + 0.26x^7 + 0.27x^9$ and
$G_Z(x) = e^{0.27x^3+0.07x^5+0.26x^7+0.27x^9-0.87}$.

Expanding the function $G_Z(x)$ gives the following first 20 coefficients A_k, as stated in Table 1.15:

Here, we have $A_0+A_1+\ldots+A_{17} = 0.9878\ldots$ and $A_0+A_1+\ldots+A_{17}+A_{25} = 0.9919\ldots$

The proposed amount of own funds would thus be $24 \times 100,000$. That is 21.05% of the rounded total face value of $114 \times 100,000$ EUR.

So we see a much higher capital ratio here. Figure 1.35 shows the coefficients A_k for the case with sector allocation (blue) and the case of independent loans (red).

However, the two results cannot be compared directly! In particular, the coefficient A_0 in the case of sector allocations depends exclusively on the parameters α_i and β_i, that is, exclusively on the parameters μ^{S_i} and σ^{S_i} of the sectors and not on the parameters of the individual loans. So a different choice of the sector parameters μ^{S_i} and σ^{S_i} would yield entirely different results despite unchanged credit parameters.

Table 1.15 First twenty coefficients of A_k

A_0	A_1	A_2	A_3	A_4	A_5	A_6	A_7	A_8	A_9
0.419	0	0	0.0293	0	0.0153	0.1089	0.0079	0.1145	0.0304
A_{10}	A_{11}	A_{12}	A_{13}	A_{14}	A_{15}	A_{16}	A_{17}	A_{18}	A_{19}
0.001	0.0382	0.0042	0.0221	0.0062	0.0298	0.0062	0.0159	0.0092	0.0009

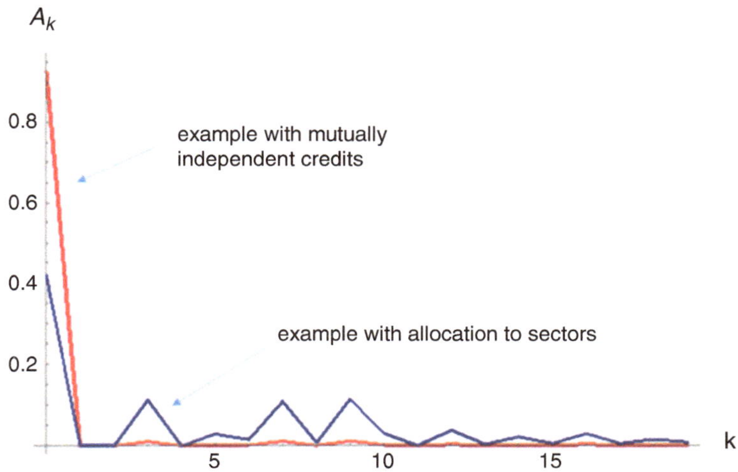

Fig. 1.35 Distribution of credit loss amounts in the example for Credit Risk+ (with sector allocation (blue), independent loans (red))

References

1. Credit Metrics. *Technical Document*. Risk Metrics Group, J.P. Morgan & Co., 2007.
2. Lucia Del Chicca and Gerhard Larcher."Hybrid Monte Carlo-Methods in Credit Risk Management". In: *Monte Carlo Methods and Applications* 20.4 (2014), pp. 245–260.
3. Credit Risk+. *A Credit Risk Management Framework*. Credit Suisse First Boston, 1997.

Optimal Investment Problems 2

Keywords

Markowitz portfolio optimization · Opportunity set · Efficient border · Sharpe ratio · Capital asset pricing model · Market portfolios · Single-index model · Optimal stochastic control theory · Hamilton-Jacobi-Bellman equations · The optimal investment and consumption problem · Merton ratio

The complex question that every investor and every asset manager is faced with in one way or another, and which we will address below—if only in broad outlines—can be summarized as follows:

We have a certain investment amount at our disposal for a certain future period of time, and a certain universe of financial assets in which that amount can be invested. How much of that investment amount should ideally be invested in which of these assets in this period?

There are various approaches to potentially solving this challenge, and these approaches have led to extremely beautiful and compelling financial mathematical theories, such as Markowitz' portfolio selection theory; Sharpe's capital asset pricing model and its extensions by Fama, French, Black, Litterman, and others; or Merton's solution of the "Optimal Investment-Consumption Problem" on the basis of stochastic control theory.

Yet as beautiful and convincing as these approaches are in theory, there is often a problem when it comes to applying them in practice: They all require precise forecasts of, for example, trends and volatilities of contemplated financial assets and of correlations between these assets. And it is hardly a secret that procuring such precise forecasts is very difficult. Another difficulty that is often encountered in seeking a relevant practical answer to this question is that the answer depends on each investor's particular context. Different types of investors will assess portfolio optimality by different criteria.

G. Larcher, *The Art of Quantitative Finance Vol. 3*, Springer Texts in Business and Economics, https://doi.org/10.1007/978-3-031-23867-3_2

Still, the methods and techniques listed above are part of the necessary basic toolbox in financial engineering, and so, for the sake of completeness, we will therefore present their fundamental principles in the following, without going into too much detail.

2.1 Classic Portfolio Optimization According to Markowitz, Part 1: Fundamental Principles

We start with the classic portfolio selection theory developed and presented by Harry M. Markowitz in 1952, for which he was awarded the Nobel Prize in Economics in 1990. As noted above, we will not go much beyond presenting the basic principles and ideas here. For readers interested in learning more about these techniques, we recommend the books [1] or [2].

The situation that we start from is as follows: We have a certain investment amount at our disposal. For simplicity, we denote this investment amount by 1. This amount is to be fully invested now, at time 0, and to be held without any changes until time T. There are n different financial assets $A_1, A_2, \ldots, A_n$ that we could invest in. We assume that we are dealing with a "reasonable, rational" investor. By that we mean that the investor prefers a higher expected return to a lower expected return and low volatility to higher volatility.

It is assumed that we have estimates $\mu_1, \mu_2, \mu_3, \ldots, \mu_n$ and $\sigma_1, \sigma_2, \sigma_3, \ldots, \sigma_n$ for the trends (i.e. the average returns) and volatilities in the price of the individual assets $A_1, A_2, \ldots, A_n$ and that we also have estimates $\rho_{i,j}$ of the correlations between the returns of the assets A_i and A_j for all $i, j = 1, 2, 3, \ldots, n$ with $i \neq j$. How we arrive at these estimates is not the issue for now (we could simply assume, for instance, that they were provided to us by analysts).

As we will frequently do in the following, we will now plot these data in a return/volatility diagram as shown in Fig. 2.1.

We start with a simple experiment to demonstrate, just very briefly, that it is in principle possible to run strategies in this setting that will perform better and worse on average, irrespective of the estimated values for the parameters of the assets $A_1, A_2, \ldots, A_n$.

To this end, we consider two large groups of individuals who play the role of investors and test two different strategies.

Strategy 1 *Every individual in the first group randomly selects (with equal probability $\frac{1}{n}$ in each case) one of the financial assets $A_1, A_2, \ldots, A_n$ and invests their entire investment in this one asset.*

Strategy 2 *Every individual in the second group invests the same amount of $\frac{1}{n}$ in each of the assets.*

On average, what characteristics do the portfolios of the individuals in each of the two test groups have?

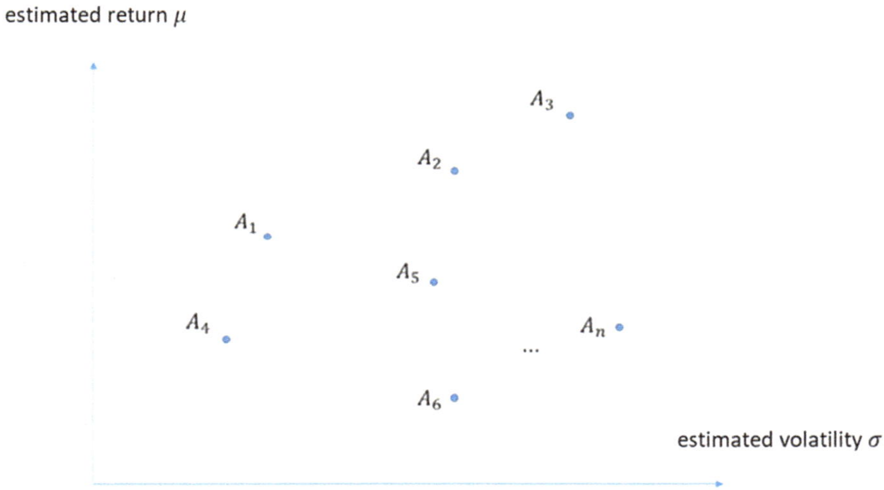

Fig. 2.1 Representation of the financial assets in a volatility/return diagram

Average Properties of Strategy 1 Since every individual chooses one of the assets at random, the average expected return $\tilde{\mu}$ (taken over all individuals in this group or, more precisely, over all portfolios traded by these individuals) is

$$\tilde{\mu} = \frac{\mu_1 + \mu_2 + \ldots + \mu_n}{n}.$$

And since every individual independently chooses one of the assets at random, the average expected variance is

$$\tilde{\sigma}^2 = \frac{\sigma_1^2 + \sigma_2^2 + \ldots + \sigma_n^2}{n}$$

Average Properties of Strategy 2 The portfolios of these individuals are all of the same type. We schematically denote the portfolios by $\frac{1}{n}A_1 + \frac{1}{n}A_2 + \frac{1}{n}A_3 + \ldots + \frac{1}{n}A_n$. (This does not mean that the individuals bought $\frac{1}{n}$ units of each asset but rather that they invested $\frac{1}{n}$ EUR in each asset!) The expected return $\hat{\mu}$ is therefore

$$\hat{\mu} = \frac{\mu_1 + \mu_2 + \ldots + \mu_n}{n}.$$

To determine the expected variance $\hat{\sigma}^2$ of the returns of this portfolio, we use the formula—well known by now—for determining the variance of the sum of dependent random variables:

$$\hat{\sigma}^2 = \left(\frac{1}{n}\right)^2 \cdot \left(\sum_{i=1}^{n} \sigma_i^2 + \sum_{\substack{i,j=1 \\ i \neq j}}^{n} \sigma_i \cdot \sigma_j \cdot \rho_{i,j}\right)$$

We reformulate the last expression somewhat and get:

$$\hat{\sigma}^2 = \frac{1}{n} \cdot \left(\frac{1}{n} \cdot \sum_{i=1}^{n} \sigma_i^2\right) + \frac{n-1}{n} \cdot \left(\frac{1}{n \cdot (n-1)} \cdot \sum_{\substack{i,j=1 \\ i \neq j}}^{n} \sigma_i \cdot \sigma_j \cdot \rho_{i,j}\right).$$

The expression $\frac{1}{n} \cdot \sum_{i=1}^{n} \sigma_i^2$ is precisely the **average variance** of the individual assets, and $\frac{1}{n \cdot (n-1)} \cdot \sum_{\substack{i,j=1 \\ i \neq j}}^{n} \sigma_i \cdot \sigma_j \cdot \rho_{i,j}$ is the **average covariance** between the assets $\left(\sigma_i \cdot \sigma_j \cdot \rho_{i,j} = cov_{i,j}\right)$.

If we assume that the number n of available assets is large, then $\frac{1}{n} \approx 0$ and $\frac{n-1}{n} \approx 1$ and thus

$$\hat{\sigma}^2 \approx \frac{1}{n \cdot (n-1)} \cdot \sum_{\substack{i,j=1 \\ i \neq j}}^{n} \sigma_i \cdot \sigma_j \cdot \rho_{i,j}.$$

We Conclude

We can expect about the same average return for both strategies. The average expected variance of returns for the first strategy is equal to the **mean of the variances** of the individual assets, while for the second strategy (assuming the number of available assets is relatively large), it is approximately equal to the **mean of the covariances** between two different assets.

Now, in Volume I Section 4.2, we already found the following: *"The average covariance between the various DAX stocks in the test period from May 2015 to May 2018 was 0.000128. The average variance of the individual DAX stocks was 0.000288 (i.e. approximately 2.3 times that value."*

In fact, whenever we have a larger number of financial assets, the average variance of the returns of these assets is practically always significantly larger than the average covariance of the returns of these assets.

Note also—and this may be even more convincing:

It is obvious, based on purely mathematical considerations, that $\tilde{\sigma}^2 = \frac{\sigma_1^2 + \sigma_2^2 + \ldots + \sigma_n^2}{n}$ must always be greater than or equal to $\hat{\sigma}^2 = \left(\frac{1}{n}\right)^2 \cdot$

$\left(\sum_{i=1}^{n} \sigma_i^2 + \sum_{\substack{i,j=1 \\ i \neq j}}^{n} \sigma_i \cdot \sigma_j \cdot \rho_{i,j}\right)$ and equality between the two variables $\tilde{\sigma}^2$ and $\hat{\sigma}^2$ can only exist (except in the trivial case of $\sigma_i = 0$ for all i) if the correlations $\rho_{i,j}$ are all equal to 1 and if, in addition, the volatilities σ_i are all equal, i.e. if $\sigma_i = \sigma_j$ holds for all i and j.

This fact results from the following:

The expression $\left(\frac{1}{n}\right)^2 \cdot \left(\sum_{i=1}^{n} \sigma_i^2 + \sum_{\substack{i,j=1 \\ i \neq j}}^{n} \sigma_i \cdot \sigma_j \cdot \rho_{i,j}\right)$ can also be written in a simpler form, namely, $\left(\frac{1}{n}\right)^2 \cdot \sum_{i,j=1}^{n} \sigma_i \cdot \sigma_j \cdot \rho_{i,j}$.

We need to demonstrate therefore that $\left(\frac{1}{n}\right)^2 \cdot \sum_{i,j=1}^{n} \sigma_i \cdot \sigma_j \cdot \rho_{i,j} \leq \frac{1}{n} \cdot \sum_{i=1}^{n} \sigma_i^2$, i.e. that $\sum_{i,j=1}^{n} \sigma_i \cdot \sigma_j \cdot \rho_{i,j} \leq n \cdot \sum_{i=1}^{n} \sigma_i^2$ is always true.

Since the volatilities σ_i, σ_j are always non-negative numbers, then certainly $\sum_{i,j=1}^{n} \sigma_i \cdot \sigma_j \cdot \rho_{i,j} \leq \sum_{i,j=1}^{n} \sigma_i \cdot \sigma_j$ and therefore the inequality $\sum_{i,j=1}^{n} \sigma_i \cdot \sigma_j \cdot \rho_{i,j} \leq n \cdot \sum_{i=1}^{n} \sigma_i^2$ is certainly always satisfied if $\sum_{i,j=1}^{n} \sigma_i \cdot \sigma_j \leq n \cdot \sum_{i=1}^{n} \sigma_i^2$ is satisfied.

In the inequality $\sum_{i,j=1}^{n} \sigma_i \cdot \sigma_j \cdot \rho_{i,j} \leq \sum_{i,j=1}^{n} \sigma_i \cdot \sigma_j$, equality holds only if all $\rho_{i,j} = 1$, except if one of the $\sigma_i = 0$ because then the correlation $\rho_{i,j}$ is not defined.

We note that for the left side of the last inequality, we have $\sum_{i,j=1}^{n} \sigma_i \cdot \sigma_j = \left(\sum_{i=1}^{n} \sigma_i\right)^2$.

So what is now left to demonstrate is that

$$\left(\sum_{i=1}^{n} \sigma_i\right)^2 \leq n \cdot \sum_{i=1}^{n} \sigma_i^2 \text{ resp. } \sum_{i=1}^{n} \sigma_i \leq \sqrt{n} \cdot \sqrt{\sum_{i=1}^{n} \sigma_i^2}$$

or, in a somewhat more complicated form:

$$\sum_{i=1}^{n} 1 \cdot \sigma_i \leq \sqrt{\sum_{i=1}^{n} 1^2} \cdot \sqrt{\sum_{i=1}^{n} \sigma_i^2}.$$

Note, however, that $\frac{\sum_{i=1}^{n} 1 \cdot \sigma_i}{\sqrt{\sum_{i=1}^{n} 1^2} \cdot \sqrt{\sum_{i=1}^{n} \sigma_i^2}}$ is simply the cosine of the angle between the n-dimensional vectors $(1, 1, \ldots, 1, 1)$ and $(\sigma_1, \sigma_2, \ldots, \sigma_{n-1}, \sigma_n)$, which is always less than or equal to 1. Equality holds here only if the vectors $(1, 1, \ldots, 1, 1)$ and $(\sigma_1, \sigma_2, \ldots, \sigma_{n-1}, \sigma_n)$ are

(continued)

parallel to each other. Thus $\sum_{i=1}^{n} 1 \cdot \sigma_i \leq \sqrt{\sum_{i=1}^{n} 1^2} \cdot \sqrt{\sum_{i=1}^{n} \sigma_i^2}$ is always true, with equality only if all σ_i have the same value. If all σ_i have the same value (not equal to 0, for obvious reasons), then, for equality to hold in the original inequality, it is required that all $\rho_{i,j} = 1$.

With this introductory simple experiment, we only wanted to illustrate the following: It is definitely possible to make "better" or "poorer" investments, in the sense that it is quite possible to reduce portfolio risk (in terms of its volatility) through suitable portfolio composition without reducing the expected return.

In the following sections, we will look at how this could be done in an optimal way.

2.2 Classic Portfolio Optimization According to Markowitz, Part 2: Two Financial Assets

We start by first looking at the case where we have two assets A_1 and A_2 that we can potentially invest in (see Fig. 2.2).

We invest an amount of x euros in asset A_1 and the remaining amount of $1 - x$ euros in asset A_2. Here (for now) $0 \leq x \leq 1$. So schematically, our portfolio $P = P(x)$ has the form $x \cdot A_1 + (1 - x) \cdot A_2$.

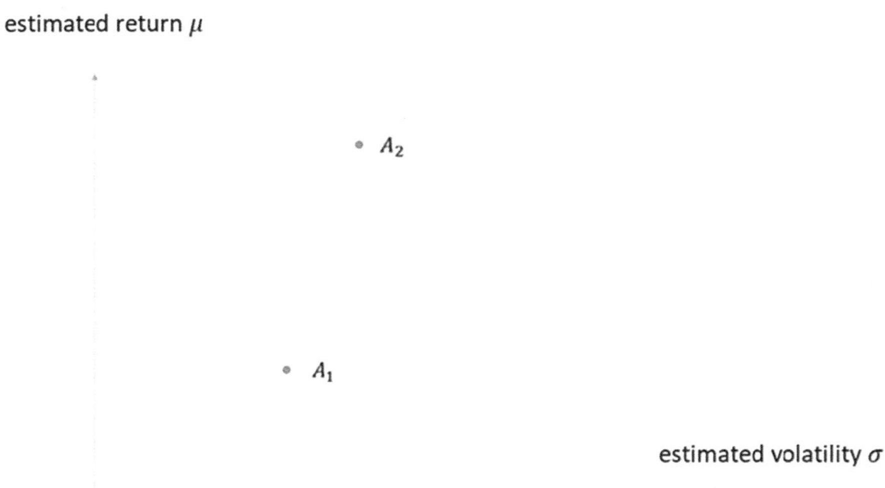

Fig. 2.2 Investment in two assets

The expected return $\mu(P)$ of the portfolio P is given by

$$\mu(P) = x \cdot \mu_1 + (1-x) \cdot \mu_2,$$

and volatility $\sigma(P)$ is given by

$$\sigma(P) = \sqrt{x^2 \cdot \sigma_1^2 + 2 \cdot x(1-x) \cdot \sigma_1 \cdot \sigma_2 \cdot \rho + (1-x)^2 \cdot \sigma_2^2}.$$

Here we write ρ as short for $\rho_{1,2}$.

We now want to determine the position of the point $P := (\sigma(P), \mu(P))$ in the volatility/return diagram. For this we distinguish between three cases.

Case 1: $\rho = 1$
Thus we have

$$\begin{aligned}\sigma(P) &= \sqrt{x^2 \cdot \sigma_1^2 + 2x(1-x) \cdot \sigma_1 \cdot \sigma_2 + (1-x)^2 \cdot \sigma_2^2} \\ &= \sqrt{(x \cdot \sigma_1 + (1-x) \cdot \sigma_2)^2} = x \cdot \sigma_1 + (1-x) \cdot \sigma_2.\end{aligned}$$

When taking the root here, we used the fact that $x \cdot \sigma_1 + (1-x) \cdot \sigma_2$ is always positive.

The point P is thus given by the coordinate representation

$$P = P(x) = (x \cdot \sigma_1 + (1-x) \cdot \sigma_2, x \cdot \mu_1 + (1-x) \cdot \mu_2).$$

For variable x between 0 and 1, this is the parameter representation of a line in the plane connecting the point (σ_1, μ_1) of asset A_1 with the point (σ_2, μ_2) of asset A_2 ($x = 1$ returns A_1 and $x = 0$ returns A_2).

This line is shown in red in Fig. 2.3.

Case 2: $\rho = -1$
Thus we have

$$\begin{aligned}\sigma(P) &= \sqrt{x^2 \cdot \sigma_1^2 - 2x(1-x) \cdot \sigma_1 \cdot \sigma_2 + (1-x)^2 \cdot \sigma_2^2} \\ &= \sqrt{(x \cdot \sigma_1 - (1-x) \cdot \sigma_2)^2} = |x \cdot \sigma_1 - (1-x) \cdot \sigma_2|.\end{aligned}$$

The point P is thus given by the coordinate representation

$$P = P(x) = (|x \cdot \sigma_1 - (1-x) \cdot \sigma_2|, x \cdot \mu_1 + (1-x) \cdot \mu_2).$$

If $x \cdot \sigma_1 - (1-x) \cdot \sigma_2 \geq 0$, that is, for $x \geq \frac{\sigma_2}{\sigma_1+\sigma_2}$, this means that $P = P(x) = (x \cdot \sigma_1 - (1-x) \cdot \sigma_2, x \cdot \mu_1 + (1-x) \cdot \mu_2)$.

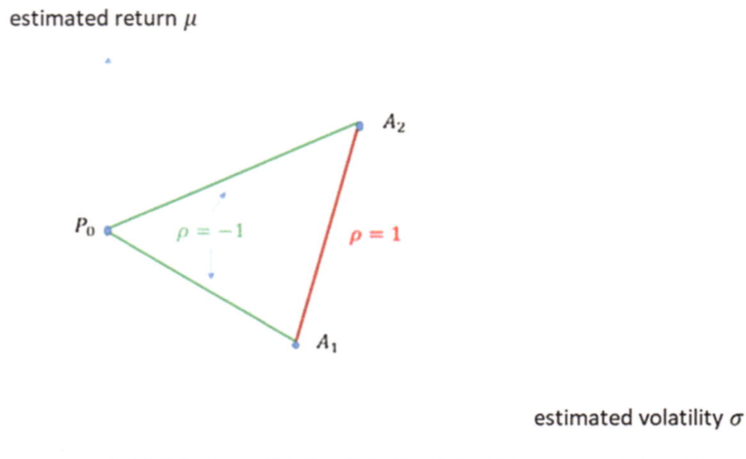

Fig. 2.3 Opportunity sets for $\rho = 1$ (red) and for $\rho = -1$ (green)

The value $x = \frac{\sigma_2}{\sigma_1+\sigma_2}$ yields the point $P_0 := \left(0, \frac{\sigma_2}{\sigma_1+\sigma_2} \cdot \mu_1 + \frac{\sigma_1}{\sigma_1+\sigma_2} \cdot \mu_2\right)$. So, for variable x between $\frac{\sigma_2}{\sigma_1+\sigma_2}$ and 1, we get a straight line segment connecting point $P_0 = \left(0, \frac{\sigma_2}{\sigma_1+\sigma_2} \cdot \mu_1 + \frac{\sigma_1}{\sigma_1+\sigma_2} \cdot \mu_2\right)$ on the y-axis with point A_1.

If $x \cdot \sigma_1 - (1-x) \cdot \sigma_2 \le 0$, that is, for $x \le \frac{\sigma_2}{\sigma_1+\sigma_2}$, this means that $P = P(x) = (-x \cdot \sigma_1 + (1-x) \cdot \sigma_2, x \cdot \mu_1 + (1-x) \cdot \mu_2)$.

The value $x = \frac{\sigma_2}{\sigma_1+\sigma_2}$ again yields the point $P_0 = \left(0, \frac{\sigma_2}{\sigma_1,\sigma_2} \cdot \mu_1 + \frac{\sigma_1}{\sigma_1+\sigma_2} \cdot \mu_2\right)$. So, for variable x between 0 and $\frac{\sigma_2}{\sigma_1+\sigma_2}$, we get a straight line segment connecting point $P_0 = \left(0, \frac{\sigma_2}{\sigma_1+\sigma_2} \cdot \mu_1 + \frac{\sigma_1}{\sigma_1+\sigma_2} \cdot \mu_2\right)$ on the y-axis with point A_2.

So, for any x between 0 and 1, we get a point somewhere on one of these two straight line segments. We note that the x value $\frac{\sigma_2}{\sigma_1+\sigma_2}$ that gives the point on the y-axis is in fact a value between 0 and 1 and that the y value $\frac{\sigma_2}{\sigma_1+\sigma_2} \cdot \mu_1 + \frac{\sigma_1}{\sigma_1+\sigma_2} \cdot \mu_2$ of this risk-free point is a value between μ_1 and μ_2.

These two lines are shown in green in Fig. 2.3.

We refer to these potential portfolio points as the respective "opportunity sets". (The red line represents the opportunity set for $\rho = 1$, and the two green lines represent the opportunity set for $\rho = -1$.)

Case 3: $-1 < \rho < 1$

For x between 0 and 1, just as in the case of correlation -1 or 1, the trend value is given by $x \cdot \mu_1 + (1-x) \cdot \mu_2$. For the volatility $\sigma(P)$, we have

$$\sigma(P) = \sqrt{x^2 \cdot \sigma_1^2 + 2x(1-x) \cdot \sigma_1 \cdot \sigma_2 \cdot \rho + (1-x)^2 \cdot \sigma_2^2}$$

and because

$$\begin{aligned} 0 &\leq x^2 \cdot \sigma_1^2 - 2x(1-x) \cdot \sigma_1 \cdot \sigma_2 + (1-x)^2 \cdot \sigma_2^2 \\ &\leq x^2 \cdot \sigma_1^2 + 2x(1-x) \cdot \sigma_1 \cdot \sigma_2 \cdot \rho + (1-x)^2 \cdot \sigma_2^2 \\ &\leq x^2 \cdot \sigma_1^2 + 2x(1-x) \cdot \sigma_1 \cdot \sigma_2 + (1-x)^2 \cdot \sigma_2^2, \end{aligned}$$

we get $|x \cdot \sigma_1 - (1-x) \cdot \sigma_2| \leq \sigma(P) \leq x \cdot \sigma_1 + (1-x) \cdot \sigma_2$.

So, given the same parameters, $\sigma(P)$ is a value between the volatility in the case of $\rho = -1$ and the volatility in the case of $\rho = 1$.

The point $P(x)$ for any x $(0 \leq x \leq 1)$ thus lies within the triangle formed by the red and green lines in Fig. 2.3. The curve formed by the points $P(x)$ for varying x between 0 and 1 again connects the points A_1 and A_2, lies within said triangle, has no jumps (both coordinate functions are continuous), and does not intersect itself (in the case of $\mu_1 \neq \mu_2$), because the coordinate function $x \cdot \mu_1 + (1-x) \cdot \mu_2$, which describes the return, is then strictly monotonic. In principle, this curve could therefore look like the one in Fig. 2.4 (blue curve), for example.

However, a simple analysis of this curve would reveal that the blue curve connects the points A_1 and A_2 in a convex curve, as shown in Fig. 2.5.

Now, what exactly do we mean by a convex curve running from point A_1 to point A_2 to the left of the red line connecting A_1 and A_2?

Such a curve is called convex if the line connecting any two points on the curve always runs completely to the right of the curve. See Fig. 2.6 for illustration.

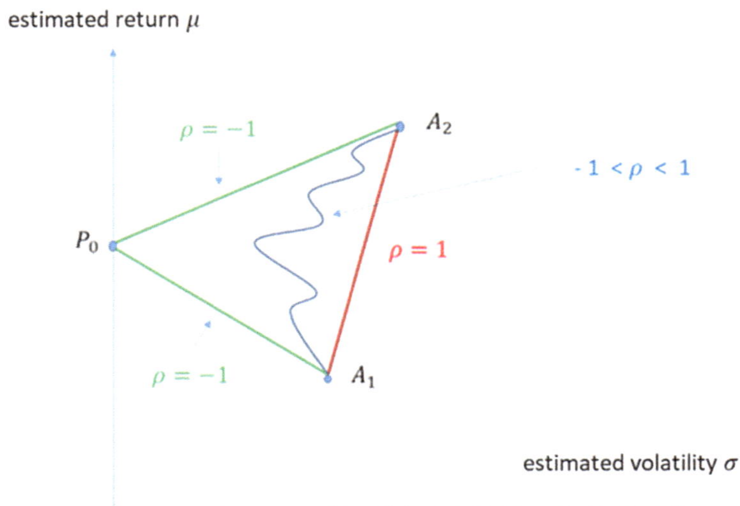

Fig. 2.4 Possible opportunity set, in principle, in the case $-1 < \rho < 1$ (blue line)?

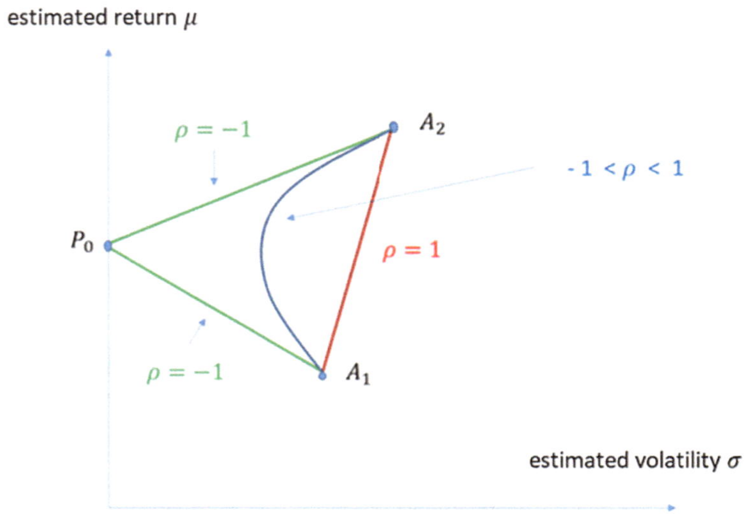

Fig. 2.5 Opportunity set in the case $-1 < \rho < 1$ (blue line)

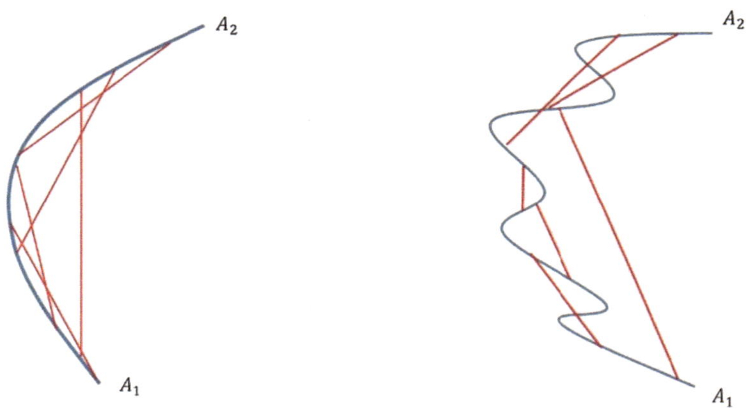

Fig. 2.6 Example of a convex curve (blue) on the left and a non-convex curve (blue) on the right

While in the left picture, every red line connecting any two points on the blue curve runs to the right of the curve, this is clearly not the case in the picture on the right.

The fact that the opportunity set in the case $-1 < \rho < 1$ (i.e. the blue curve in Fig. 2.4 or in Fig. 2.5) must indeed be convex is easily verified by curve analysis, as mentioned, or can also be understood from the following considerations:

Suppose for once that the opportunity set is not convex. Then there would certainly be points Q_1 and Q_2 on this opportunity set where the line connecting these points would lie entirely to the left of that opportunity set.

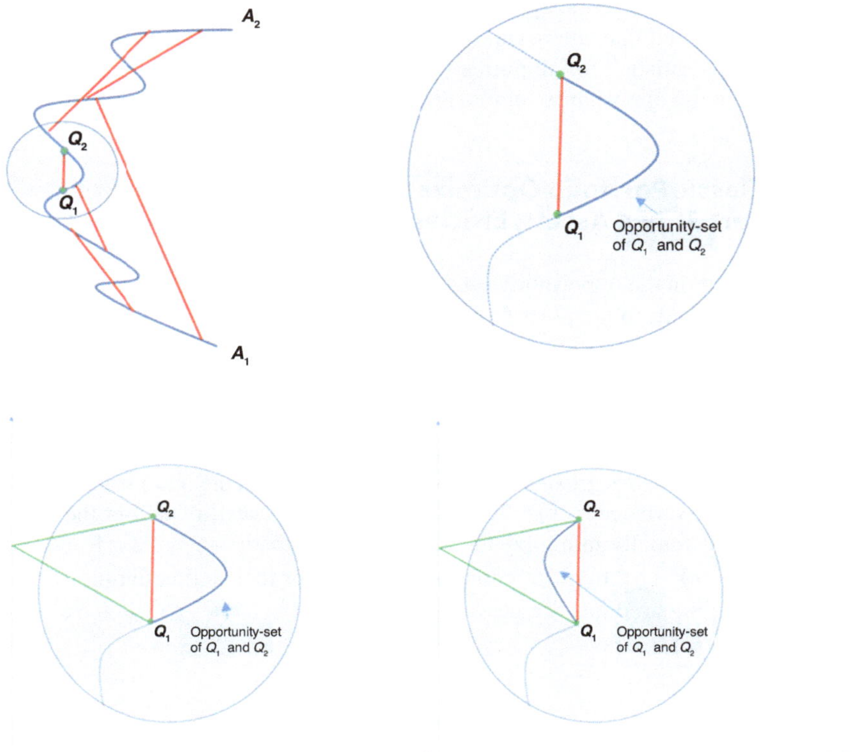

Fig. 2.7 The curve of the opportunity set is convex

Let us take any two points from this opportunity set, say, points Q_1 and Q_2 in Fig. 2.7, top left.

These two points Q_1 and Q_2 represent two specific portfolios built with the assets $A_1, A_2, \ldots, A_n$.

We could also invest our money in combinations of Q_1 and Q_2. Each such combination is then again a combination of $A_1, A_2, \ldots, A_n$ and thus lies on the blue line (the opportunity set consisting of $A_1, A_2, \ldots, A_n$).

The opportunity set of the two portfolios Q_1 and Q_2 forms a continuous curve that connects Q_1 and Q_2 seamlessly. The opportunity set of Q_1 and Q_2 is therefore precisely the part of the opportunity set of $A_1, A_2, \ldots, A_n$ that connects Q_1 and Q_2. In our top-right Fig. 2.7, that would be the blue curve segment from Q_1 to Q_2 to the right of the red line connecting Q_1 and Q_2.

However, for the opportunity set of Q_1 and Q_2, it can be shown, using exactly the same arguments as above, that it must lie between the red line connecting Q_1 and Q_2 and a green segment of two lines, from Q_1 to the y-axis and from there to Q_2 (Fig. 2.7, bottom left). This contradicts our assumption!

The curve segment of the opportunity set between Q_1 and Q_2 must definitely lie to the left of the red line, approximately as shown in Fig. 2.7, bottom right.

Since this is satisfied for all points Q_1 and Q_2 on the curve between A_1 and A_2, it follows that the opportunity set must be convex.

2.3 Classic Portfolio Optimization According to Markowitz, Part 3: Two Assets, Efficient Border

A special point on the opportunity set seems of particular interest, namely, the point P_0 that corresponds to the portfolio, that is, to the combination of A_1 and A_2, with minimum volatility σ (see Fig. 2.8).

The coordinates of the portfolio P_0 are easily calculated by minimizing the volatility function $\sigma = \sigma(x) = \sqrt{x^2 \cdot \sigma_1^2 + 2x(1-x) \cdot \sigma_1 \cdot \sigma_2 \cdot \rho + (1-x)^2 \cdot \sigma_2^2}$ with respect to x.

Since $\sigma(x)$ is always greater than or equal to 0, minimizing $\sigma(x)$ is equivalent to minimizing the variance $\sigma(x)^2$. To spare ourselves the need of taking the root for $\sigma(x)^2$, we will actually minimize the function $\sigma(x)^2 = x^2 \cdot \sigma_1^2 + 2x(1-x) \cdot \sigma_1 \cdot \sigma_2 \cdot \rho + (1-x)^2 \cdot \sigma_2^2$ by differentiating with respect to x and equating to zero in the following. Squared out, we get $\sigma(x)^2 = x^2 \cdot \left(\sigma_1^2 - 2 \cdot \sigma_1 \cdot \sigma_2 \cdot \rho + \sigma_2^2\right) + 2x \cdot \left(\sigma_1 \cdot \sigma_2 \cdot \rho - \sigma_2^2\right) + \sigma_2^2$.

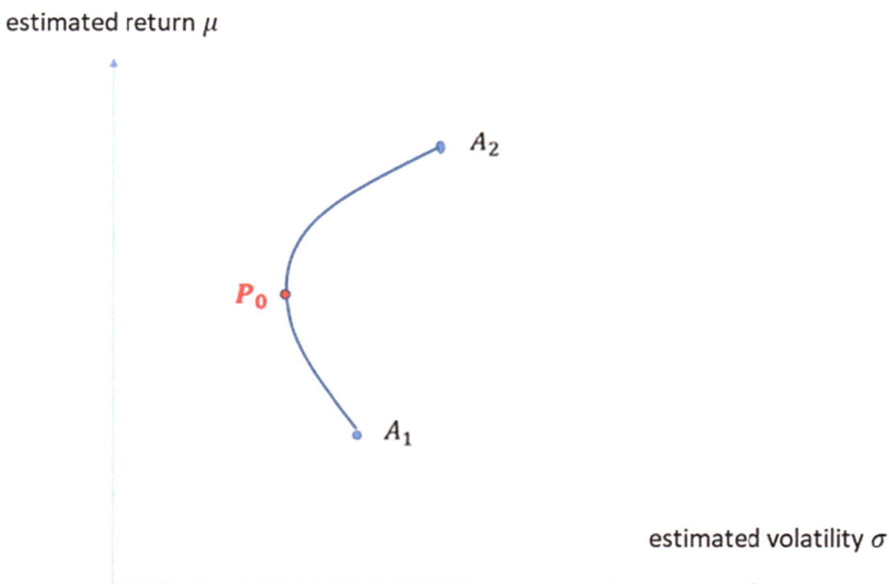

Fig. 2.8 Portfolio P_0 with lowest volatility

The coefficient $\sigma_1^2 - 2 \cdot \sigma_1 \cdot \sigma_2 \cdot \rho + \sigma_2^2$ is always positive (except in trivial cases where $\rho = 1$ and $\sigma_1 = \sigma_2$ or where $\sigma_1 = \sigma_2 = 0$), and so this quadratic function has exactly one extremum, namely, a minimum.

We have $\left(\sigma(x)^2\right)' = 2x\left(\sigma_1^2 - 2 \cdot \sigma_1 \cdot \sigma_2 \cdot \rho + \sigma_2^2\right) + 2 \cdot \left(\sigma_1 \cdot \sigma_2 \cdot \rho - \sigma_2^2\right)$, and thus we get a minimum for $x = \frac{\sigma_2^2 - \sigma_1 \cdot \sigma_2 \cdot \rho}{\sigma_1^2 - 2 \cdot \sigma_1 \cdot \sigma_2 \cdot \rho + \sigma_2^2}$. By substituting this value x in $\sigma(x) = \sqrt{x^2 \cdot \sigma_1^2 + 2x(1-x) \cdot \sigma_1 \cdot \sigma_2 \cdot \rho + (1-x)^2 \cdot \sigma_2^2}$, we get the minimum volatility value as $\sigma_{min} = \frac{\sigma_1 \sigma_2 \sqrt{1-\rho^2}}{\sqrt{\sigma_1^2 - 2\sigma_1 \cdot \sigma_2 \cdot \rho + \sigma_2^2}}$.

It remains to be verified, however, under which circumstances this value x actually lies between 0 and 1.This is not necessarily a given. In principle, the situations illustrated in Fig. 2.9 can also occur. In the upper chart, minimum volatility would be attained for a value of $x < 0$, and in the situation shown in the lower chart, it would be attained for a value of $x > 1$. Since these values are not permissible in our current setting, the actual portfolio with minimum volatility in the upper chart would be the portfolio that consists only of asset A_2, and in the lower chart, it would be the portfolio consisting only of asset A_1.

So the question now is: When is $0 \leq \frac{\sigma_2^2 - \sigma_1 \cdot \sigma_2 \cdot \rho}{\sigma_1^2 - 2 \cdot \sigma_1 \cdot \sigma_2 \cdot \rho + \sigma_2^2} \leq 1$ satisfied?

Since the denominator is positive (except in "pathological" cases), we can multiply by this denominator and obtain the equivalent condition

$$0 \leq \sigma_2^2 - \sigma_1 \cdot \sigma_2 \cdot \rho \leq \sigma_1^2 - 2 \cdot \sigma_1 \cdot \sigma_2 \cdot \rho + \sigma_2^2.$$

The left side of this inequality yields the condition $\rho \leq \frac{\sigma_2}{\sigma_1}$ (otherwise $x < 0$) and the right side of the inequality yields the condition $\rho \leq \frac{\sigma_1}{\sigma_2}$ (otherwise $x > 1$).

If $\sigma_1 \leq \sigma_2$ (e.g. Fig. 2.9, bottom), then $\rho \leq \frac{\sigma_2}{\sigma_1}$ is satisfied in any case. And if $\rho \leq \frac{\sigma_1}{\sigma_2}$, then the point P_0 corresponds to a possible portfolio. But if $\rho > \frac{\sigma_1}{\sigma_2}$, then we are in the situation shown in the lower chart of Fig. 2.9. The possible portfolio with the smallest volatility is the portfolio consisting only of A_1.

If $\sigma_2 \leq \sigma_1$ (e.g. Fig. 2.9, top), then $\rho \leq \frac{\sigma_1}{\sigma_2}$ is satisfied in any case. And if $\rho \leq \frac{\sigma_2}{\sigma_1}$, then the point P_0 corresponds to a possible portfolio. But if $\rho > \frac{\sigma_2}{\sigma_1}$, then we are in the situation shown in the upper chart of Fig. 2.9. The possible portfolio with the smallest volatility is the portfolio consisting only of A_2.

So we see: Reducing volatility through diversification is possible only if the correlation between the two assets A_1 and A_2 is not too high. More precisely, that correlation needs to be smaller than the minimum of $\frac{\sigma_1}{\sigma_2}$ and of $\frac{\sigma_2}{\sigma_1}$. So, if our goal in the investment process is to reduce risk through diversification, we need to choose the assets A_1 and A_2 such that their correlation meets this condition.

Also important to note: The value of the minimum volatility is independent of the estimates μ_1 and μ_2 for the expected returns of the two assets!

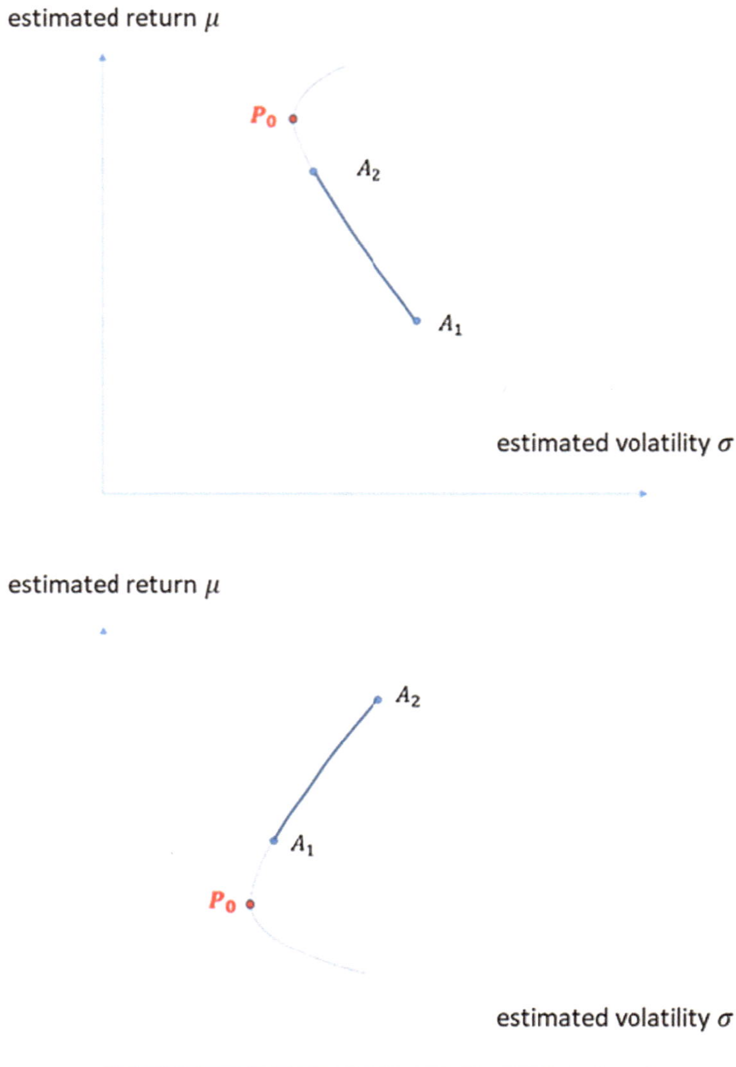

Fig. 2.9 Portfolio P_0 with lowest volatility outside the opportunity set

In the following, we will refer to the situation illustrated in Figs. 2.8 and 2.10 as the "normal case" for the location of point P_0.

Let us now consider (for this normal case) the portfolios the representatives of which are on the opportunity set but below point P_0. For example, the portfolio or point Q in Fig. 2.10. Would you invest in portfolio Q? And if not, why not?

The answer to both questions is given in the upper chart of Fig. 2.10:

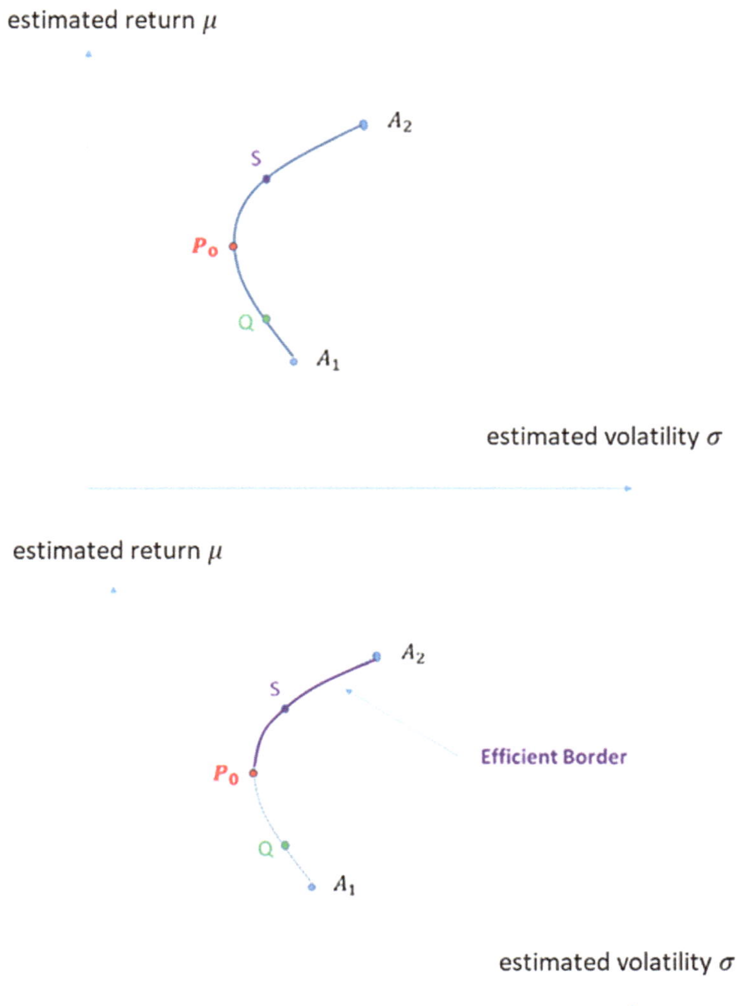

Fig. 2.10 Opportunity set and efficient border

For every point on the opportunity set below the point P_0, there is a point that promises a higher expected return at the same (or lower) expected volatility. For example, for the point Q, there is a point S vertically above it on the opportunity set that has the same estimated volatility but a higher estimated return. Any rational investor (as defined above) will therefore prefer portfolio S to portfolio Q! Meaning: A rational investor will only consider investments in portfolios on the opportunity set going from portfolio P_0 (the portfolio with minimal volatility) towards A_2. Only these portfolios should be considered, and we refer to the curve segment containing these portfolios as the **efficient border**. For our example, this efficient border is shown in purple in the lower chart of Fig. 2.10.

A rational investor will not be able to find justifiable preferences between the portfolios on the efficient border by just looking at and comparing expected return and expected volatility. For each pair of portfolios on the efficient border, one portfolio has a higher return and the other has lower volatility.

In the special cases shown in Fig. 2.9, and again in Fig. 2.11, the efficient borders are also clearly defined.

In the upper picture, the portfolio consisting only of asset A_2 dominates all other portfolios both in terms of expected return and in terms of low volatility. Thus, the efficient border consists only of the portfolio $P = A_2$.

In the bottom picture, none of the possible portfolios dominates any other portfolio in both return and low volatility. The entire opportunity set is therefore also efficient border.

Finally, we want to explicitly point out a fact by another graphic (Fig. 2.12):

Asset A_2 is far superior to asset A_1 in the situation depicted in Fig. 2.12 in terms of both return and volatility. Nevertheless, instead of investing solely in A_2, it would be well worth considering investing in portfolio S, which promises a slightly lower return than A_2 but at a significantly lower volatility than A_2. Yet an investment in the portfolio S also means a partial investment in asset A_1. So, in a nutshell: Although all parameters speak against asset A_1 compared to asset A_2, a partial investment in A_1 may well make sense.

2.4 Classic Portfolio Optimization According to Markowitz, Part 4: Two Assets, with Short Selling

Under certain conditions—in particular in frictionless markets (!)—it is also possible to short sell (without limitation) the financial assets available for investment in our portfolio problem. That is, we can create portfolios of the form $x_1 A_1 + x_2 A_2$ with $x_1 + x_2 = 1$ and with x_1 and x_2 arbitrarily from real numbers.

$A_n x_1$ with $x_1 < 0$ means short selling asset $A_1, and a_n x_2$ with $x_2 < 0$ means short selling asset A_2.

The shape of the opportunity set does not change on the curve segment between A_1 and A_2, but it then continues beyond A_1 and A_2. We see the typical shape of the opportunity set in the case of unlimited short selling in Fig. 2.13.

The efficient border in this case is the curve segment from P_0 upwards. In this case (as shown in Fig. 2.13), one can achieve arbitrarily high expected return by short selling asset A_1 and buying asset A_2.

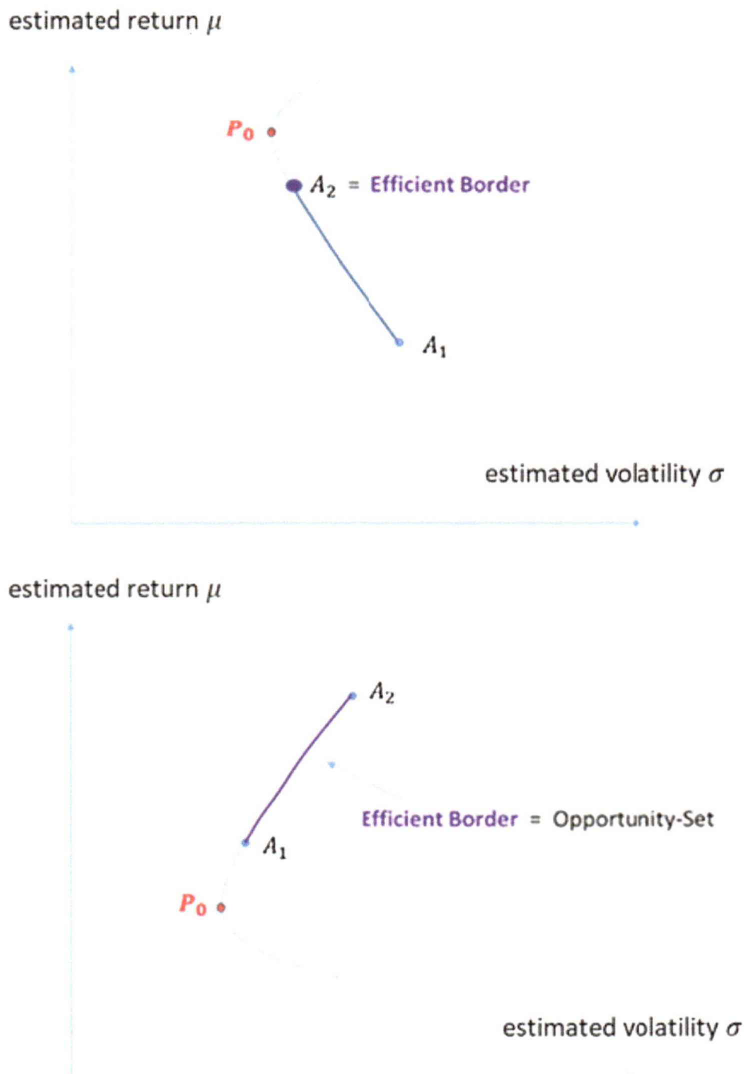

Fig. 2.11 Efficient border in special cases

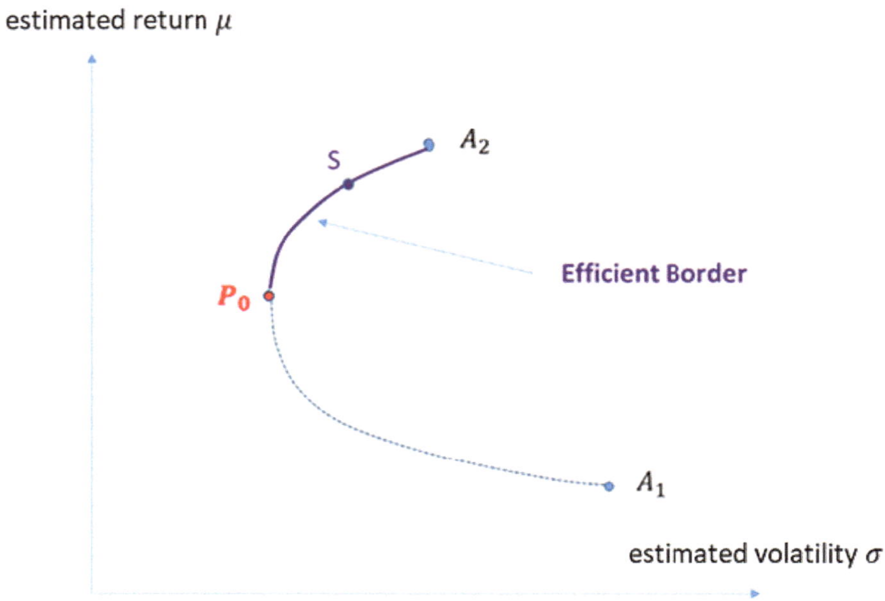

Fig. 2.12 Another possible situation

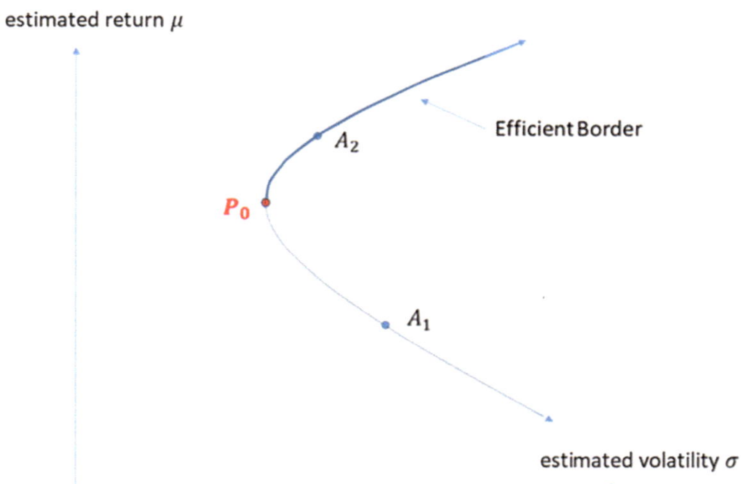

Fig. 2.13 Opportunity set and efficient border in the case of unlimited short selling

2.5 Classic Portfolio Optimization According to Markowitz, Part 5: Two Risky Assets and One Risk-Free Asset. Portfolios with Maximal Sharpe Ratio

In addition to the two risky assets A_1 and A_2, we now have an additional risk-free asset A_0. The risky assets are characterized by the fact that the volatilities σ_1 and σ_2 are both positive.

The risk-free asset A_0 has volatility $\sigma_0 = 0$ and return $\mu_0 = r$, i.e. equal to the risk-free interest rate. A_0 has the coordinates $(0, r)$ and sits on the y-axis of the volatility/return diagram. Our first question here is: What is the opportunity set of all possible combinations $x_0A_0 + x_1A_1 + x_2A_2$ with $x_0 + x_1 + x_2 = 1$ of these three objects? We get these combinations by choosing an arbitrary combination Q of A_1 and A_2 from the opportunity set of A_1 and A_2 and investing an amount of y_1 in Q and an amount of $y_0 = 1 - y_1$ in A_0.

But where are the possible combinations $y_0A_0 + y_1Q$ located?

We illustrate the situation and answer this question with the charts in Figs. 2.14, 2.15, and 2.16 for the case without short selling (i.e. $0 \leq x_0, x_1, x_2, y_0, y_1 \leq 1$) and in Figs. 2.18 and 2.19 for the case with unlimited short selling.

In the **case without short selling**, the portfolio Q is on the opportunity curve between A_1 and A_2.

Let us denote the parameters of portfolio Q by (σ_Q, μ_Q).

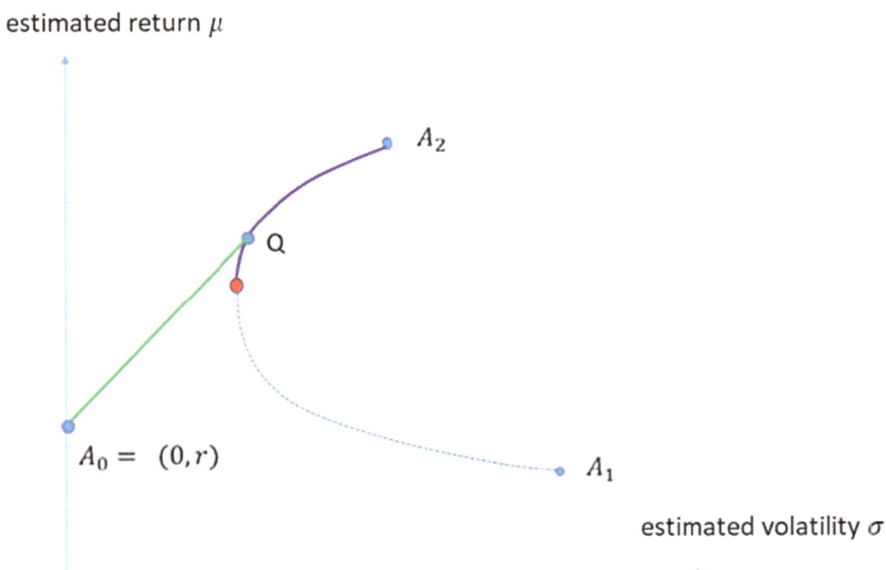

Fig. 2.14 Combinations of a portfolio Q with a risk-free asset A_0 (without short selling)

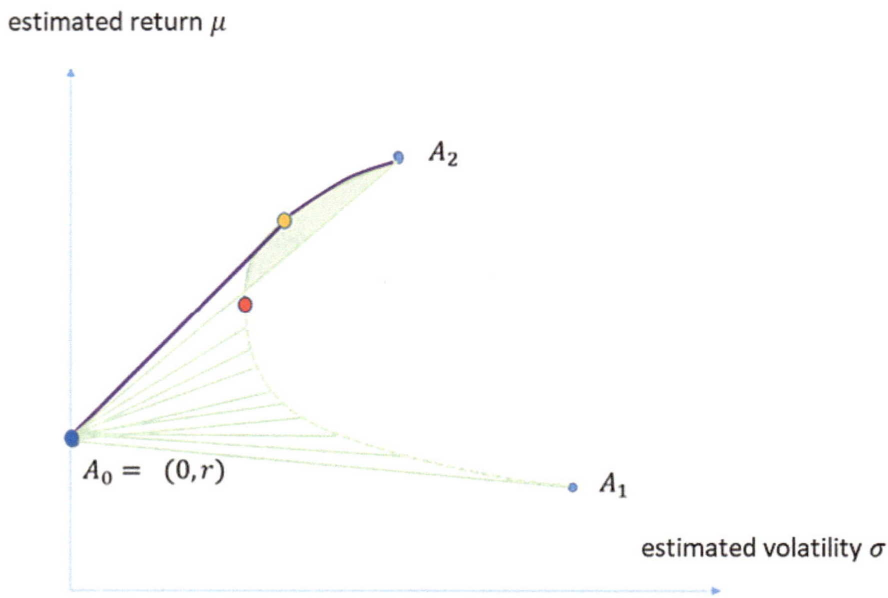

Fig. 2.15 Opportunity set and efficient border of the combinations of two risky assets and one risk-free asset, case 1

The covariance $cov(A_0, Q)$ between A_0 and Q is 0, since $cov(A_0, Q) = E(\mu(A_0)) - E(\mu(A_0)) \cdot (\mu(Q) - E(\mu(Q))) = E((r-r) \cdot (\mu(Q))) - E(\mu(Q)) = 0$.

A combination $(1-y) \cdot A_0 + y \cdot Q$ therefore has expected return $(1-y) \cdot r + y \cdot \mu_Q$ and expected standard deviation $\sqrt{y^2 \cdot \sigma_Q^2} = |y| \cdot \sigma_Q$. Thus, for y between 0 and 1, we have the linear parameter representation $(y \cdot \sigma_Q, (1-y) \cdot r + y \cdot \mu_Q)$ with $0 \leq y \leq 1$, that is, we get the line connecting A_0 and Q as the locus of all possible combinations (with no short selling) of A_0 and Q. This line is shown in green in Fig. 2.14.

If we now connect point A_0 with all possible portfolios Q on the opportunity set of A_1 and A_2, we get all possible portfolios of A_0, A_1, and A_2, i.e. the opportunity set of A_0, A_1, and A_2 (without short selling). This opportunity set is the green shaded area bordered in purple in Fig. 2.15. The purple line is (based on the same considerations as above) the efficient border of the combinations of A_0, A_1, and A_2.

The orange point plays a special role. It denotes the point of contact of the upper tangent from A_0 to the opportunity set of A_1 and A_2, if such a point exists (case 1). If it does not exist (case 2), however, then the opportunity set of A_0, A_1, and A_2 is simply formed by the line connecting A_0 and A_2, and this connecting line is then also the efficient border of the combinations of A_0, A_1, and A_2 (see Fig. 2.16).

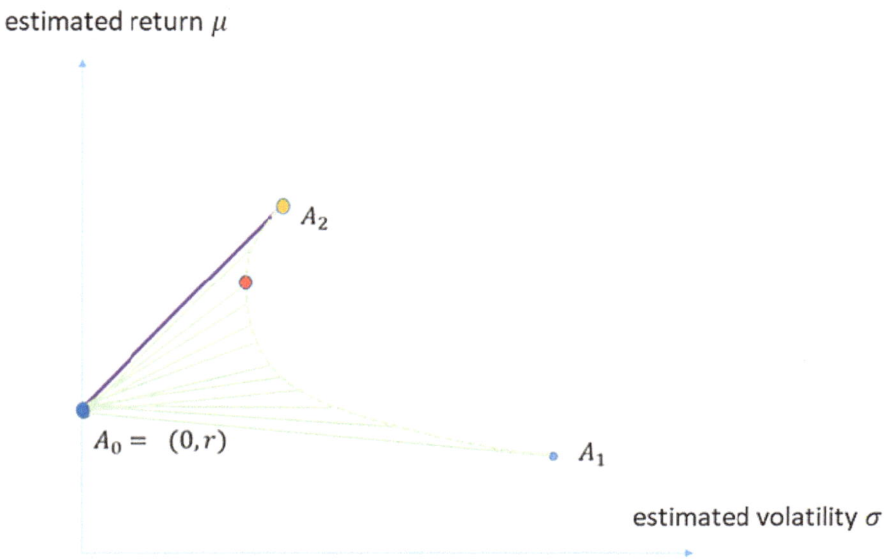

Fig. 2.16 Opportunity set and efficient border of the combinations of two risk assets and one risk-free asset, case 2

The special relevance of the orange point results from the following:

Let us revisit Fig. 2.14. The slope of the straight lines through A_0 and the portfolio Q is given by $\frac{\mu_Q - r}{\sigma_Q}$ (see Fig. 2.17). However, this value is precisely the Sharpe ratio of the portfolio Q.

The orange point is now clearly on the line that has the largest slope of all the lines between A_0 and a portfolio Q on the opportunity set of A_1 and A_2 (this is true both in the situation of Fig. 2.15 (case 1) and in the situation of Fig. 2.16 (case 2)). The orange point thus denotes the portfolio Q of A_1 and A_2 that is expected to have the highest Sharpe ratio.

We Summarize, for the Case Without Short Selling

If A_2 (the asset with the higher expected return) is the permissible combination of A_1 and A_2 with the highest Sharpe ratio, then the only investment that makes sense is in a permissible combination of A_2 and the risk-free investment A_0. How much of the investment should be invested in A_2 and how much of the investment should be invested in A_0 depend solely on the investor's risk appetite.

If portfolio SR is the permissible combination of A_1 and A_2 with the highest Sharpe ratio (orange point in Fig. 2.15) and if SR is different from A_2 (the asset with the higher expected return), then the only investment that makes sense is in a permissible combination of SR and the risk-free investment A_0 or in a permissible combination of SR and A_2.

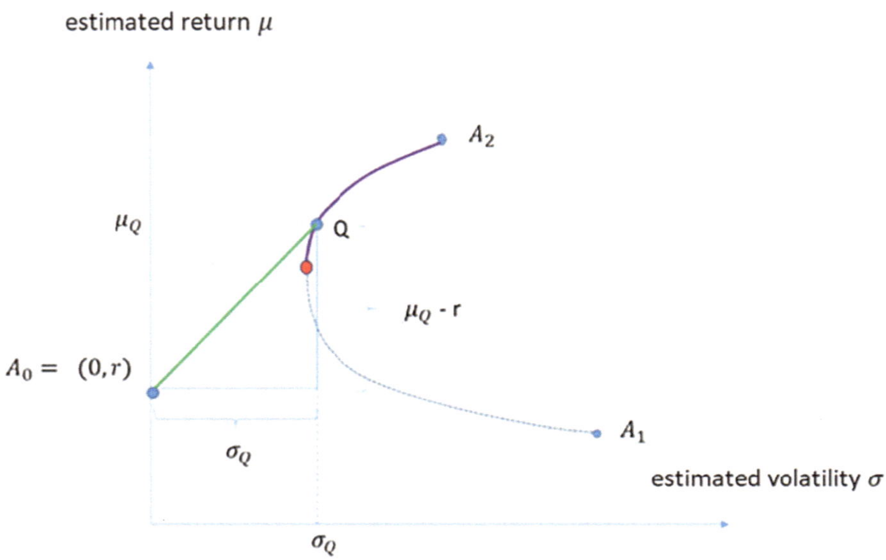

Fig. 2.17 Slope of the line between A_0 and Q (= Sharpe ratio of Q)

It is an easy task to analyse under which conditions case 1 and under which conditions case 2 occurs.

In the **case with short selling**, the portfolios Q lie on the opportunity curve between A_1 and A_2 and also beyond A_1 and A_2.

We first consider again where the permissible portfolios of Q and A_0 lie in the case of unlimited short selling.

As noted above, the parameter representation of the combinations is $\left(|y| \cdot \sigma_Q, (1-y) \cdot r + y \cdot \mu_Q\right)$, with y now running through the real numbers arbitrarily. Thus, for $y \geq 0$, we obtain the parameter representation $\left(y \cdot \sigma_Q, (1-y) \cdot r + y \cdot \mu_Q\right)$, and this gives us the half-line starting at point A_0 and going through Q (for $y = 1$) and beyond. For $y \leq 0$ we get $\left(-y \cdot \sigma_Q, (1-y) \cdot r + y \cdot \mu_Q\right)$ and this gives the half-line that also starts at point A_0 (for $y = 0$) and goes through point $\widetilde{Q} := \left(\sigma_Q, 2r - \mu_Q\right)$ and beyond. This second half-line is exactly the reflection of the first half-line on the line $y = r$. This is illustrated in Fig. 2.18. The two green half-lines denote all combinations of A_0 and Q.

If we now construct the set of all possible combinations of A_0, A_1, and A_2 by generating these two half-lines for each Q on the opportunity set of A_1 and A_2, then, in principle, the following four different cases can occur, illustrated in Fig. 2.19.

Case 1 (Top Left) *There exists a unique upper tangent from A_0 to the opportunity curve of A_1 and A_2, and the reflection of this tangent is consistently below the opportunity curve of A_1 and A_2.*

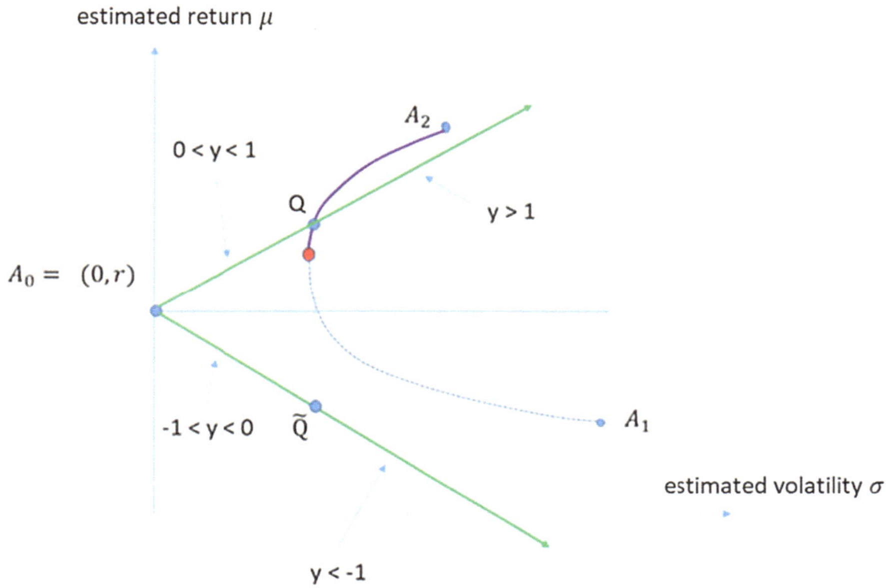

Fig. 2.18 Combinations of a portfolio Q with a risk-free asset A_0 (with short selling)

Then there is a unique point $Q(= SR)$ with maximum Sharpe ratio on the upper branch of the opportunity curve from A_1 to A_2. Only investments in combinations of SR and A_0 make sense. A positive portion is invested in SR. It is also possible to short sell the risk-free product (= borrowing) and fully invest the initial investment and the loan amount in SR.

The **light green triangle, which is open on the right and closed at the top and bottom, is the opportunity set**, and the **upper green boundary line is the efficient border**.

Case 2 (Top Right) *There exists an upper asymptote from A_0 to the opportunity curve of A_1 and A_2, and the reflection of this asymptote is consistently below the opportunity curve of A_1 and A_2.*

The Sharpe ratio of the portfolios on the upper branch of the opportunity curve of A_1 and A_2 increases as μ increases. The **light green triangle, which is open on the right and at the top and bottom, is the opportunity set**. There is no definitive efficient border. Investments as close as possible to the upper asymptote make the most sense. Meaning: Combinations of A_0 and of portfolios Q of A_1 and A_2 as far to the top right as possible on the opportunity curve of A_1 and A_2. In these combinations, a positive portion of the investment is invested in Q (possibly even a portion greater than 1 if a loan is taken out). **We still refer to this upper asymptote as efficient border** (even though the points on this asymptote do not correspond to any portfolios).

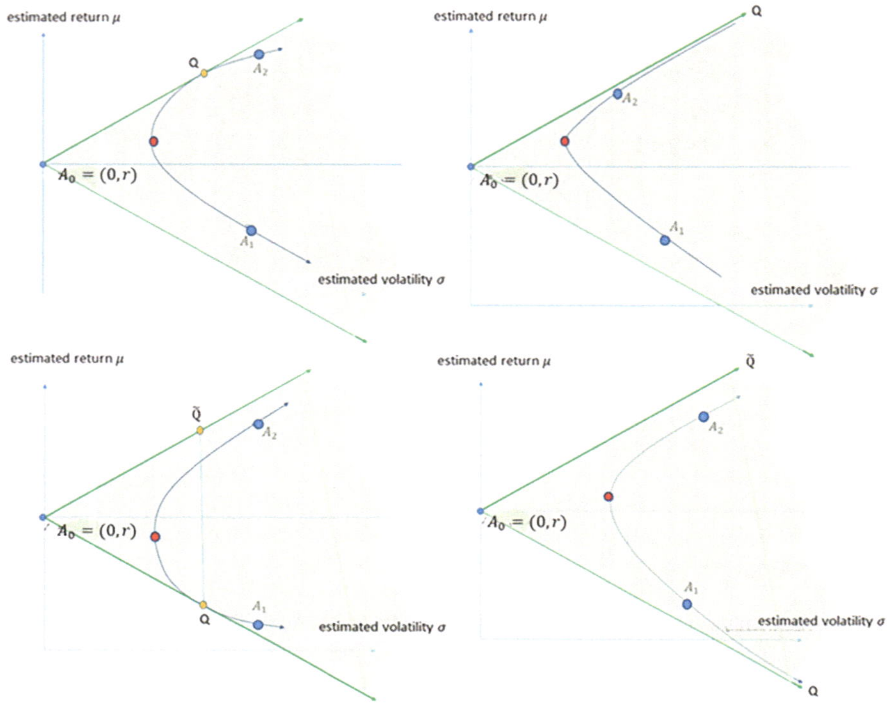

Fig. 2.19 Possible forms of the opportunity set in the case of unlimited short selling, four cases

Case 3 (Bottom Left) *There exists a unique lower tangent from A_0 to the opportunity curve of A_1 and A_2, and the reflection of this tangent is consistently above the opportunity curve of A_1 and A_2.*

Then there is a unique point Q with minimum Sharpe ratio on the lower branch of the opportunity curve from A_1 to A_2. Only investments in combinations of A_0 and the reflection $\widetilde{Q}$ of Q make sense. $\widetilde{Q}$ in this case means short selling one unit of the portfolio Q. $\widetilde{Q}$ is the portfolio of A_1 and A_2 with maximum Sharpe ratio. Thus, an optimal portfolio consists of a combination of A_0 and a negative investment in (short selling of) Q. The **light green triangle, which is open on the right and closed at the top and bottom, is the opportunity set**, and the **upper green boundary is the efficient border**.

Case 4 (Bottom Right) *There exists a lower asymptote from A_0 to the opportunity curve of A_1 and A_2, and the reflection of this asymptote is consistently above the opportunity curve of A_1 and A_2.*

The Sharpe ratio of the portfolios on the lower branch of the opportunity curve of A_1 and A_2 decreases as μ decreases. The **light green triangle, which is open on the right and at the top and bottom, is the opportunity set**. There is no definitive efficient border. Investments as close as possible to the upper asymptote make the most sense. Meaning: Combinations of A_0 and of short positions in portfolios Q of A_1 and A_2 that are as far to the bottom right as possible on the opportunity curve of A_1 and A_2. In these combinations, a negative portion of the investment is invested in Q. **We still refer to the upper asymptote as efficient border** (even though the points on this asymptote do not correspond to any portfolios).

Again, it could be easily analysed under which circumstances which of the four cases occurs and how to calculate the orange point, i.e. the **portfolio with maximum Sharpe ratio (if it exists).** This **will be carried out in detail in subsequent chapters for the general case of any number of financial assets $A_1, A_2, \ldots, A_n$.**

2.6 Classic Portfolio Optimization According to Markowitz, Part 6: Two Risky Assets and One Risk-Free Asset: Practical Example

All previous illustrations of the Markowitz model were sketches and did not correspond to any specific numerical example. In this section we will therefore give an illustrative numerical example of the opportunity set and the efficient border of combinations consisting of two stocks and one risk-free investment. As already indicated, the main problem lies in the model's approach, that is, in the estimation of the necessary parameters. Our focus in the following is not on this estimation issue; instead, we will simply use an estimate based on historical daily closing prices (knowing full well that this is a very debatable approach).

We consider an investment of 1 (= 1 million) USD for the period of 1 year from today (7 May 2019 to 7 May 2020) in the following assets:

A_1 = Ford stock
A_2 = Apple stock

as well as in

A_0 = a 1-year fixed-rate investment at an interest rate of 2%.

We estimate the expected per-annum returns μ_1 and μ_2 and the expected per-annum standard deviations σ_1 and σ_2 and the correlation ρ between the returns of these two stocks based on the daily closing prices of the immediately preceding trading year and obtain:

$\mu_1 = 0.1627$ (corresponding to 16.27%)
$\sigma_1 = 0.2948$ (corresponding to 29.48%)

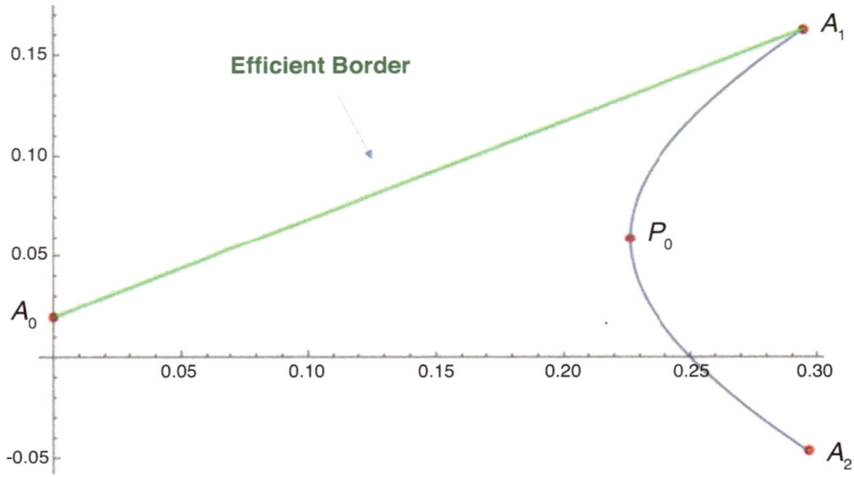

Fig. 2.20 Example: Opportunity set for A_1 and A_2 and efficient border for A_0, A_1, and A_2 without short selling

$\mu_2 = -0.0465$ (corresponding to -4.65%)
$\sigma_2 = 0.2962$ (corresponding to 29.62%)
$\rho = 0.1728$

We first treat the case **without short selling**:

The portfolio P_0 consisting of A_1 and A_2 with minimum standard deviation is obtained using the formula derived earlier, which gives us $P_0 = x \cdot A_1 + (1-x) \cdot A_2$ with $x = 0.5029$ and standard deviation 0.2263 and expected return 0.0587.

The opportunity set is obtained using the parameter representation derived above and is the blue curve shown in Fig. 2.20. It is clear to see that there is no tangent from A_0 to the opportunity set. The efficient border for combinations of A_0, A_1, and A_2 therefore consists of the green straight line connecting A_0 and A_1. Thus, the efficient portfolios consist of combinations of A_0 and A_1.

With Short Selling

The opportunity set now has the form as shown in Fig. 2.21. Point P_0 with minimum standard deviation remains unchanged.

Using the general formula—which we will derive in the following sections—for the portfolio with maximal Sharpe ratio in the case of unlimited short selling (see Sect. 2.13), it can be shown that, in our case, we do in fact get a portfolio SR of A_1 and A_2 with maximum Sharpe ratio on the upper branch of the opportunity set of A_1 and A_2 and thus an upper tangent to the upper branch of the opportunity set of A_1 and A_2. The reflection of this tangent lies entirely below the opportunity set of A_1

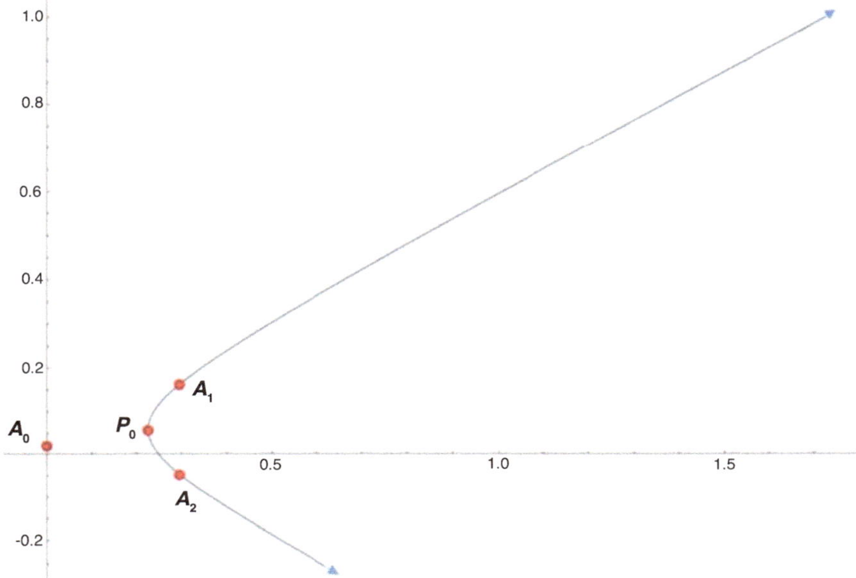

Fig. 2.21 Example: Opportunity set with short selling

and A_2. The portfolio SR has the form $SR = 2.42A_1 - 1.42A_2$ and the coordinates (0.763, 0.46), given in the Fig. 2.22:

2.7 Classic Portfolio Optimization According to Markowitz, Part 7: Sketch of a Sensitivity Test

The following graphs are meant to serve only as a superficial indication of how sensitive the opportunity set and efficient border react to changing parameter estimates. In these graphs, we have again assumed that we generated opportunity sets for the parameters of the above numerical example.

Figure 2.23 shows how the shape of the opportunity set changes as we increase the parameter μ_1 four times successively by 1%. The darkest curve represents the opportunity set for the original parameters. The brightest curve represents the opportunity set for the return parameter $\mu_1 + 4\%$ for the first asset A_1.

Figure 2.24 shows how the shape of the opportunity set changes as we increase the parameter σ_1 four times successively by 2%. The darkest curve represents the opportunity set for the original parameters. The brightest curve represents the opportunity set for the volatility parameter $\sigma_1 + 8\%$ for the first asset A_1.

Figure 2.25 shows how the shape of the opportunity set changes as we increase the parameter ρ four times successively by 0.05. The darkest curve represents

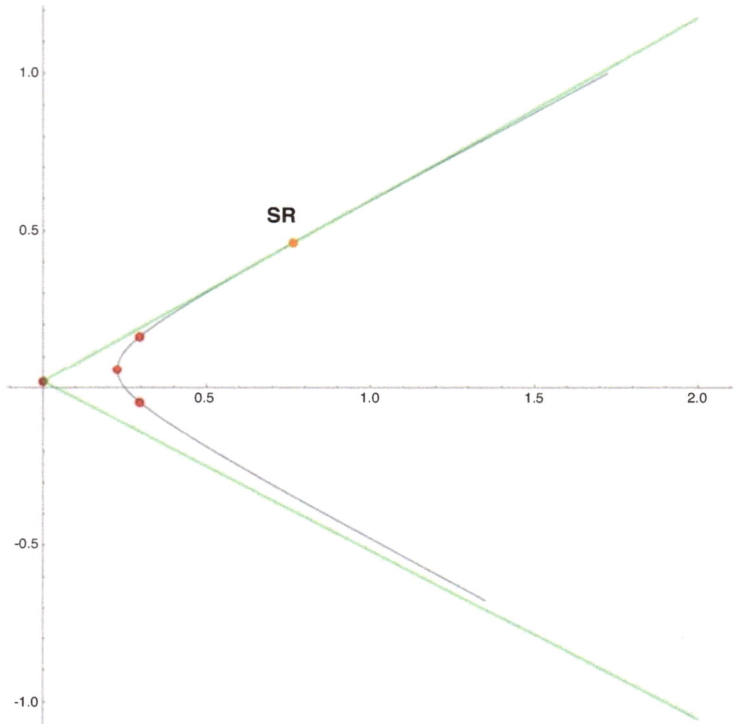

Fig. 2.22 Example: Opportunity set and efficient border as well as portfolio SR with maximal Sharpe ratio in the case with short selling

the opportunity set for the original parameters. The brightest curve represents the opportunity set for the correlation parameter $\rho + 0.2$.

2.8 Classic Portfolio Optimization According to Markowitz, Part 8: Arbitrary Number of Assets, Principal Form of Opportunity Sets

If we not only have two risky (and possibly another risk-free) assets that we can invest in but also n assets $A_1, A_2, \ldots, A_n$ (and possibly one risk-free asset A_0) with expected returns $\mu_1, \mu_2, \ldots, \mu_n$ (resp. r) and with estimated standard deviations $\sigma_1, \sigma_2, \ldots, \sigma_n$ (resp. 0) and with correlations $\rho_{i,j}$, then the opportunity curve becomes an opportunity set of all permissible combinations $A_1x_1 + A_2x_2 + \ldots + A_nx_n$ from $A_1, A_2, \ldots, A_n$.

The typical shape of such an opportunity set can be illustrated using a Monte Carlo method. We generate a large sample (consisting of, e.g. 2,000,000 scenarios, see Fig. 2.26) of random vectors $(y_1, y_2, \ldots, y_n)$ with $0 \leq y_i \leq 1$, which

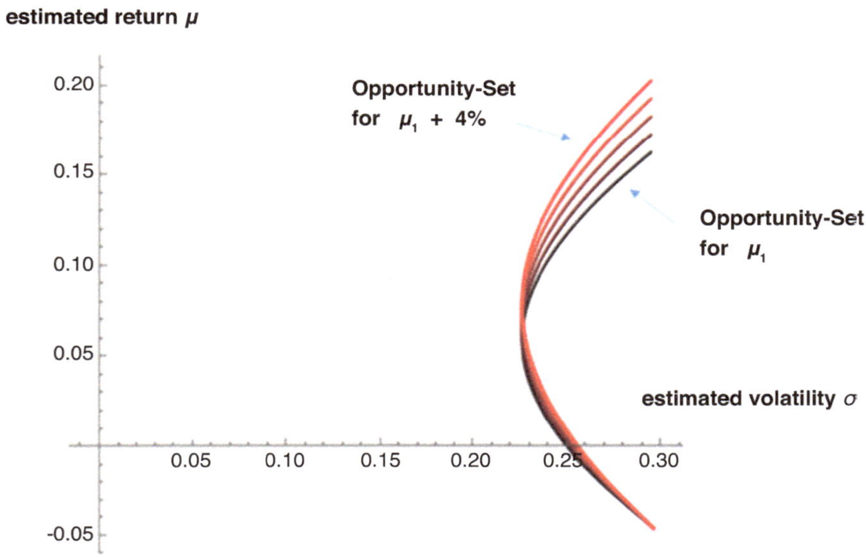

Fig. 2.23 Example: Sensitivity of the opportunity set when the return parameter changes by 1%, respectively

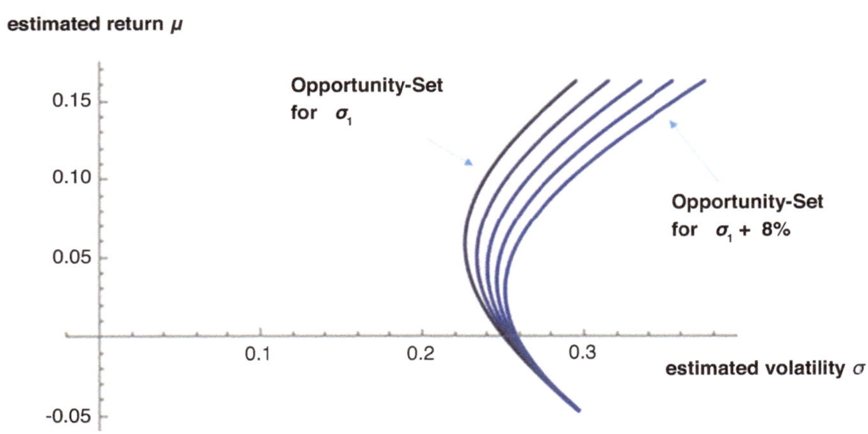

Fig. 2.24 Sensitivity of the opportunity set when the volatility parameter changes by 2%, respectively

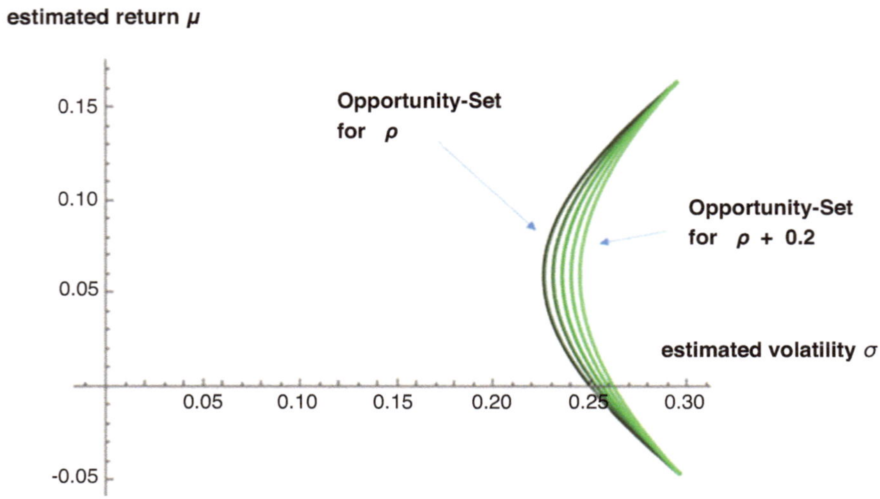

Fig. 2.25 Sensitivity of the opportunity set when the correlation changes by 0.05, respectively

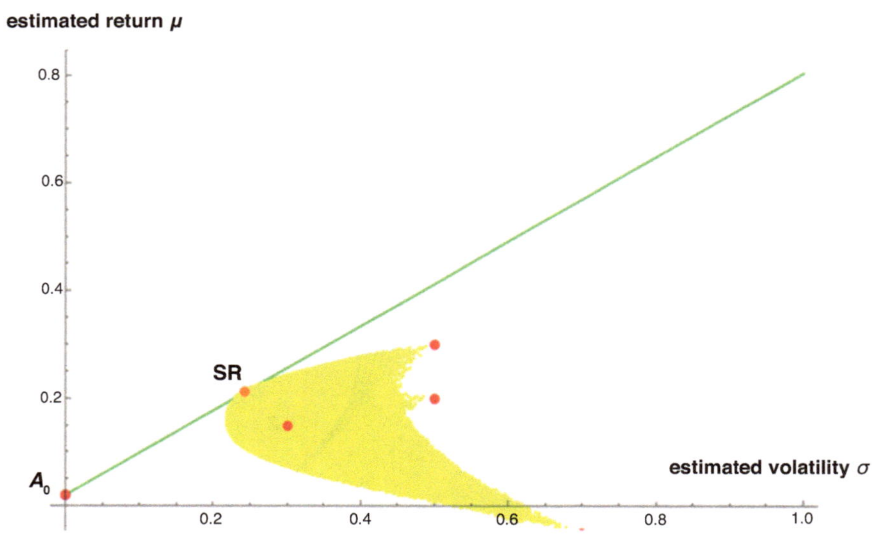

Fig. 2.26 Approximation of the opportunity set and portfolio with maximum Sharpe ratio based on four assets using Monte Carlo simulation, without short selling

are normalized to vectors $(x_1, x_2, \ldots, x_n)$ using $x_i = \frac{y_i}{y_1+y_2+\ldots+y_n}$, such that $x_1+x_2+\ldots+x_n = 1$. For each of these normalized random vectors $(x_1, x_2, \ldots, x_n)$, expected return and volatility are calculated for the associated portfolio $P = A_1x_1 + A_2x_2 + \ldots + A_nx_n$ using the formulas $\mu(P) = \mu_1x_1 + \mu_2x_2 + \ldots + \mu_nx_n$ and $\sigma(P) = \sqrt{\sum_{i,j=1}^{n} x_ix_j\sigma_i\sigma_j\rho_{i,j}}$, and the point $P = (\sigma(P), \mu(P))$ is plotted in the volatility/return diagram. The resulting point cloud (yellow area in Fig. 2.26) gives an approximation for the opportunity set of $A_1, A_2, \ldots, A_n$.

For each of these points, we can also calculate the Sharpe ratio $\frac{\mu(P)-r}{\sigma(P)}$ and select the point SR, i.e. the one with the highest Sharpe ratio among all scenarios, as the portfolio with an approximate maximal Sharpe ratio (orange point in Fig. 2.26). The green connecting line between A_0 and SR in Fig. 2.26 contains all combinations of A_0 and SR.

The determinants for the example shown in Figs. 2.26, 2.27, and 2.28 are four assets with the following (volatility, return) coordinates:

(50%, 20%)
(50%, 30%)
(70%, −5%)
(30%, 15%)

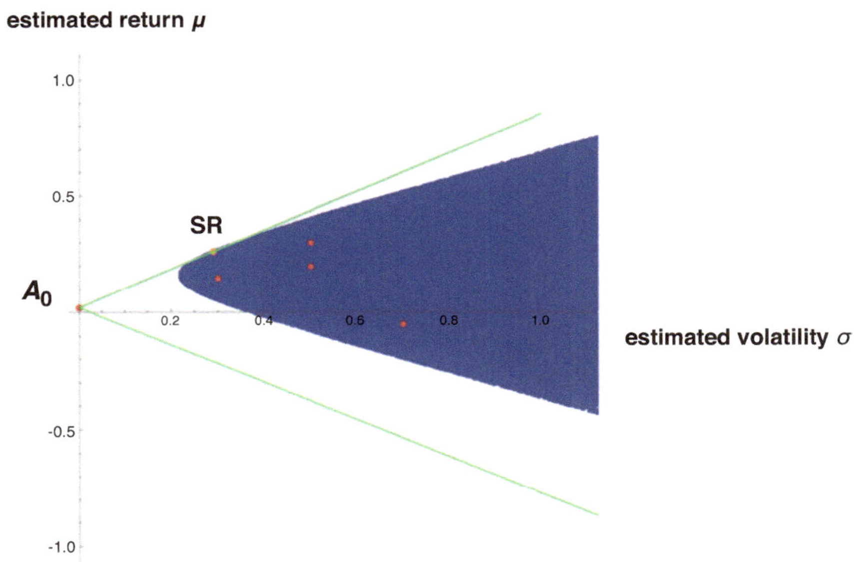

Fig. 2.27 Approximation of the opportunity set and portfolio with maximum Sharpe ratio based on four assets using Monte Carlo simulation, with unlimited short selling

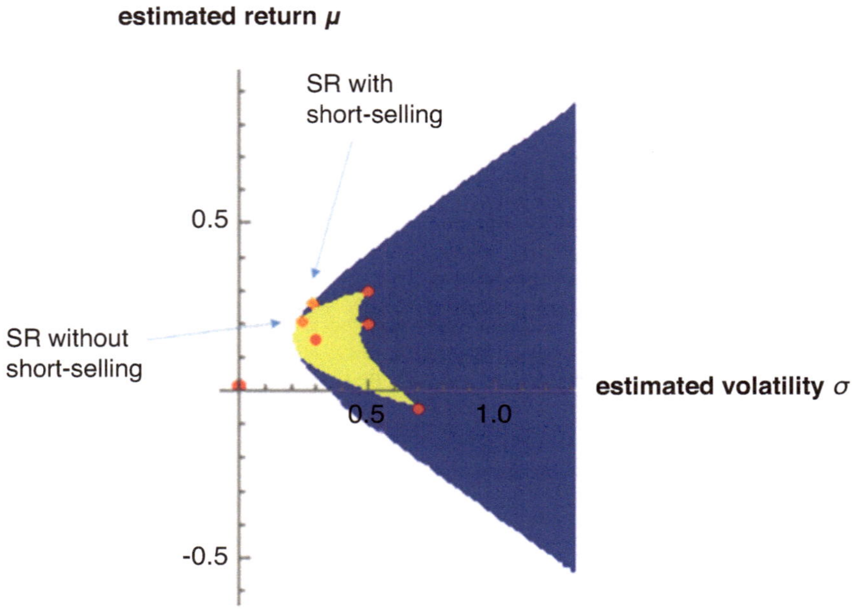

Fig. 2.28 Comparison of the size of opportunity sets with short selling (blue) and without short selling (yellow) and of the position of portfolios with approximately maximal Sharpe ratio (orange)

and with correlation matrix

$$\begin{pmatrix} 1 & 0 & 0.2 & -0.1 \\ 0 & 1 & 0.3 & 0.1 \\ 0.2 & 0.3 & 1 & -0.2 \\ -0.1 & 0.1 & -0.2 & 1 \end{pmatrix}$$

risk-free interest rate $r = 2\%$

We can proceed analogously for the case of unlimited short selling. The only difference is that we do not start with random vectors $(y_1, y_2, \ldots, y_n)$ with $0 \leq y_i \leq 1$, but with arbitrary y_i. To determine the opportunity set and efficient border approximately, it suffices in practice that we choose y_i in the range $-3 \leq y_i \leq 5$, for example, to get an idea of the basic shape of the opportunity set, which is now unbounded to the right, however.

In Fig. 2.27 we see the shape of a portion of the opportunity set, which in reality is unbounded to the right (blue area), as well as the resulting portfolio SR (orange point) with the approximately maximum Sharpe ratio, and again the connecting line between A_0 and SR as an approximate representation of the efficient border.

The graph in Fig. 2.28 shows the ratio of opportunity set in the case with short selling (blue, unbounded to the right) to the size of the opportunity set in the case without short selling (yellow) and also the respective position of the portfolio with (approximately) maximal Sharpe ratio. This approximately maximal Sharpe ratio is 0.789 in the case without short selling and 0.843 in the case with short selling.

The **efficient border and the portfolio with maximal Sharpe ratio** (if it exists) **can be determined explicitly** in the **case with unlimited short selling** by simple extreme-value calculation, and this is in fact what we will do in the next sections.

In the case without short selling or with limited short selling, we can no longer use conventional extreme-value calculation. Using convex optimization methods, we could determine the efficient border and the portfolio with maximal Sharpe ratio more precisely than by the Monte Carlo simulation performed above. However, Monte Carlo simulation is much more accessible than the optimization method and provides results that are sufficiently accurate (considering that the model's real Achilles heel is the reliability of the required parameter estimates). We will therefore not address these optimization methods for the portfolio selection case without short selling.

2.9 Classic Portfolio Optimization According to Markowitz, Part 9: Arbitrary Number of Assets, Explicit Calculation of Efficient Border, and Portfolio with Maximal Sharpe Ratio: Planning the Procedure

In the following sections, we will explicitly compute the efficient border of the portfolios of n assets $A_1, A_2, \ldots, A_n$ as well as the portfolio SR with maximal Sharpe ratio (if it exists) for the case with unlimited short selling.

In this section, we use the illustration in Fig. 2.29 to plan how we want to proceed.

2.10 Classic Portfolio Optimization According to Markowitz, Part 10: Arbitrary Number of Assets, Explicit Calculation of Efficient Border, and Portfolio with Maximal Sharpe Ratio

We will now run the program in detail that we sketched above to calculate the efficient border and the portfolio with maximal Sharpe ratio. Let us first recap all the required notations.

The assets available for investment are $A_1, A_2, \ldots, A_n$.

We denote the estimated returns by $\mu_1, \mu_2, \ldots, \mu_n$ and the estimated volatilities by $\sigma_1, \sigma_2, \ldots, \sigma_n$.

The covariances between the returns of A_i and A_j are denoted by c_{ij}. We further set $c_{ii} := \sigma_i^2$. The covariance matrix M is then given by $M := (c_{ij})_{i,j,=1,2,\ldots,n}$.

We have $c_{ij} = \sigma_i \cdot \sigma_j \cdot \rho_{i,j}$, where $\rho_{i,j}$ denotes the correlation between A_i and A_j.

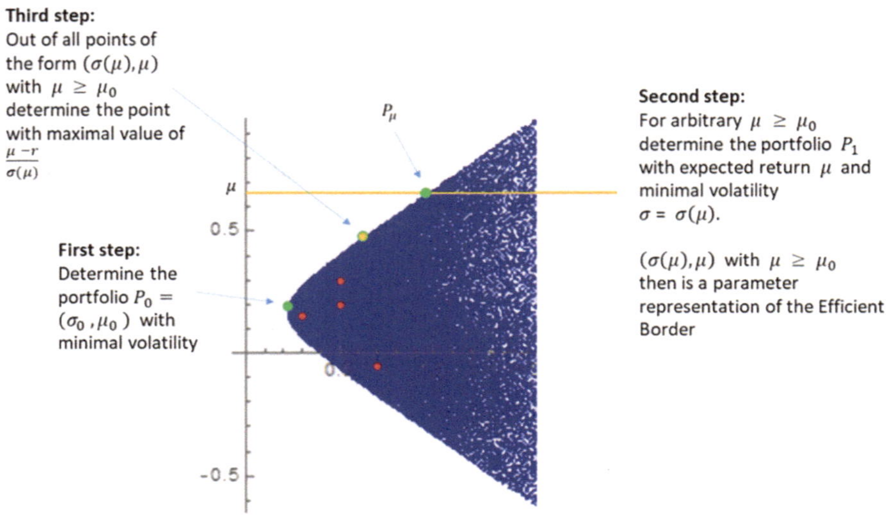

Fig. 2.29 Program for explicit determination of efficient border and portfolio with maximal Sharpe ratio in the case of unlimited short selling

We consider portfolios P of the form $P = x_1A_1 + x_2A_2 + \ldots + x_nA_n$, where $x_1 + x_2 + \ldots + x_n = 1$. This schematic representation means:

"x_i EUR are invested in asset A_i".

The expected return $\mu(P)$ is given by

$$\mu(P) = \mu_1 x_1 + \mu_2 x_2 + \ldots + \mu_n x_n$$

and the expected volatility $\sigma(P)$ of the portfolio P satisfies

$$\sigma(P) = \sqrt{\sum_{i,j=1}^{n} x_i \cdot x_j \cdot c_{ij}}.$$

(We note that for indices $i = j$, we get summands of the form $x_i^2 \cdot \sigma_i^2$. Furthermore, we note that $c_{ij} = c_{ji}$ and that for $i \neq j$, each summand of the form $x_i \cdot x_j \cdot c_{ij}$ thus occurs twice.)

If necessary, instead of $\mu(P)$ or $\sigma(P)$, we will occasionally write

$$\mu(x_1, \ldots, x_n) \text{ or } \sigma(x_1, \ldots, x_n).$$

We assume in the following that $\sigma_i > 0$ holds for all i (i.e. that none of our available assets is risk-free). Furthermore, we assume that we do not obtain a risk-free asset through combinations of $A_1, A_2, \ldots, A_n$, which means that $\sigma(P) > 0$ always holds.

Step 1

Our first task is to determine the portfolio P_0, i.e. the portfolio with the smallest volatility. Since $\sigma(x_1, \ldots, x_n) > 0$, this task is effectively equivalent to minimizing the square of the volatility, i.e. of $(\sigma(x_1, \ldots, x_n))^2$, which saves us manipulations of the root in the definition of $\sigma(x_1, \ldots, x_n)$. We also have to take into account the constraint $x_1 + x_2 + \ldots + x_n = 1$. Such extreme-value determinations for functions in several variables with one or more constraints for the variables in the form of equations can often be solved by applying the Lagrange multiplier method.

Excursus (Lagrange Multiplier)

The Lagrange multiplier method can help us find local extrema of functions $F(x_1, \ldots, x_n)$ in several variables subject to one or more constraints of the form $G_1(x_1, \ldots, x_n) = 0, G_2(x_1, \ldots, x_n) = 0, \ldots, G_s(x_1, \ldots, x_n) = 0$. We will present this method only briefly here and not discuss it further. In the applications we have in mind below, unique and exclusive extrema will always exist a priori, and the Lagrangian method will give us what we need in a straightforward way. The method consists of generating the so-called Lagrange function

$$H_{\lambda_1,\lambda_2,\ldots,\lambda_s}(x_1, x_2, \ldots, x_n) := F(x_1, \ldots, x_n) - \sum_{j=1}^{s} G_j(x_1, x_2, \ldots, x_n) \cdot \lambda_j$$

For this function, the n partial derivatives $\frac{dH_{\lambda_1,\lambda_2,\ldots,\lambda_s}(x_1,x_2,\ldots,x_n)}{dx_k}$ for $k = 1, 2, \ldots, n$ are generated and set equal to 0.

These n equations $\frac{dH_{\lambda_1,\lambda_2,\ldots,\lambda_s}(x_1,x_2,\ldots,x_n)}{dx_k} = 0$ for $k = 1, 2, \ldots, n$ together with the s equations $G_j(x_1, \ldots, x_n) = 0$ for $j = 1, 2, \ldots, s$ yield a system of equations (*) of $n + s$ equations in the $n + s$ variables $x_1, x_2, \ldots, x_n$ and $\lambda_1, \lambda_2, \ldots, \lambda_s$.

Under certain smoothness conditions, every local extremum $x_1, x_2, \ldots, x_n$ of the function F that satisfies all the constraints must be the initial part of a solution $(x_1, x_2, \ldots, x_n, \lambda_1, \lambda_2, \ldots, \lambda_s)$ of the system of equations (*).

In our following sample applications, the situation will always be such that the respective system of equations (*) has exactly one solution in each case.

Thus, the Lagrange function for our problem is

$$H_\lambda(x_1, x_2, \ldots, x_n) = \sum_{i,j=1}^{n} x_i \cdot x_j \cdot c_{ij} - \lambda \cdot (x_1 + x_2 + \ldots + x_n - 1).$$

We differentiate $H_\lambda(x_1, x_2, \ldots, x_n)$ with respect to x_k:

$$\frac{dH_\lambda(x_1, x_2, \ldots, x_n)}{dx_k} = 2x_k c_{kk} + 2\sum_{i \neq k} x_i c_{ik} - \lambda.$$

We set all of these derivatives equal to 0 and thus obtain the system of equations

$$\begin{aligned} x_1 c_{11} + x_2 c_{21} + \ldots + x_n c_{n1} &= \frac{\lambda}{2} \\ x_1 c_{12} + x_2 c_{22} + \ldots + x_n c_{n2} &= \frac{\lambda}{2} \\ \ldots \\ x_1 c_{1n} + x_2 c_{2n} + \ldots + x_n c_{nn} &= \frac{\lambda}{2}. \end{aligned}$$

In matrix notation this is simply

$$M \cdot \begin{pmatrix} x_1 \\ x_2 \\ \vdots \\ x_n \end{pmatrix} = \lambda \cdot \begin{pmatrix} \frac{1}{2} \\ \frac{1}{2} \\ \vdots \\ \frac{1}{2} \end{pmatrix} \text{ and therefore } \begin{pmatrix} x_1 \\ x_2 \\ \vdots \\ x_n \end{pmatrix} = \lambda \cdot M^{-1} \begin{pmatrix} \frac{1}{2} \\ \frac{1}{2} \\ \vdots \\ \frac{1}{2} \end{pmatrix}.$$

Here, M^{-1} denotes the inverse matrix to the covariance matrix M.

We denote the vector $M^{-1} \begin{pmatrix} \frac{1}{2} \\ \frac{1}{2} \\ \vdots \\ \frac{1}{2} \end{pmatrix}$ by $\begin{pmatrix} a_1 \\ a_2 \\ \vdots \\ a_n \end{pmatrix}$.

Since $x_1 + x_2 + \ldots + x_n = 1$, this means that $\lambda a_1 + \lambda a_2 + \ldots + \lambda a_n = 1$, so $\lambda = \frac{1}{a_1 + a_2 + \ldots + a_n}$. Consequently,

$$\boldsymbol{x_i = \frac{a_i}{a_1 + a_2 + \ldots + a_n}} \quad \text{for all } i = 1, 2, \ldots, n$$

and we have thus found portfolio P_0.

Substituting these values for $x_1, x_2, \ldots, x_n$ in $\sigma(x_1, \ldots, x_n)$ resp. $\mu(x_1, \ldots, x_n)$, we get the coordinates σ_0 resp. μ_0 of P_0.

Step 2

Now comes the second step in our program. We choose an arbitrary value $\mu > \mu_0$ and now minimize the function $(\sigma(x_1, \ldots, x_n))^2$ subject to the two constraints $x_1 + x_2 + \ldots + x_n = 1$ and $\mu_1 x_1 + \mu_2 x_2 + \ldots + \mu_n x_n = \mu$.

Again we use the Lagrange multiplier method. The Lagrange function $H_{\kappa,\tau}$ here is

$$H_{\kappa,\tau}(x_1, x_2, \ldots, x_n) = \sum_{i,j=1}^{n} x_i \cdot x_j \cdot c_{i,j} - \kappa \cdot (x_1 + x_2 + \ldots + x_n - 1) - $$
$$- \tau \cdot (\mu_1 x_1 + \mu_2 x_2 + \ldots + \mu_n x_n - \mu)$$

We differentiate $H_{\kappa,\tau}(x_1, x_2, \ldots, x_n)$ with respect to x_k:

$$\frac{dH_{\kappa,\tau}(x_1, x_2, \ldots, x_n)}{dx_k} = 2x_k c_{kk} + 2\sum_{i \neq k} x_i \cdot c_{ik} - \kappa - \tau \cdot \mu_k.$$

Again we set these equations equal to 0 and thus get the system of equations (presented here in matrix notation right away):

$$M \cdot \begin{pmatrix} x_1 \\ x_2 \\ \vdots \\ x_n \end{pmatrix} = \kappa \cdot \begin{pmatrix} \frac{1}{2} \\ \frac{1}{2} \\ \vdots \\ \frac{1}{2} \end{pmatrix} + \tau \cdot \begin{pmatrix} \frac{\mu_1}{2} \\ \frac{\mu_2}{2} \\ \vdots \\ \frac{\mu_n}{2} \end{pmatrix}, \quad \text{thus}$$

$$\begin{pmatrix} x_1 \\ x_2 \\ \vdots \\ x_n \end{pmatrix} = \kappa \cdot \begin{pmatrix} a_1 \\ a_2 \\ \vdots \\ a_n \end{pmatrix} + \tau \cdot \begin{pmatrix} b_1 \\ b_2 \\ \vdots \\ b_n \end{pmatrix},$$

where $\begin{pmatrix} b_1 \\ b_2 \\ \vdots \\ b_n \end{pmatrix}$ denotes the vector $\begin{pmatrix} b_1 \\ b_2 \\ \vdots \\ b_n \end{pmatrix} := M^{-1} \cdot \begin{pmatrix} \frac{\mu_1}{2} \\ \frac{\mu_2}{2} \\ \vdots \\ \frac{\mu_n}{2} \end{pmatrix}$.

We now introduce further notations (don't worry: at the end of this section, we will provide a compiled overview of all the notations that are actually needed to determine the efficient border and the portfolio with maximal Sharpe ratio):

$$\alpha := a_1 + a_2 + \ldots + a_n$$
$$\beta := b_1 + b_2 + \ldots + b_n$$
$$\gamma := \mu_1 a_1 + \mu_2 a_2 + \ldots + \mu_n a_n$$
$$\delta := \mu_1 b_1 + \mu_2 b_2 + \ldots + \mu_n b_n$$

The two constraints $x_1 + x_2 + \ldots + x_n = 1$ and $\mu_1 x_1 + \mu_2 x_2 + \ldots + \mu_n x_n = \mu$ then lead to the system of equations

$$\alpha \cdot \kappa + \beta \cdot \tau = 1$$
$$\gamma \cdot \kappa + \delta \cdot \tau = \mu$$

with the solution

$$\kappa = \frac{\beta \cdot \mu - \delta}{\beta \cdot \gamma - \alpha \cdot \delta} \text{ and } \tau = \frac{\alpha \cdot \mu - \gamma}{\alpha \cdot \delta - \beta \cdot \gamma}.$$

We now set (more new notations):

$$A := \frac{\beta}{\beta \cdot \gamma - \alpha \cdot \delta}, \; B := \frac{-\delta}{\beta \cdot \gamma - \alpha \cdot \delta},$$
$$C := \frac{\alpha}{\alpha \cdot \delta - \beta \cdot \gamma} \text{ and } D := \frac{-\gamma}{\alpha \cdot \delta - \beta \cdot \gamma},$$

then $\kappa = A \cdot \mu + B$ and $\tau = C \cdot \mu + D$ and thus

$$\begin{pmatrix} x_1 \\ x_2 \\ \vdots \\ x_n \end{pmatrix} = \mu \cdot \left(A \cdot \begin{pmatrix} a_1 \\ a_2 \\ \vdots \\ a_n \end{pmatrix} + C \cdot \begin{pmatrix} b_1 \\ b_2 \\ \vdots \\ b_n \end{pmatrix} \right) + \left(B \cdot \begin{pmatrix} a_1 \\ a_2 \\ \vdots \\ a_n \end{pmatrix} + D \cdot \begin{pmatrix} b_1 \\ b_2 \\ \vdots \\ b_n \end{pmatrix} \right), \text{ hence}$$

$$\begin{pmatrix} x_1 \\ x_2 \\ \vdots \\ x_n \end{pmatrix} = \begin{pmatrix} \mu \cdot \xi_1 + \eta_1 \\ \mu \cdot \xi_2 + \eta_2 \\ \vdots \\ \mu \cdot \xi_n + \eta_n \end{pmatrix} \quad \text{where} \quad \xi_i := A \cdot a_i + C \cdot b_i \quad \text{and}$$

$\eta_i := B \cdot a_i + D \cdot b_i$.

And with this we have determined, for a given μ, the portfolio P_μ with expected return μ and (under this assumption) minimal volatility $\sigma(\mu)$.

Substituting $\begin{pmatrix} x_1 \\ x_2 \\ \vdots \\ x_n \end{pmatrix} = \begin{pmatrix} \mu \cdot \xi_1 + \eta_1 \\ \mu \cdot \xi_2 + \eta_2 \\ \vdots \\ \mu \cdot \xi_n + \eta_n \end{pmatrix}$ in $\sigma(x_1, \ldots, x_n)$ yields

$$\sigma(\mu) = \sigma(x_1, \ldots, x_n) = \sqrt{\sum_{i,j=1}^{n} (\mu \cdot \xi_i + \eta_i) \cdot (\mu \cdot \xi_j + \eta_j) \cdot c_{i,j}}$$

$$= \sqrt{\mu^2 \cdot \sum_{i,j=1}^{n} c_{i,j} \cdot \xi_i \cdot \xi_j + \mu \cdot \sum_{i,j=1}^{n} c_{i,j} \cdot (\xi_i \cdot \eta_j + \xi_j \cdot \eta_i) + \sum_{i,j=1}^{n} c_{i,j} \cdot \eta_i \cdot \eta_j}$$

$$=: \sqrt{\mu^2 \cdot E + \mu \cdot F + E}, \text{ where we set}$$

$E := \sum_{i,j=1}^{n} c_{i,j} \cdot \xi_i \cdot \xi_j$,
$F := \sum_{i,j=1}^{n} c_{i,j} \cdot (\xi_i \cdot \eta_j + \xi_j \cdot \eta_i)$ and
$G := \sum_{i,j=1}^{n} c_{i,j} \cdot \eta_i \cdot \eta_j$

The **efficient border** is now been given through the **parameter representation**

$$(\sigma(\mu), \mu) = \left(\sqrt{\mu^2 \cdot E + \mu \cdot F + E}, \mu\right) \text{ for } \mu \geq \mu_0.$$

Step 3
The last step in our program is to determine the portfolio with maximal Sharpe ratio $\frac{\mu - r}{\sigma}$. Since we can restrict our search to portfolios that lie on the efficient border, we just need to find the one μ for which $\frac{\mu - r}{\sigma(\mu)}$ is maximal.

Being that only portfolios with positive Sharpe ratios can be considered optimal, we can restrict ourselves to $\mu > r$ and to maximizing $\frac{(\mu-r)^2}{(\sigma(\mu))^2} = \frac{(\mu-r)^2}{\mu^2 \cdot E + \mu \cdot F + G}$ for $\mu > r$.

We recall that, by premise, the denominator $(\sigma(\mu))^2 = \mu^2 \cdot E + \mu \cdot F + G$ is strictly greater than 0 for all μ.

For later purposes, we will now go a bit further and not only determine the maximum of the function $f(\mu) := \frac{(\mu-r)^2}{\mu^2 \cdot E + \mu \cdot F + G}$ but also discuss the function f in some detail.

So we know that the function f is always positive for $\mu > r$. For $\mu = r$, it has the value 0.

If μ tends to $+\infty$, then $f(\mu)$ tends to $\frac{1}{E}$.

Since $\mu^2 \cdot E + \mu \cdot F + G$ is always positive, E and thus $\frac{1}{E}$ have to be positive.

We now differentiate the function $f(\mu)$ with respect to μ and get the derivative

$$f'(\mu) = \frac{\mu - r}{(\mu^2 \cdot E + \mu \cdot F + G)^2} \cdot (\mu \cdot (F + 2r \cdot E) + (2G + r \cdot F)).$$

The first factor is always positive for $\mu > r$. The second factor is monotonically increasing if $F + 2r \cdot E \geq 0$ and monotonically decreasing if $F + 2r \cdot E \leq 0$.

Note that the derivative of $(\sigma(\mu))^2$ is $(\mu^2 \cdot E + \mu \cdot F + G)' = 2\mu \cdot E + F$. So $F + 2r \cdot E$ is exactly the derivative of $(\sigma(\mu))^2$ at $\mu = r$.

The second factor $(\mu \cdot (F + 2r \cdot E) + (2G + r \cdot F))$ in the derivative of $f(\mu)$ is equal to 0 for $\mu = \tilde{\mu} := -\frac{2G + r \cdot F}{F + 2r \cdot E}$.

So if a local extremum of the function $f(\mu)$ exists, then for this $\tilde{\mu}$.

What is relevant now is under which circumstances this $\tilde{\mu}$ is larger than r, that is, when do we have (*) $\quad -\frac{2G + r \cdot F}{F + 2r \cdot E} > r$.

For this we distinguish between three cases:

Case 1 $F + 2r \cdot E = ((\sigma(r))^2)' > 0$

Then $(*) \Leftrightarrow -2G - r \cdot F > r \cdot F + 2r^2 \cdot E \Leftrightarrow r^2 \cdot E + r \cdot F + G < 0$. However, as we know, this cannot be the case.

Thus: If $((\sigma(r))^2)' > 0$, then the only $\tilde{\mu}$ that qualifies for a local extremum is less than or equal to r. Moreover, the factor $(\mu \cdot (F + 2r \cdot E) + (2G + r \cdot F))$ in $f'(\mu)$ is monotonically increasing, that is, positive for $\mu > r$, and thus the function f is monotonically increasing for $\mu \geq r$. So the maximal Sharpe ratio is only attained approximately for $\mu \to \infty$ and tends to $\sqrt{\frac{1}{E}}$. We are therefore dealing with a situation as illustrated in Fig. 2.30.

Case 2 $F + 2r \cdot E = ((\sigma(r))^2)' < 0$

Then $(*) \Leftrightarrow -2G - r \cdot F < r \cdot F + 2r^2 \cdot E \Leftrightarrow r^2 \cdot E + r \cdot F + G > 0$, and that is, by premise, always satisfied.

Thus: In the case $((\sigma(r))^2)' < 0$, we have a local extremum in $\tilde{\mu} > r$, and the dependence of the evolution of the Sharpe ratio on μ must have the form as shown in Fig. 2.31. The maximal Sharpe ratio is therefore attained in $\mu = \tilde{\mu}$ and is greater than $\sqrt{\frac{1}{E}}$.

Case 3 $F + 2r \cdot E = ((\sigma(r))^2)' = 0$

As we will see in a moment, when we discuss the illustrative meaning of the conditions of the three cases for $((\sigma(r))^2)'$, this case means that $\mu_0 = r$ must hold. The derivative of the function f is then $\frac{\mu - r}{(\mu^2 \cdot E + \mu \cdot F + G)^2} \cdot (2G + r \cdot F)$.

If $2G + r \cdot F \neq 0$, then there is no local maximum for $\mu > r$, which means we are inevitably back to the situation of Fig. 2.30.

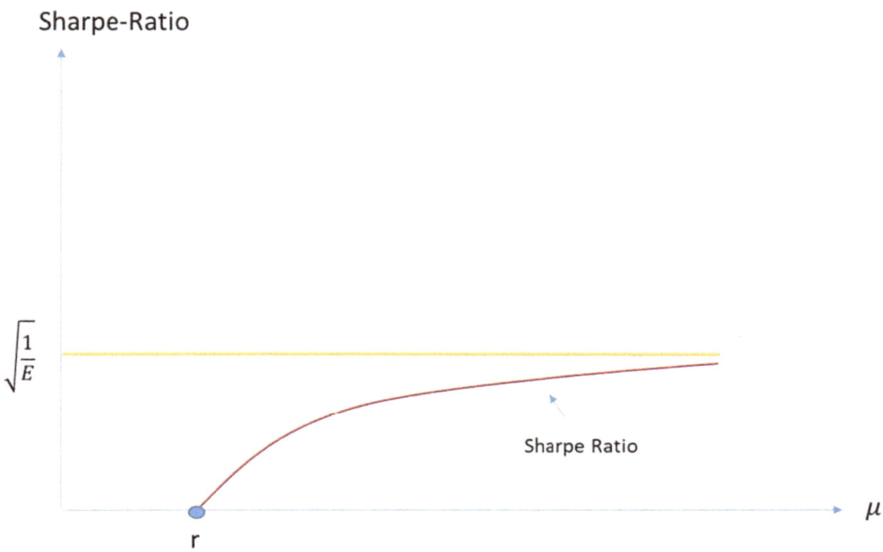

Fig. 2.30 Evolution of Sharpe ratio for $\mu \geq r$ in the case $((\sigma(r))^2)' > 0$

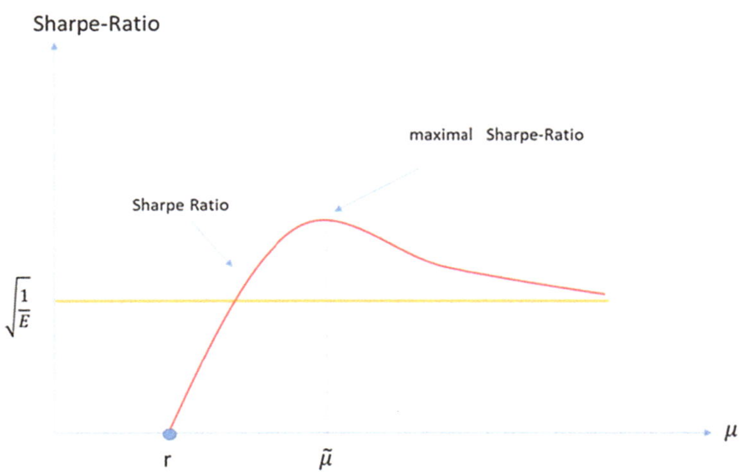

Fig. 2.31 Evolution of Sharpe ratio for $\mu \geq r$ in the case $((\sigma(r))^2)' < 0$

If $2G + r \cdot F = 0$, then, since $F + 2r \cdot E = 0$, we automatically have (as is easily calculated) $r^2 \cdot E + r \cdot F + G = 0$, which contradicts our assumption.

In Other Words If and only if $((\sigma(r))^2)' < 0$, there is one $\tilde{\mu} > r$ in which the maximal Sharpe ratio is attained.

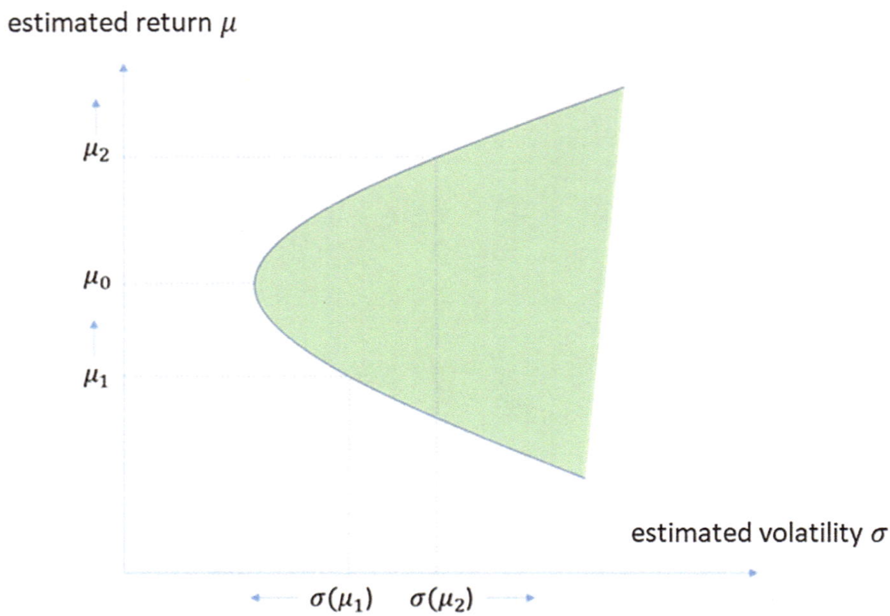

Fig. 2.32 Illustrative meaning of the condition $((\sigma(r))^2)' < 0$

But what exactly does the condition $((\sigma(r))^2)' < 0$ mean, in more illustrative terms? We explain this meaning and the resulting conclusion graphically by means of Fig. 2.32.

By μ_0 we again denote the expected return of the portfolio P_0 with minimal volatility.

If μ_1 is a return smaller than μ_0 and μ_1 continues to increase locally, then $\sigma(\mu_1)$ obviously becomes smaller locally. Therefore, for $\mu_1 < \mu_0$, the volatility $\sigma(\mu_1)$ and hence its square $(\sigma(\mu_1))^2$ is strictly monotonically decreasing and therefore $((\sigma(\mu_1))^2)' < 0$. Especially if $r < \mu_0$, then $((\sigma(r))^2)' < 0$.

If μ_2 is a return greater than μ_0 and μ_2 continues to increase locally, then $\sigma(\mu_2)$ obviously becomes greater locally. Therefore, for $\mu_2 > \mu_0$, the volatility $\sigma(\mu_2)$ and hence its square $(\sigma(\mu_2))^2$ is strictly monotonically increasing and therefore $((\sigma(\mu_2))^2)' > 0$. Especially if $r > \mu_0$ then $((\sigma(r))^2)' > 0$.

If $r = \mu_0$ then $((\sigma(r))^2)' = 0$.

Consequently

- If and only if $r < \mu_0$, we have $((\sigma(r))^2)' < 0$, and then there exists a uniquely determined value $\tilde{\mu}$, such that the portfolio on the efficient border with expected return $\tilde{\mu}$ is the portfolio with maximal Sharpe ratio.

- If $r \geq \mu_0$, then $((\sigma(r))^2)' \geq 0$ and the Sharpe ratio becomes larger and larger as μ increases for portfolios on the efficient border, increasingly approaching $\sqrt{\frac{1}{E}}$.

We summarize all necessary notations below for ease of reference, so that the efficient border and the portfolio with maximal Sharpe ratio can be easily represented with the help of these data.

Summary

- The **portfolio $P_0 = (\sigma_0, \mu_0)$ with minimal volatility** is obtained through $P_0 = x_1 A_1 + x_2 A_2 + \ldots + x_n A_n$ where $x_i = \frac{a_i}{a_1+a_2+\ldots+a_n}$ for all $i = 1, 2, \ldots, n$ and where $\begin{pmatrix} a_1 \\ a_2 \\ \vdots \\ a_n \end{pmatrix} = M^{-1} \begin{pmatrix} \frac{1}{2} \\ \frac{1}{2} \\ \vdots \\ \frac{1}{2} \end{pmatrix}$ and M is the covariance matrix of the investment assets.
  Furthermore, $\mu_0 = x_1\mu_1 + x_2\mu_2 + \ldots + x_n\mu_n$ and $\sigma_0 = \sqrt{\sum_{i,j=1}^n x_i \cdot x_j \cdot c_{i,j}}$.
- The efficient border is given by the parameter representation
  $(\sigma(\mu), \mu) = \left(\sqrt{\mu^2 \cdot E + \mu \cdot F + G}, \mu\right)$ for $\mu \geq \mu_0$, where
  $E := \sum_{i,j=1}^n c_{i,j} \cdot \xi_i \cdot \xi_j$,
  $F := \sum_{i,j=1}^n c_{i,j} \cdot (\xi_i \cdot \eta_j + \xi_j \cdot \eta_i)$,
  $G := \sum_{i,j=1}^n c_{i,j} \cdot \eta_i \cdot \eta_j$,
  where $\xi_i := A \cdot a_i + C \cdot b_i$ and $\eta_i := B \cdot a_i + D \cdot b_i$,
  where $A := \frac{\beta}{\beta\cdot\gamma-\alpha\cdot\delta}$, $B := \frac{-\delta}{\beta\cdot\gamma-\alpha\cdot\delta}$, $C := \frac{\alpha}{\alpha\cdot\delta-\beta\cdot\gamma}$ and $D := \frac{-\gamma}{\alpha\cdot\delta-\beta\cdot\gamma}$,
  where
  $\alpha := a_1 + a_2 + \ldots + a_n$
  $\beta := b_1 + b_2 + \ldots + b_n$
  $\gamma := \mu_1 a_1 + \mu_2 a_2 + \ldots + \mu_n a_n$
  $\delta := \mu_1 b_1 + \mu_2 b_2 + \ldots + \mu_n b_n$, and where
  $$\begin{pmatrix} b_1 \\ b_2 \\ \vdots \\ b_n \end{pmatrix} = M^{-1} \cdot \begin{pmatrix} \frac{\mu_1}{2} \\ \frac{\mu_2}{2} \\ \vdots \\ \frac{\mu_n}{2} \end{pmatrix}.$$
- If $r \geq \mu_0$, then the Sharpe ratio becomes larger and larger as μ increases for portfolios on the efficient border, increasingly approaching the value $\sqrt{\frac{1}{E}}$.
  If $r < \mu_0$, then the portfolio SR with maximal Sharpe ratio has coordinates $SR = \left(\sqrt{\tilde{\mu}^2 \cdot E + \tilde{\mu} \cdot F + G}, \tilde{\mu}\right)$, where $\tilde{\mu} = -\frac{2G+r\cdot F}{(F+2r\cdot E)}$.
  In this case, the portfolio SR has the form $x_1 A_1 + x_2 A_2 + \ldots + x_n A_n$ with $x_i = \tilde{\mu} \cdot \xi_i + \eta_i$ for $i = 1, 2, \ldots, n$.

2.11 Classic Portfolio Optimization According to Markowitz, Part 11: Arbitrary Number of Assets, Explicit Calculation of Efficient Border, and Portfolio with Maximal Sharpe Ratio: Practical Example

In Sect. 2.8 we illustrated approximately, using Monte Carlo simulation, the opportunity set and thus the efficient border as well as the portfolio with maximal Sharpe ratio in the case of unlimited short selling for four assets with the following (volatility, return) coordinates:

$$(\sigma_1, \mu_1) = (50\%, 20\%)$$
$$(\sigma_2, \mu_2) = (50\%, 30\%)$$
$$(\sigma_3, \mu_3) = (70\%, -5\%)$$
$$(\sigma_4, \mu_4) = (30\%, 15\%)$$

and with correlation matrix

$$\begin{pmatrix} 1 & 0 & 0.2 & -0.1 \\ 0 & 1 & 0.3 & 0.1 \\ 0.2 & 0.3 & 1 & -0.2 \\ -0.1 & 0.1 & -0.2 & 1 \end{pmatrix}$$

and with the risk-free interest rate $r = 2\%$.

For the following example, we are now going to calculate exactly, based on the formulas given in the summary at the end of the previous section, the portfolio with minimal volatility, the efficient border, and the portfolio with maximal Sharpe ratio in the case of unlimited short selling, and at the end we will graphically compare the exact result with the approximate result.

We get the following values:

The covariance matrix M is calculated from the correlation matrix and the volatilities as

$$M = \begin{pmatrix} 0.25 & 0 & 0.07 & -0.015 \\ 0 & 0.25 & 0.105 & 0.015 \\ 0.07 & 0.105 & 0.49 & -0.042 \\ -0.015 & 0.015 & -0.042 & 0.09 \end{pmatrix}$$

The inverse M^{-1} of the covariance matrix is given by

$$M^{-1} = \begin{pmatrix} 4.195\ldots & 0.237\ldots & -0.618\ldots & 0.371\ldots \\ 0.237\ldots & 4.541\ldots & -1.113\ldots & -1.236\ldots \\ -0.618\ldots & -1.113\ldots & 2.473\ldots & 1.236\ldots \\ 0.371\ldots & -1.236\ldots & 1.236\ldots & 11.956\ldots \end{pmatrix}.$$

Furthermore we get:

$$a = \begin{pmatrix} 2.092\ldots \\ 1.214\ldots \\ 0.989\ldots \\ 6.163\ldots \end{pmatrix},$$

and thus the **portfolio P_0 with minimal volatility** is given by the weights

$$\begin{pmatrix} x_1 \\ x_2 \\ x_3 \\ x_4 \end{pmatrix} = \begin{pmatrix} 0.200\ldots \\ 0.116\ldots \\ 0.094\ldots \\ 0.590\ldots \end{pmatrix}$$

and has expected return $\mu_0 = 0.158\ldots$ and volatility $\sigma_0 = 0.218\ldots$.

The vector b is given by

$$b = \begin{pmatrix} 0.498\ldots \\ 0.640\ldots \\ -0.197\ldots \\ 0.717\ldots \end{pmatrix}$$

and thus we get the values

$$\begin{aligned} \alpha &= 10.460\ldots \\ \beta &= 1.658\ldots \\ \gamma &= 1.658\ldots \\ \delta &= 0.409\ldots \end{aligned}$$

and subsequently

$$\begin{aligned} A &= -1.082\ldots \\ B &= 0.267\ldots \\ C &= 6.829\ldots \\ D &= -1.082\ldots \end{aligned}$$

We use this to calculate the vectors $\xi = \begin{pmatrix} \xi_1 \\ \xi_2 \\ \xi_3 \\ \xi_4 \end{pmatrix}$ and $\eta = \begin{pmatrix} \eta_1 \\ \eta_2 \\ \eta_3 \\ \eta_4 \end{pmatrix}$ and get

$$\xi = \begin{pmatrix} 1.138\ldots \\ 3.056\ldots \\ -2.422\ldots \\ -1.772\ldots \end{pmatrix} \text{ and } \eta = \begin{pmatrix} 0.019\ldots \\ -0.368\ldots \\ 0.478\ldots \\ 0.870\ldots \end{pmatrix}.$$

The resulting values for E, F and G are then $E = 3.414\ldots$, $F = -1.082\ldots$ and $G = 0.133\ldots$.

The **efficient border** thus has the parameter representation $\left(\sqrt{3.414\ldots \cdot \mu^2 - 1.082\ldots \cdot \mu + 0.133\ldots}, \mu\right)$ with $\mu \geq 0.158\ldots$

The portfolio SR with maximal Sharpe ratio is given by the allocation

$$\begin{pmatrix} x_1 \\ x_2 \\ x_3 \\ x_4 \end{pmatrix} = \begin{pmatrix} 0.315\ldots \\ 0.425\ldots \\ -0.150\ldots \\ 0.410\ldots \end{pmatrix} \text{ and}$$

has expected return 0.287..., volatility 0.259..., and hence Sharpe ratio 1.03.

In Fig. 2.33 we see once again (this time in yellow) the opportunity set with unlimited short selling that was approximated by Monte Carlo simulation. We see that the curve obtained through exact calculation borders the approximated opportunity set on the left. We have plotted the entire boundary curve here, not

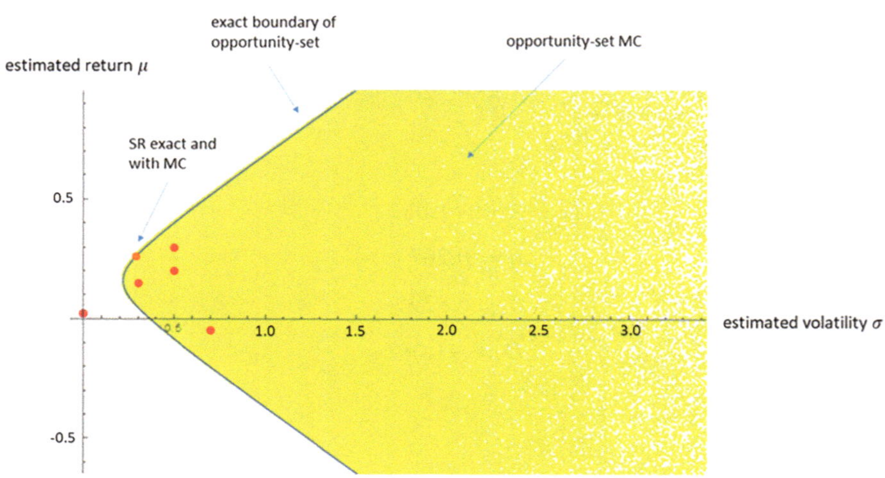

Fig. 2.33 Opportunity set with unlimited short selling approximated with Monte Carlo and boundary and portfolio SR with maximal Sharpe ratio (exact calculation)

just the efficient border from P_0 upward. Additionally, we have also plotted the two portfolios that mark the portfolio with maximal Sharpe ratio in the Monte Carlo simulation and the one in the exact calculation in orange. The two portfolios are almost exactly on top of each other, so that no difference can be detected between the two.

The software on our website provides the entire array of possibilities for determining efficient borders and portfolios with maximal Sharpe ratio for any investment assets. See https://app.lsqf.org/portfolioselektion.

2.12 The "Market Portfolio": The Portfolio with Maximal Sharpe Ratio

A rational investor following the approach of the classic Markowitz portfolio selection theory would invest only in a combination of risk-free investment and portfolio $\boldsymbol{SR}$ with maximal Sharpe ratio (if it exists. Otherwise, a portfolio with approximately maximal Sharpe ratio should be chosen as an approximation. In the following, we are going to assume that $\boldsymbol{SR}$ exists). This portfolio is often referred to as the "**market portfolio**".

If we now proceed, very intuitively and heuristically, from the assumption of an efficient market hypothesis and the assumption that the vast majority of rational investors in the market use very good analytical tools and if, for simplicity, we restrict ourselves to the globally traded stocks as the risky assets in a Markowitz portfolio selection, then with regard to this market portfolio, we could argue—rather generously—more or less as follows:

"The totality of essentially rational market participants, whose transactions on the market determine the prices of (exchange-traded) financial assets, relies on average in those transactions on having the correct parameters (after all, it is their transactions that actually define these parameters!). So, on average, the rational market participant trades the portfolio $\boldsymbol{SR}$ *worldwide with maximal Sharpe ratio based on the 'correct' parameters. Of course, there are also less rational investors out there with deviating trading behavior, but overall these deviations balance each other out.*

Let us now denote the tradable stocks on international stock exchanges by $A_1, A_2, \ldots, A_n$, *their tradable quantities by* $L_1, L_2, \ldots, L_n$, *and their current prices by* $K_1, K_2, \ldots, K_n$ *(for the sake of our generous speculation, let us forego a discussion of the correct use of different currencies here and, for simplicity, just use 'EUR' throughout in the following) and let us assume that the portfolio* $\boldsymbol{SR}$ *is given by* $z_1 A_1 + z_2 A_2 + \ldots + z_n A_n$. *This means: Out of an investment of 1 EUR,* z_i *EUR is invested in stock* A_i. *So* $\frac{z_i}{K_i}$ *shares of stock* A_i *are held in the portfolio* $\boldsymbol{SR}$.

By the above line of reasoning, the totality of stocks held worldwide by (mostly and on average rational) investors would then turn out to be essentially a multiple of the portfolio $\boldsymbol{SR}$.

Let the total number of stocks held worldwide by (mostly and on average rational) investors be given by, for instance, M *units of the portfolio* $\boldsymbol{SR}$. *The capital invested*

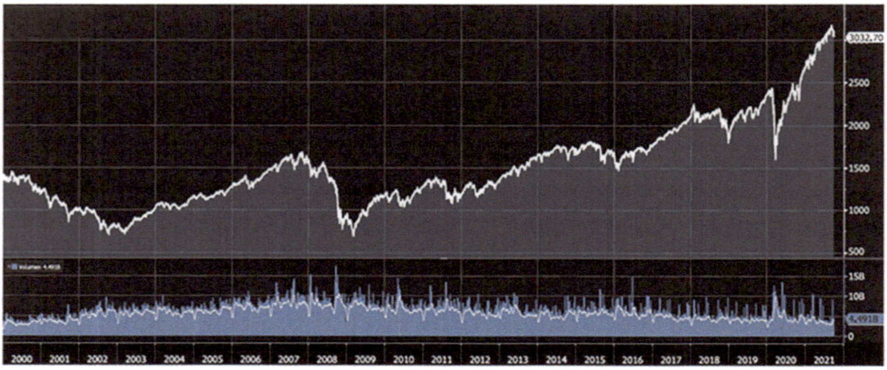

Fig. 2.34 MSCI world index 1990 to October 2021

in stocks worldwide is therefore $L_1K_1 + L_2K_2 + \ldots + L_nK_n$ *on the one hand and* $M \cdot z_1 + M \cdot z_2 + \ldots + M \cdot z_n = M \cdot (z_1 + z_2 + \ldots + z_n) = M$ *on the other hand. Thus we have* $M = L_1K_1 + L_2K_2 + \ldots + L_nK_n$.
The total number L_i *of shares of stock* A_i *on the market is also given by* $M \cdot \frac{z_i}{K_i}$ *(= number of traded* $\boldsymbol{SR}$ *multiplied by the number of shares of* A_i in $\boldsymbol{SR}$*).*
It follows that $L_i = M \cdot \frac{z_i}{K_i}$ *and thus* $z_i = \frac{K_iL_i}{M} = \frac{K_iL_i}{K_1L_1+K_2L_2+\ldots+K_nL_n}$.
If we were to follow this line of reasoning, this would give us the form of the portfolio $\boldsymbol{SR}$ *with maximal Sharpe ratio: We would have to invest just the amount in each stock* A_i *as corresponds to the proportion of market capitalization of stock* A_i *relative to total market capitalization.*
This is why the portfolio $\boldsymbol{SR}$ *with maximal Sharpe ratio is also referred to as the* ***'Market Portfolio'***.”

Certain stock indexes attempt to approximately replicate such a universal market portfolio, from a variety of angles. The best known and probably most important example of such an index is the MSCI World Index, which reflects the performance of 1644 stocks (across regions (industrialized countries) and sectors) that have been selected as representative and weighted according to their market capitalization.

If we follow our reasoning above, then this MSCI World Index could thus be seen as an approximation of a universal market portfolio. The development of the index from 1990 to October 2021 can be found in the Fig. 2.34:

2.13 Portfolio Selection Based on a Single-Index Model

The compellingly beautiful theory of the Markowitz portfolio selection model has one major flaw, which we have already pointed out several times: The usefulness of the model is critically dependent on the underlying parameter estimates for the investment assets under consideration. For each asset, a trend and a volatility must be estimated in a reliable way, and for every two different assets, a correlation estimate is needed.

Apart from the question of stability and "correctness" of these estimates, the task alone of estimating correlations, for example, can quickly reach its limits. For a universe of 1000 assets, for instance, we would have to estimate $\frac{1000{,}999}{2} = 499{,}500$ correlations.

Many extensions as well as simplifications of the model try to overcome the model's parameter dependency, or to make the model practicable. Examples of such approaches are single-index and multi-index models or the Black-Litterman model approach. In the following we will only give an idea of the direction in which these efforts are heading, and we do so by briefly sketching the concept of a single-index model:

We start again with assets $A_1, A_2, \ldots, A_n$ from a certain stock universe (e.g. stocks included in the DAX or stocks of the S&P500, ...). The first assumption we make is as follows: The performance of each asset depends on two factors which are essentially independent of each other, namely, the **general market situation** (within this universe) on the one hand and **purely company-specific criteria** on the other hand. The second assumption is: The first factor is directly represented by the performance of the stock index representing the universe in question.

Explicitly, this dynamic should be of the following form:

We denote by

r_i the return of the stock A_i (random variable)
r_m the return of the corresponding index (market return) (random variable)
r the risk-free interest rate (constant)

Assume then a relationship between these two returns r_i and r_m of the form $r_i = \beta_{i,1} \cdot r_m + \beta_{i,0} + \varepsilon_i$.

Here, $\beta_{i,1}$ and $\beta_{i,0}$ are constants (which are to be estimated), and ε_i is a random variable independent of r_m for which we assume an expected value of 0 ($\beta_{i,0}$ is chosen accordingly). The standard deviation of ε_i is denoted by σ_{ε_i}. The $\varepsilon_i, \varepsilon_j$ are mutually independent for $i \neq j$.

The $\beta_{i,0} + \varepsilon_i$ part in the dynamics of the return of stock A_i reflects the influence of company-specific criteria.

The $\beta_i \cdot r_m$ part reflects the dependence of the return of stock A_i on the general market situation, represented by the return of the market index. A homogeneous linear dependence is assumed.

Assuming the model $r_i = \beta_{i,1} \cdot r_m + \beta_{i,0} + \varepsilon_i$, the parameters $\beta_{i,1}$ and $\beta_{i,0}$ and the standard deviation σ_{ε_i} can be estimated by linear regression. (We will explain the procedure for this below with a practical example.)

Let us now further assume that we have estimates for μ_m and σ_m, the expected value and the standard deviation of r_m, i.e. for the trend and the volatility of the market index; then the expected values μ_i and the standard deviations σ_i of the individual returns r_i as well as the covariances $\sigma_{i,j}$ between the returns of the stocks A_i and A_j can be estimated as follows:

We have

$$r_i = \beta_{i,1} \cdot r_m + \beta_{i,0} + \varepsilon_i$$

and therefore (because of $E(\varepsilon_i) = 0$)

$$\boldsymbol{\mu_i = \beta_{i,1} \cdot \mu_m + \beta_{i,0}}.$$

Furthermore (because of the independence of r_m and ε_i)

$$\boldsymbol{{\sigma_i}^2 = {\beta_{i,1}}^2 \cdot {\sigma_m}^2 + {\sigma_{\varepsilon_i}}^2}.$$

For the covariances, we obtain for $i \neq j$ (because of the independence of ε_i and ε_j from each other and from r_m)

$$\begin{aligned}\boldsymbol{\sigma_{i,j}} &= E((r_i - \mu_i) \cdot (r_j - \mu_j)) = \\ &= E((\beta_{i,1} \cdot (r_m - \mu_m) + \varepsilon_i) \cdot (\beta_{j,1} \cdot (r_m - \mu_m) + \varepsilon_j)) = \\ &= \boldsymbol{\beta_{i,1} \cdot \beta_{j,1} \cdot {\sigma_m}^2}.\end{aligned}$$

We thus have all the necessary parameters to perform portfolio selection according to Markowitz.

If we now work with, say, 1000 assets, it is no longer necessary to estimate 499,500 correlations. All we need to do is estimate the parameters μ_m and σ_m of a representative market index and for each i the values $\beta_{j,0}$, $\beta_{j,1}$ and σ_{ε_i} (thus a total of 3002 parameters).

To illustrate this, we use the example given in Sect. 2.6 and consider an investment of 1 (= 1 million) USD for the period of 1 year from today (7 May 2019 to 7 May 2020) in the following assets:

A_1 = Ford stock
A_2 = Apple stock

as well as in

A_0 = a 1-year fixed-rate investment at an interest rate of 2%.

In Sect. 2.6 we estimated the expected per-annum returns μ_1 and μ_2 and the expected per-annum standard deviations σ_1 and σ_2 and the correlation ρ between the returns of these two stocks based on the daily closing prices of the immediately preceding trading year and got:

$\mu_1 = 0.1627$ (corresponding to 16.27%)
$\sigma_1 = 0.2948$ (corresponding to 29.48%)
$\mu_2 = -0.0465$ (corresponding to −4.65%)
$\sigma_2 = 0.2962$ (corresponding to 29.62%)
$\rho = 0.1728$

We now do an alternative estimation of these parameters based on the single-factor model set out above, using the S&P500 as the market index. We again base the analysis on daily data from 7 May 2018 to 7 May 2019.

In Figs. 2.35 and 2.36, we see the pairs of points (x_i, y_i) and (x_i, z_i), respectively, with x_i for $i = 1, 2, \ldots, n$ denoting the daily returns of the SPX, y_i for $i = 1, 2, \ldots, n$ denoting the daily returns of the Apple stock, and z_i for $i = 1, 2, \ldots, n$ denoting the daily returns of the Ford stock in the trading year from 7 May 2018 to 7 May 2019.

In addition, the two regression lines $y = b_{a,0} + b_{a,1} \cdot x$ and $y = b_{f,0} + b_{f,1} \cdot x$ for the two point clouds are plotted in red in both figures.

The regression line $y = b_{a,0} + b_{a,1} \cdot x$ is the line for which the sum of the squared perpendicular distances of the points (x_i, y_i) from the line is minimal. (Analogously, of course, $y = b_{f,0} + b_{f,1} \cdot x$ is the line for which the sum of the squared perpendicular distances of the points (x_i, z_i) from the line is minimal.)

The values $b_{a,0}$ or $b_{a,1}$ are obtained by means of

$$b_{a,1} = \frac{\sum_{i=1}^{n} (x_i - \bar{x}) \cdot (y_i - \bar{y})}{\sum_{i=1}^{n} (x_i - \bar{x})^2} \quad \text{and} \quad b_{a,0} = \bar{y} - b_{a,1} \cdot \bar{x},$$

where $\bar{x}$ denotes the average value of x_i and $\bar{y}$ denotes the average value of y_i.

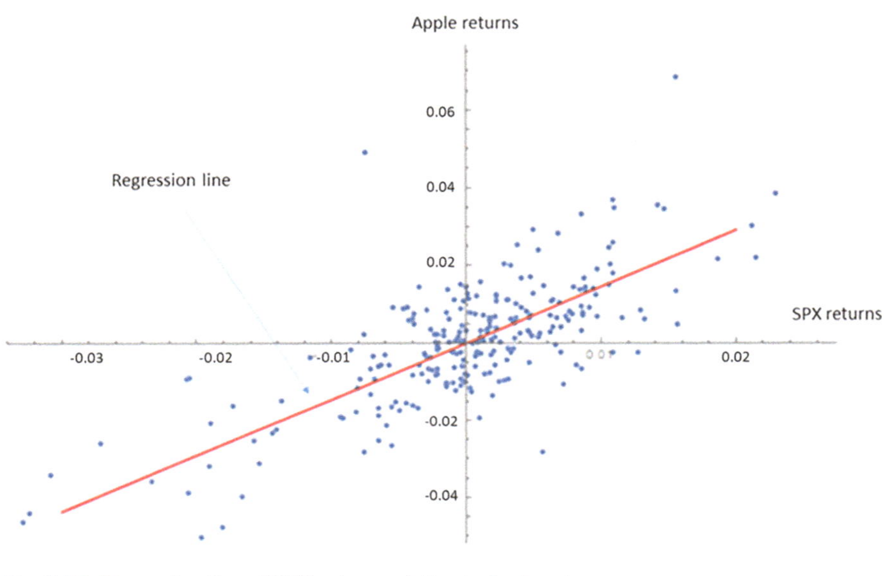

Fig. 2.35 Regression line of SPX returns relative to Apple returns

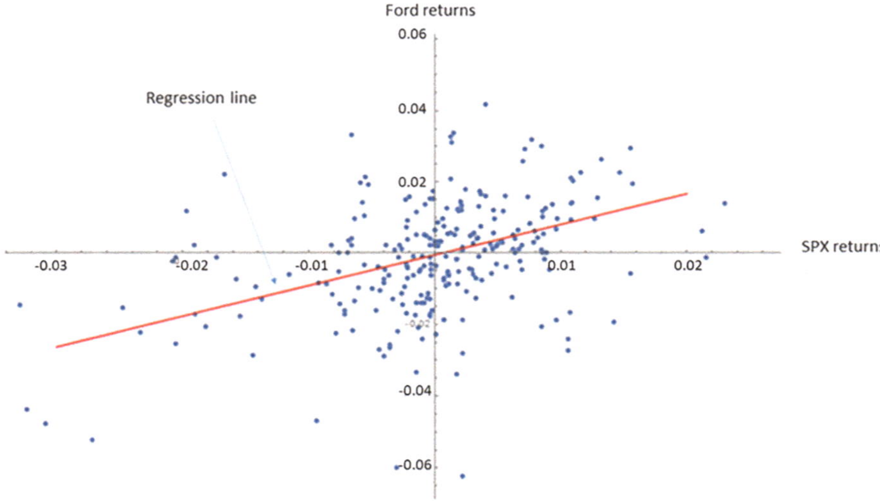

Fig. 2.36 Regression line of SPX returns relative to Ford returns

Note that the form of the parameters b_0 and b_1 (we omit the index a and f here) follows from the following simple extreme-value problem:

We minimize the sum of squared distances of the points along the y-axis, that is, we seek to find b_0 and b_1 such that $\sum_{i=1}^{n}(y_i - (b_0 + b_1 \cdot x_i))^2$ becomes minimal. We differentiate the expression with respect to b_0 and b_1 and set the two partial derivatives equal to 0. We get

$$\frac{d}{db_0} = -2 \cdot \sum_{i=1}^{n}(y_i - (b_0 + b_1 \cdot x_i)) = -2 \cdot n \cdot (\bar{y} - b_0 - b_1 \cdot \bar{x}) = 0$$

$$\frac{d}{db_1} = -2 \cdot \sum_{i=1}^{n} x_i \cdot (y_i - (b_0 + b_1 \cdot x_i)) =$$

$$= -2 \cdot \sum_{i=1}^{n} x_i \cdot y_i + 2 \cdot b_0 \cdot n \cdot \bar{x} + 2 \cdot b_1 \cdot \sum_{i=1}^{n} x_i^2 = 0$$

(continued)

From the first equation, we already get $b_0 = \bar{y} - b_1 \cdot \bar{x}$. Substituting this expression for b_0 in the second equation, we get

$$-\sum_{i=1}^{n} x_i \cdot y_i + n \cdot \bar{x} \cdot \bar{y} - b_1 \cdot n \cdot \bar{x}^2 + b_1 \cdot \sum_{i=1}^{n} x_i^2 = 0 \text{ and thus}$$

$$b_1 = \frac{\sum_{i=1}^{n} x_i \cdot y_i - n \cdot \bar{x} \cdot \bar{y}}{\sum_{i=1}^{n} x_i^2 - n \cdot \bar{x}^2}.$$

This last expression—as one can easily calculate—is precisely the above $\frac{\sum_{i=1}^{n}(x_i-\bar{x})\cdot(y_i-\bar{y})}{\sum_{i=1}^{n}(x_i-\bar{x})^2}$.

The straight lines generated by the linear regression $y = b_{a,0} + b_{a,1} \cdot x$ resp. $y = b_{f,0} + b_{f,1} \cdot x$ now represent the parts $\beta_{a,1} \cdot r_m + \beta_{a,0}$ resp. $\beta_{f,1} \cdot r_m + \beta_{f,0}$ in the representations $r_a = \beta_{a,1} \cdot r_m + \beta_{a,0} + \varepsilon_a$ resp. $r_f = \beta_{f,1} \cdot r_m + \beta_{f,0} + \varepsilon_f$ of the returns r_a and r_f of the Apple and Ford stock, respectively. We can therefore set $b_{a,1} = \beta_{a,1}$, $b_{a,0} = \beta_{a,0}$, $b_{f,1} = \beta_{f,1}$, and $b_{f,0} = \beta_{f,0}$.

If we substitute our data for the SPX, the Apple stock, and the Ford stock, we get the values

$$b_{a,1} = \beta_{a,1} = 1.4576$$
$$b_{a,0} = \beta_{a,0} = 0.000044$$
$$b_{f,1} = \beta_{f,1} = 0.8574$$
$$b_{f,0} = \beta_{f,0} = -0.00054$$

Now we determine the standard deviations σ_{ε_a} and σ_{ε_f} of the disturbance terms ε_a and ε_f in the case of Apple stock and Ford stock, respectively. These can be estimated by the empirical standard deviations of the differences $y_i - (b_{a,0} + b_{a,1} \cdot x_i)$ and $z_i - (b_{f,0} + b_{f,1} \cdot x_i)$, respectively, for $i = 1, 2, \ldots, n$. With our values we get:

$\sigma_{\varepsilon_a} = 0.01258$ and thus ${\sigma_{\varepsilon_a}}^2 = 0.000158$
$\sigma_{\varepsilon_f} = 0.01687$ and thus ${\sigma_{\varepsilon_f}}^2 = 0.000285$

Now we only need estimates for μ_m and σ_m, the expected value and the standard deviation of the SPX returns. Again, we estimate these values simply from the values of the SPX in the time period 7 May 2018 to 7 May 2019 and get

$$\mu_m = 0.000415$$
$$\sigma_m = 0.0094$$

Remember that in all of the above, we worked only with daily parameters!

Now, with the help of the formulas derived above

$$\mu_i = \beta_{i,1} \cdot \mu_m + \beta_{i,0}$$
$$\sigma_i{}^2 = \beta_{i,1}{}^2 \cdot \sigma_m{}^2 + \sigma_{\varepsilon_i}{}^2$$

and

$$\sigma_{i,j} = \beta_{i,1} \cdot \beta_{j,1} \cdot \sigma_m{}^2$$

we can obtain the estimates for $\mu_a, \mu_f, \sigma_a, \sigma_f$ and for the covariance $\sigma_{a,f}$. For the time being, we continue to use the daily values and only annualize them at the end.

We get

$\mu_a = \beta_{a,1} \cdot \mu_m + \beta_{a,0} = 0.0006489$, annualized 0.16547
$\mu_f = \beta_{f,1} \cdot \mu_m + \beta_{f,0} = -0.000184$, annualized -0.0469
$\sigma_a^2 = \beta_{a,1}{}^2 \cdot \sigma_m{}^2 + \sigma_{\varepsilon_a}{}^2 = 0.0003457$,
so that $\sigma_a = 0.01859$, annualized 0.2969
$\sigma_f{}^2 = \beta_{f,1}{}^2 \cdot \sigma_m{}^2 + \sigma_{\varepsilon_f}{}^2 = 0.0003499$,
so that $\sigma_f = 0.01871$, annualized 0.2987

and finally
$\sigma_{a,f} = \beta_{a,1} \cdot \beta_{f,1} \cdot \sigma_m{}^2 = 0.0001104$, which corresponds to a correlation of $\rho = \frac{\sigma_{a,f}}{\sigma_a \cdot \sigma_f} = 0.317$.

We see that the parameters estimated in this way are very similar to those originally estimated purely from historical data. Only the correlation of 0.317 is now significantly higher than the 0.1728 estimated earlier.

Figure 2.37 compares the opportunity sets and the efficient borders for the Apple and Ford stock example when parameters are estimated directly from historical data

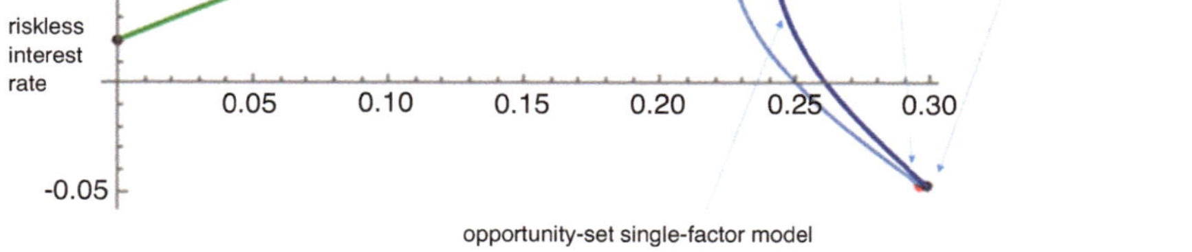

Fig. 2.37 Comparison of opportunity sets and efficient borders by direct parameter estimation and by parameter estimation in the single-factor model

(as we did in Sect. 2.6) and when parameters are estimated in the single-factor model using the SPX. The efficient border remains virtually unchanged; the opportunity set is slightly less curved in the single-factor model.

In our portfolio selection theory software, parameter estimation can also be done on the basis of a single-factor model. See https://app.lsqf.org/portfolioselektion.

2.14 The Optimal Investment and Consumption Problem: Introduction and Formulation of the Problem

The last topic that we will address in this introduction to the basic principles of quantitative finance and financial engineering—before we proceed to look at select case studies—relates to the "classic" optimal investment and consumption problem. In the context of this problem (first solved by Robert Merton), we will also present an introduction (which will again be highly intuitive and heuristic) to a basic technique of stochastic control theory, the Hamilton-Jacobi-Bellman equation.

Informally speaking, our issue can be described as follows (we will specify it in more detail right afterwards):

We want to invest a certain amount 1 over an upcoming period $[0, T]$.

The available investment options are a risk-free investment B *(a bond with a fixed interest rate* r*) and a risky portfolio* S*. (For example,* S *may be an individual stock, but in light of what has been said in the preceding sections, we could also take* S *to mean the market portfolio.) Furthermore, it is also possible to use parts of the investment (or any investment profits) for consumption during the investment period. The question (in very informal terms) is: What is the optimal way to dynamically (!) allocate the available investment to* B, S *and consumption over the investment period?*

To clarify that question in more precise terms, we are going to proceed step by step. In principle, we will use stochastic notation.

So, we have a bond B with price $B(t)$ and the dynamics

$$dB = r \cdot B \cdot dt$$

and a portfolio S with price $S(t)$, for which we assume the dynamics

$$dS(t) = \alpha \cdot S(t) \cdot dt + \beta \cdot S(t) \cdot dW$$

with a Brownian motion W and (for simplicity) constants α and β. S thus follows a geometric Brownian motion.

We model the possible consumption over the investment period $[0, T]$ by a time-dependent non-negative consumption rate $c(t)$.

$c(t)$ is a random variable for every t. How much we are going to consume at a given time t in the future depends on the performance of our assets, i.e. on our trading strategy and the performance of our portfolio S until time t. Thus, we assume that the stochastic process $(c(t))_{t\in[0,T]}$ is adapted to the filtration of

Brownian motion. $c(t)$ denotes a consumption rate per annum. This means: In a tiny time range $[t, t + \Delta t]$, we consume approximately $c(t) \cdot \Delta t$ EUR. (As usual, Δt is expressed in years.)

By $X(t)$ we denote the total value of the investment at time t.

At time t we hold a share of $1 - u(t)$ of the total value $X(t)$ in the bond and a share of $u(t)$ of the total value $X(t)$ in the portfolio S. Expressed in EUR, we hold bond shares worth $(1 - u(t)) \cdot X(t)$ EUR and shares in the portfolio S worth $u(t) \cdot X(t)$ EUR at time t. Expressed in number of units, we hold, at time t, a total of $\frac{(1-u(t))\cdot X(t)}{B(t)}$ units of bond B and $\frac{u(t)\cdot X(t)}{S(t)}$ units of the portfolio S.

The dynamics of the total value $X(t)$ (we will refer to $X(t)$ as the value process in the following) is then

$$\begin{aligned} dX(t) &= \frac{u(t) \cdot X(t)}{S(t)} \cdot dS(t) + \frac{(1 - u(t)) \cdot X(t)}{B(t)} \cdot dB(t) - c(t) \cdot dt = \\ &= X(t) \cdot (u(t) \cdot \alpha + (1 - u(t)) \cdot r)dt - c(t)dt + X(t) \cdot u(t) \cdot \beta \cdot dW(t). \end{aligned}$$

So we see that in our investment process we can freely choose at any time t the consumption rate $c(t)$ and the share $u(t)$ of the value process $X(t)$ that is invested in the portfolio S.

The actual performance of the value process $X(t)$ is thus essentially dependent on c and on u. We will emphasize this dependence (where essential) in the following by using the more explicit notation $X_{c,u}(t)$ instead of $X(t)$.

Our task now is to come up with a strategy for choosing c and u such that the investment and consumption process in the time period $[0, T]$ yields an "optimal result". For this purpose, however, we first need to clarify how we want to measure the "quality of the investment and consumption process in the time period $[0, T]$".

In principle—in Merton's approach—this "quality" is assessed on the basis of two components.

First, on the basis of the consumption $c(t)$ over the period $[0, T]$. This is done by means of an expression of the form $\int_0^T F(c(t))dt$, which should be as large as possible, with a function F, which in principle can be chosen arbitrarily but is typically chosen in a specific form that we will discuss in more detail in the following. F will definitely be a positive, monotonically increasing function, since, in general, the result of a consumption (of a purchased good) is perceived as positive and thus the result of higher consumption is perceived as more positive than the result of lower consumption.

The second component of the "quality of the investment and consumption process in the period $[0, T]$" is generally made dependent on how much of the investment is still available at time T, i.e. after investment and consumption in the period $[0, T]$. This is done by means of an expression of the form $\Phi(X_{c,u}(T))$, which should also be as large as possible. The function Φ (which can be chosen arbitrarily in principle) is also monotonically increasing. Higher terminal wealth $X(T)$ is deemed more positive than lower terminal value.

Thus, the "quality of a specific outcome (path) of the investment and consumption process in the time period $[0, T]$" is measured by an expression of the form

$$\int_0^T F(c(t))dt + \Phi(X_{c,u}(T)).$$

However, due to the stochasticity of $S(t)$, this measure is a priori a random variable (even for deterministically chosen c and u). Thus we obviously wish for the **expected value** of this measure to turn out as large as possible.

The goal then is to determine the consumption strategy c and the trading strategy u such that the expected value of the above measure is maximized. The task is therefore: Determine

$$\max_{c,u} E\left(\int_0^T F(c(t))dt + \Phi(X_{c,u}(T))\right),$$

taking the maximum over all allowable consumption strategies c and trading strategies u. We will discuss which strategies are "admissible" here in more detail later. So far we have one admissibility condition, namely, that $c(t) \geq 0$ needs to be satisfied at all times.

The two parts of the measure, $\int_0^T F(c(t))dt$ and $\Phi(X_{c,u}(T))$, obviously work "against each other". A consistently high consumption rate (and thus a high value of $\int_0^T F(c(t))dt$) implies a rather low terminal wealth (and thus a lower value of $\Phi(X_{c,u}(T))$) and vice versa.

A "large" or "strongly increasing" function F relative to the Φ function values large consumption more positively and as more important than high terminal wealth, and vice versa. The relation of the two deterministically chosen functions F and Φ to each other therefore depends on the respective investor's individual preferences (does that investor attach more importance to successful consumption or to "safety" in the form of high terminal wealth).

In the literature and in practical applications, the chosen function F is often a concave function and, more specifically, a function of the form $F(c(t)) = c(t)^\gamma$ with a value γ with $0 < \gamma < 1$ or else a logarithmic function $F(c(t)) = \log(1+c(t))$. We refer to such a function F as a "utility function". It is used to measure the "utility" (usefulness) of a given consumption rate $c(t)$ at a given time t.

Figure 2.38 shows some of these functions. The picture might be a bit misleading, however, so to be clear: Each of the functions $c(t)^\gamma$ with $0 < \gamma < 1$ eventually grows faster towards infinity than the function $\log(1 + c(t))$, even if for small γ the function $c(t)^\gamma$ overtakes the function $\log(1 + c(t))$ only at a very late point (as can be seen, e.g. in Fig. 2.39).

What is the reason for choosing concave functions as utility functions? In general, a given increment of consumed goods increases utility more when the consumer has acquired only a few goods so far rather than many goods. For instance, the purchase of the first car for the household brings more benefit to a (private) consumer than the purchase of the fifth car, and even the purchase of the fifth

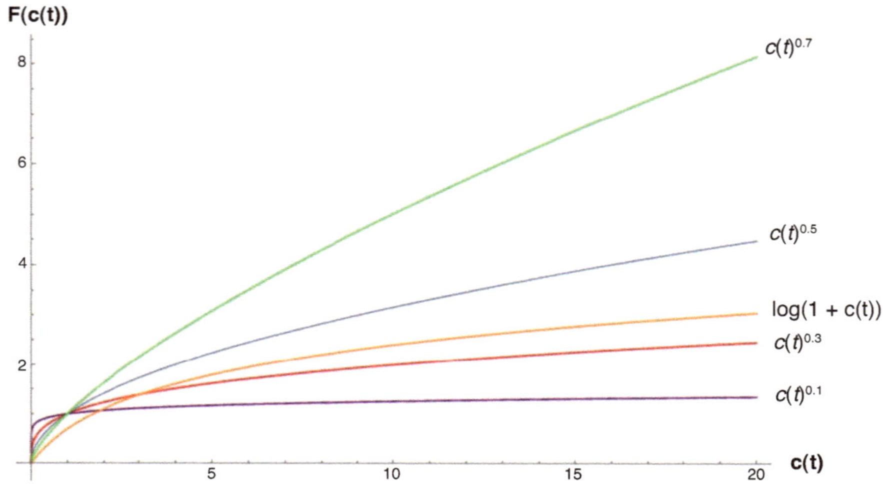

Fig. 2.38 Concave functions (utility functions)

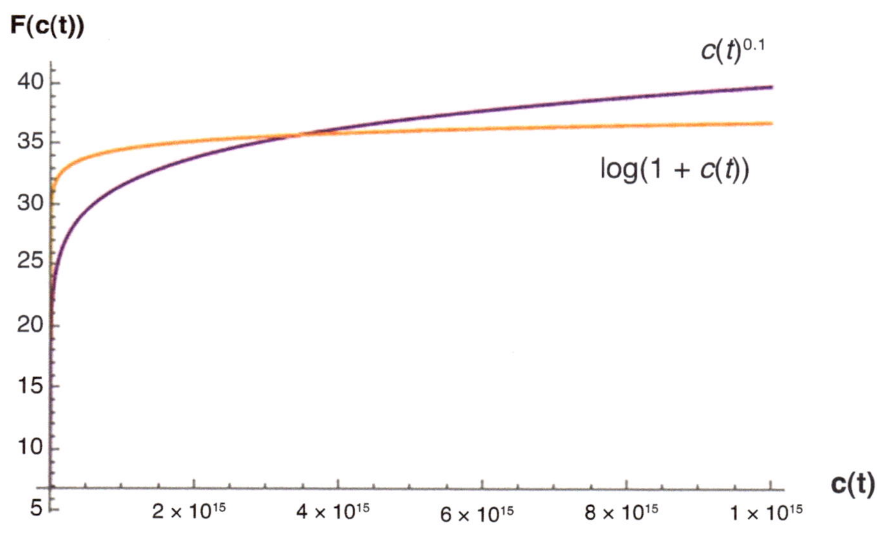

Fig. 2.39 Concave functions (utility functions)

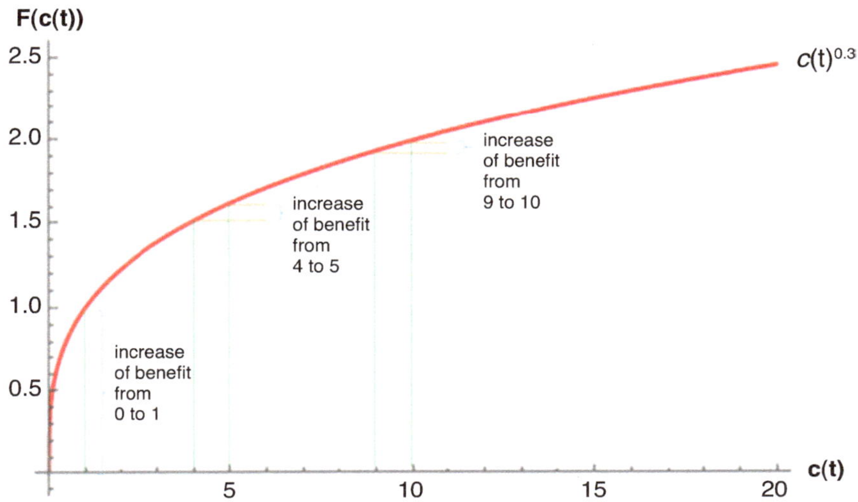

Fig. 2.40 Decrease in incremental utility as $c(t)$ increases

car brings more benefit than the purchase of the tenth car (there are already nine cars anyway, so it is very questionable whether a tenth car is of much additional use). And that is precisely why concave utility functions make sense, because they account for this fact, as shown in Fig. 2.40.

In actual applications, the utility function F is often defined in a somewhat more complex form, as also depending directly on time t, so in the form $F(t, c(t))$ instead of $F(c(t))$.

A frequently chosen version (which is also the one we will work with in the following) is:

$$F(t, c(t)) = e^{-\delta \cdot t} \cdot c(t)^{\gamma}$$

Here, δ is a given positive constant. One possible justification for this choice could be: Earlier consumption represents higher utility because it can be used over a longer period of time. The utility of a later consumption (large t) is reduced in value by a factor $e^{-\delta \cdot t}$.

So the task that we are going to solve in a later section is:

Determine

$$\max_{c,u} E\left(\int_0^T e^{-\delta \cdot t} \cdot c(t)^{\gamma} dt + \Phi(X_{c,u}(T))\right),$$

taking the maximum over all allowable consumption strategies c and trading strategies u.

Another admissibility condition that we will require (in addition to $c \geq 0$) is that the value process $X_{c,u}$ is always greater than or equal to 0. If $X_{c,u}$ becomes 0 at any one point, investments and consumption are no longer possible and the process ends.

Furthermore, we also need to determine the optimal admissible strategies c and u of course.

A variant of this task will be the following: We again consider the time interval $[0, T]$ as the investment interval and introduce a random variable τ. This random variable τ depends on the values of the random variable $X_{c,u}$ and is defined as "the first point in time t at which $X_{c,u}(t)$ is equal to 0 (or $\tau = T$ if $X_{c,u}$ remains positive at all times)". In this variant we then determine

$$\max_{c,u} E\left(\int_0^\tau e^{-\delta \cdot t} \cdot c(t)^\gamma \, dt\right).$$

Thus, the upper bound in the integral is a random variable. The condition $X_{c,u} \geq 0$ can then be ignored, since it is already included in this definition of the problem.

And of course we also need to determine the optimal admissible strategies c and u again.

In the next sections, we will show, heuristically and intuitively, how such types of problems can be solved in principle using techniques from stochastic control theory, and in the process, we will derive (again using heuristic reasoning) the Hamilton-Jacobi-Bellman equations.

2.15 Fundamentals of Stochastic Control Theory, the HJB Equations

We restate the stochastic control problem—which we defined above in the context of the optimal consumption investment problem—in more general terms.

In a general stochastic control problem, the task is to determine the following value

$$\max_{\boldsymbol{u}} E\left(\int_0^T F(t, X^{(\boldsymbol{u})}(t), \boldsymbol{u}(t, X^{(\boldsymbol{u})}(t)))dt + \Phi(X^{(\boldsymbol{u})}(T))\right)$$

or

$$\max_{\boldsymbol{u}} E\left(\int_0^\tau F(t, X^{(\boldsymbol{u})}(t), \boldsymbol{u}(t, X^{(\boldsymbol{u})}(t)))dt\right).$$

Here,

- $\boldsymbol{u} = \boldsymbol{u}(t, x)$ denotes a k-dimensional stochastic process;
  $\boldsymbol{u}$ is called control process.

- $X^{(\boldsymbol{u})}(t)$ denotes an n-dimensional stochastic process with dynamics of the form

$$dX^{(\boldsymbol{u})}(t) = \mu(t, X^{(\boldsymbol{u})}(t), \boldsymbol{u})dt + \sigma(t, X^{(\boldsymbol{u})}(t), \boldsymbol{u})d\boldsymbol{W}(t),$$

where $\boldsymbol{W}$ is a d-dimensional Brownian motion and with functions

$$\mu : [0, T] \times \mathbb{R}^n \times \mathbb{R}^k \to \mathbb{R}^n$$

$$\sigma : [0, T] \times \mathbb{R}^n \times \mathbb{R}^k \to \mathbb{R}^{n\times d}$$

$X(0) = x_0$ is given.
- F denotes any function $F : [0, T] \times \mathbb{R}^n \times \mathbb{R}^k \to \mathbb{R}$.
- Φ denotes any function $\Phi : \mathbb{R}^n \to \mathbb{R}$.
- τ denotes any stopping time.

For the control process $\boldsymbol{u}$, an admissibility condition $\boldsymbol{u}(t, x) \in U_t$ is often required for all $t \in [0, T]$ and $x \in R^n$, where U_t is a particular subset of $\mathbb{R}^k$ for each t.

(For example, in the optimal consumption investment problem above, $c(t) \geq 0$ was required for all t, so that the U_t for this control process was $U_t = \mathbb{R}_0^+$ for all t.)

For all occurring processes, we assume unrestricted Ito integrability in the following, and for all occurring functions, we assume the existence and continuity of all required derivatives. Of course, in a mathematically exact derivation of the following steps, all arguments need to be verified exactly.

In classic extreme-value problems, the challenge is to select certain real parameters $x, y, z, \ldots$ occurring in a function F in such a way that the function attains maximum or minimum values. The preferred choice for solving these problems is to differentiate the function F in a suitable way and set these derivatives to zero.

For problems of the classic calculus of variation, one is supposed to maximize or minimize a function F (or rather a functional F) that depends on other functions $f, g, h, \ldots$. So what we are looking for are the functions $f, g, h, \ldots$ for which F attains maximal values. The central tool for solving such problems is the Euler-Lagrange equation. In a stochastic control theory problem, the fundamental tool for solving it is the Hamilton-Jacobi-Bellman (HJB) equation, which we will now derive heuristically.

For this purpose we define the functions

$$J_0(\boldsymbol{u}) := E\left(\int_0^T F(t, X^{(\boldsymbol{u})}(t), \boldsymbol{u}(t, X^{(\boldsymbol{u})}(t)))dt + \Phi(X^{(\boldsymbol{u})}(T))\right)$$

and

$$J(t, x, \boldsymbol{u}) := E_{t,x}\left(\int_t^T F(s, X^{(\boldsymbol{u})}(s), \boldsymbol{u}(s, X^{(\boldsymbol{u})}(s)))ds + \Phi(X^{(\boldsymbol{u})}(T))\right).$$

(continued)

Here we denote by $E_{t,x}$ the expected value from the perspective of time t and assuming that $X^{(\boldsymbol{u})}(t) = x$. Also, $J_0(\boldsymbol{u}) = J(0, x_0, \boldsymbol{u})$.

Further, we define $V(t, x) := \max_{\boldsymbol{u}} J(t, x, \boldsymbol{u})$.

We refer to the problem of determining the variable $V(t, x)$ and the associated optimal control process(es) $\boldsymbol{u}$ as $Pro(t, x)$.

Our goal is to determine $V(0, x_0)$ and the $\boldsymbol{u}$ for which the maximum is attained in $J(0, x_0, \boldsymbol{u})$, i.e. to solve $Pro(0, x_0)$.

V is a function $V : [0, T] \times \mathbb{R}^n \to \mathbb{R}$. In the following we also assume continuous differentiability for V, if needed.

We further assume that, for each choice of t and x, there is indeed an optimal $\boldsymbol{u}$ for the problem $Pro(t, x)$.

We now fix an arbitrary time/location point (t, x) and consider the time/location point $(t + h, X^{(\boldsymbol{u})}(t + h))$ for a small positive value h and an **arbitrarily selected** control function $\boldsymbol{u}$ within the time period $[t, t + h]$. It is assumed here that the condition $X^{(\boldsymbol{u})}(t) = x$ holds for the process $X^{(u)}$. The exact value $X^{(\boldsymbol{u})}(t+h)$ is not yet known from the point of view of time t, i.e. it is a random variable and can take any possible value z. For $(t+h, X^{(\boldsymbol{u})}(t+h))$, we therefore write $(t + h, z)$ for short, with a random variable z.

We consider the problem $Pro(t+h, z)$ from the point of view of a possible time/location point $(t + h, z)$ and assume that the solution to this problem is the control process $\hat{u}^{(z)}$.

We now use the following control process u^* in the time range $[t, x]$ to $[T, X^{(u^*)}(T)]$:

$$u^*(s, y) = \begin{cases} \boldsymbol{u}(s, y) & \text{for } t \le s \le t + h \\ \hat{u}^{(z)} & \text{for } t + h \le s \le T \end{cases}$$

where $z := X^{(\boldsymbol{u})}(t + h)$, and **u** is an arbitrary control process.

The situation is sketched in Fig. 2.41.

Finally, let $\hat{\hat{u}}$ denote the optimal control process for the $Pro(t, x)$. Then $V(t, x) = J(t, x, \hat{\hat{u}}) \ge J(t, x, u^*)$, and if $\hat{\hat{u}} = u^*$, then equality holds.

It follows that:

$$V(t, x) \ge J(t, x, u^*) = E_{t,x}\left(\int_t^{t+h} F(s, X^{(\boldsymbol{u})}(s), \boldsymbol{u}(t, X^{(\boldsymbol{u})}(t)))ds + \right.$$
$$\left. + V(t + h, X^{(\boldsymbol{u})}(t + h))\right) \tag{2.1}$$

We now apply the Ito formula to the expression $V := V(t + h, X^{(\boldsymbol{u})}(t + h))$. For simplicity, we are going to assume a one-dimensional Brownian motion W and will therefore use only the one-dimensional Ito formula. The approach

(continued)

in the multidimensional case is analogous but is more complex in terms of writing.

We recall: $dX^{(\boldsymbol{u})}(t) = \mu(t, X^{(\boldsymbol{u})}(t), \boldsymbol{u})dt + \sigma(t, X^{(\boldsymbol{u})}(t), \boldsymbol{u})d\boldsymbol{W}(t)$

In shorthand we therefore have:

$$dV = (V_s + \mu \cdot V_x + \frac{\sigma^2}{2} \cdot V_{xx}) \cdot dt + \sigma \cdot V_x \cdot dW$$

and thus, in integral notation, starting in (t, x):

$$V(t+h, X^{(\boldsymbol{u})}(t+h)) = V(t,x) + \int_t^{t+h} \left(V_s + \mu \cdot V_x + \frac{\sigma^2}{2} \cdot V_{xx}\right) ds + \\ + \int_t^{t+h} \sigma \cdot V_x dW.$$

It follows that

$$E_{t,x}(V(t+h), X^{(\boldsymbol{u})}(t+h)) = V(t,x) + \\ + E_{t,x}\left(\int_t^{t+h} \left(V_s + \mu \cdot V_x + \frac{\sigma^2}{2} \cdot V_{xx}\right) ds\right)$$

and substituting this equality in Formula (2.1), we get

$$0 \geq E_{t,x}\left(\int_t^{t+h} \left(F + V_s + \mu \cdot V_x + \frac{\sigma^2}{2} \cdot V_{xx}\right) ds\right).$$

For our next steps, we write this last inequality again in long form:

$$0 \geq E_{t,x}\Bigg(\int_t^{t+h} \Big(F(s, X^{(\boldsymbol{u})}(s), \boldsymbol{u}(s, X^{(\boldsymbol{u})}(s))) + V_s(s, X^{(\boldsymbol{u})}(s)) + \\ + \mu(s, X^{(\boldsymbol{u})}(s), \boldsymbol{u}) \cdot V_x(s, X^{(\boldsymbol{u})}(s)) + \\ + \frac{\sigma(s, X^{(\boldsymbol{u})}(s), \boldsymbol{u})^2}{2} \cdot V_x(s, X^{(\boldsymbol{u})}(s))\Big) ds\Bigg)$$

Here, we now divide by (the positive) h and let h converge to zero. This gives us:

$$0 \geq E_{t,x}\Bigg(F(t, x, \boldsymbol{u}(t,x)) + V_t(t,x) + \mu(t, x, \boldsymbol{u}(t,x)) \cdot V_x(t,x) + \\ + \frac{\sigma(t, x, \boldsymbol{u}(t,x))^2}{2} \cdot V_{xx}(t,x)\Bigg).$$

(continued)

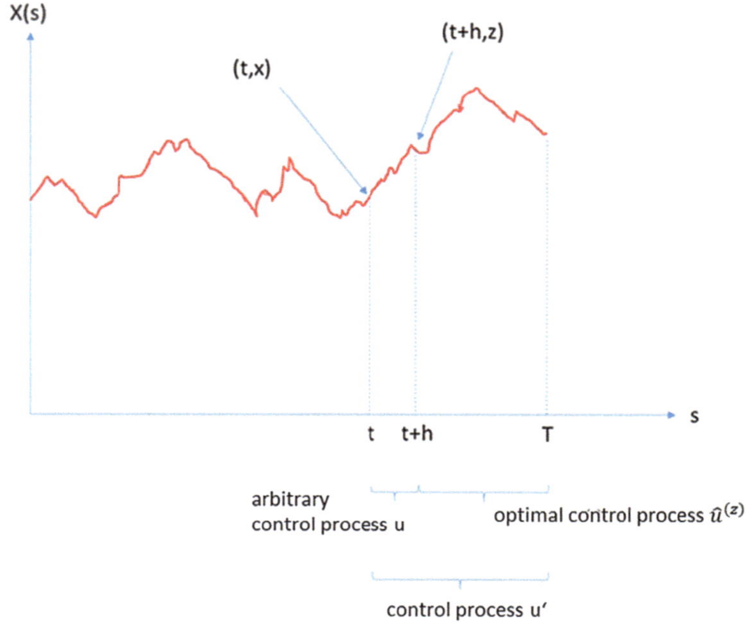

Fig. 2.41 Sketch for deriving the HJB equations

In this inequality the expectation value $E_{t,x}$ is no longer needed, and we get

$$0 \geq F(t,x,\boldsymbol{u}(t,x)) + V_t(t,x) + \mu(t,x,\boldsymbol{u}(t,x)) \cdot V_x(t,x) + \frac{\sigma(t,x,\boldsymbol{u}(t,x))^2}{2} \cdot V_{xx}(t,x),$$

where equality holds if $\boldsymbol{u}(t,x) = \hat{\hat{u}}(t,x)$. (Remember: $\hat{\hat{u}}$ is the optimal control process for the problem $Pro(t,x)$!)

So for all t and x, we have:

$$0 = V_t(t,x) + \sup_u \left[F(t,x,u) + \mu(t,x,u) \cdot V_x(t,x) + \frac{\sigma(t,x,u)^2}{2} \cdot V_{xx}(t,x) \right]$$

and for $u = \hat{\hat{u}}(t,x)$, the supremum is attained.

We have thus proved the following theorem, if not exactly, but at least plausibly:

Theorem 2.1 (Hamilton-Jacobi-Bellman Equation) *If, for all $t \in [0, T]$, the stochastic control problem*

$$J(t, x, \boldsymbol{u}) := E_{t,x}\left(\int_t^T F(s, X^{(\boldsymbol{u})}(s), \boldsymbol{u}(s, X^{(\boldsymbol{u})}(s)))ds + \Phi(X^{(\boldsymbol{u})}(T))\right) \rightarrow \max$$

with

$$dX^{(\boldsymbol{u})}(s) = \mu(s, X^{(\boldsymbol{u})}(s), \boldsymbol{u})ds + \sigma(s, X^{(\boldsymbol{u})}(s), \boldsymbol{u})dW(s) \text{ and } X^{(\boldsymbol{u})}(s) = x$$

and with $\boldsymbol{u}(t, x) \in U_t$ for all $t \in [0, T]$ and $x \in \mathbb{R}^n$ has a solution $\hat{\boldsymbol{u}}$ and if $V(t, x) := J(t, x, \hat{\boldsymbol{u}}) \in C^{1,2}$, then it follows that:

$$0 = V_t(t, x) + \sup_{\boldsymbol{u}}\left[F(t, x, \boldsymbol{u}) + \mu(t, x, \boldsymbol{u}) \cdot V_x(t, x) + \frac{\sigma(t, x, \boldsymbol{u})^2}{2} \cdot V_{xx}(t, x)\right],$$

where the following constraint must be satisfied for V: $V(T, x) = 0$ for all $x \in \mathbb{R}^n$.

For all $(t, x) \in [0, T] \times R^n$, this supremum is attained for $\boldsymbol{u} = \hat{\boldsymbol{u}}(t, x)$.

If we are to determine the minimum instead of the maximum of $J(t, x, \boldsymbol{u})$, then the same equation applies but with an infimum instead of the supremum.

Just as in the case of a simple classic extreme-value problem, we obviously should check here too whether the solutions $\boldsymbol{u}$ and V of the above HJB equation are indeed solutions of the control problem. There is also a "verification theorem" for this, which states specific technical conditions under which this is always the case. We will not discuss this verification theorem any further here. More details can be found in [3] or [4].

In the following section, we will demonstrate with a simple example how the HJB equation can be applied.

2.16 An Application Example for the HJB Equation: The Linear Regulator

Before we use the HJB equations to solve the optimal consumption investment problem, let us apply the HJB equations to a simple example, for demonstration purposes.

The starting point of this problem is a one-dimensional stochastic process X following an Orenstein-Uhlenbeck model. Thus:

$$dX(t) = a \cdot X(t)dt + \sigma \cdot dW(t) \text{ and } X(0) = 0$$

Time t runs from 0 to T.

We now want the process $X(t)$ to move, on average, as closely as possible to the t-axis over that period of time.

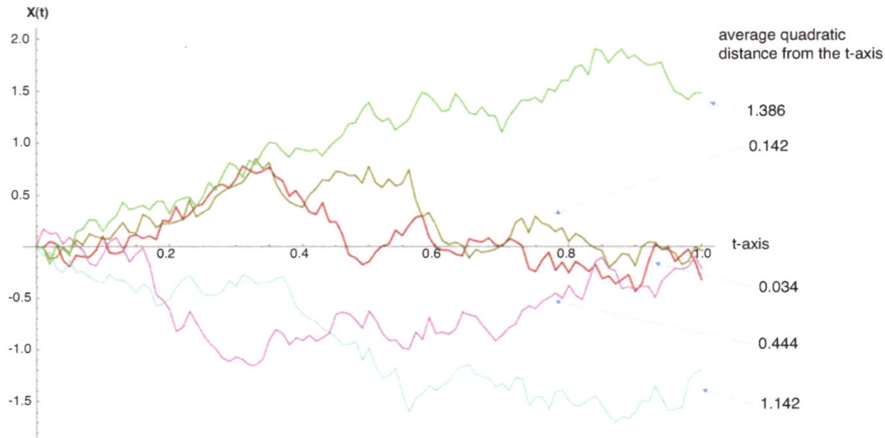

Fig. 2.42 Linear regulator: five unregulated paths with average squared distances from the t-axis

More precisely: We look at average **squared** distances and want to measure this average squared distance using $\int_0^T X(t)^2 dt$. However, the actual evolution of $X(t)$ is purely random, and so we can only wait and see what happens and what average distance results for individual realizations. In Fig. 2.42 we plotted a few realizations of the process

$$dX(t) = X(t)dt + dW(t) \text{ and } X(0) = 0$$

on the interval $[0, 1]$ and noted the respective average distance from the t-axis.

So in Fig. 2.42, the red path would be the one that would most closely match what we want. However, as noted above, this is a purely random evolution, and we can only hope in each case that a specific realization will have a shape more like the red path than the green path.

We will now assume, however, that we can intervene and influence the shape of the stochastic process by means of a control process $u(t, x)$ (for instance, by "slowing down" or "steering" the trend of process X).

We then obtain a new controlled process, which we denote by $X^{(u)}$.

More precisely: We assume that we can intervene in the trend of process X and adapt it at each point in time such that the controlled process $X^{(u)}$ has the dynamics

$$dX^{(u)}(t) = (a \cdot X^{(u)}(t) + u(t, X^{(u)}(t))dt + \sigma \cdot dW(t) \text{ and } X^{(u)}(0) = 0$$

The average squared distance we obtain for such a process is then given by $\int_0^T (X^{(u)}(t))^2 dt$.

However, intervening in the process X requires energy. Let the energy expended at time t be exactly proportional to $(u(t, X^{(u)}(t)))^2$, i.e. the energy expended in a tiny time period from t to $t + \Delta t$ is approximately

$(u(t, X^{(u)}(t)))^2 \cdot \Delta t$. Therefore, the total energy expended during the time period $[0, T]$ is $\int_0^T (u(t, X^{(u)}(t)))^2 dt$.

The aim now is to minimize average squared distance plus expended energy. So, the task at hand is:
Determine

$$\min_u E\left(\int_0^T ((X^{(u)}(t))^2 + (u(t, X^{(u)}(t)))^2)dt\right)$$

and we are therefore looking at a typical stochastic control problem. The function $F(t, x, u)$ is given by $F(t, x, u) = x^2 + u^2$. The corresponding HJB equation is therefore (in short form)

$$0 = V_t + \inf_u \left(x^2 + u^2 + (ax + u) \cdot V_x + \frac{\sigma^2}{2} \cdot V_{xx}\right).$$

First, we simply solve the minimization problem

$$\inf_u \left(x^2 + u^2 + (ax + u) \cdot V_x + \frac{\sigma^2}{2} \cdot V_{xx}\right)$$

in this equation by ordinary differentiation with respect to u and setting the result to zero. This yields the equation $2 \cdot u + V_x = 0$ with the solution $u = -\frac{V_x}{2}$, which in fact gives the unique minimum.

We substitute $u = -\frac{V_x}{2}$ in the HJB equation and obtain the partial differential equation (PDE)

$$0 = V_t - \frac{{V_x}^2}{4} + a \cdot x \cdot V_x + \frac{\sigma^2}{2} \cdot V_{xx} + x^2$$

with the constraint $V(T, x) = 0$ (since $\int_T^T ((X^{(u)}(t))^2 + (u(t, X^{(u)}(t)))^2)dt$ is zero for any u).

This PDE can be solved using an approach of the form $V(t, x) := p(t) \cdot x^2 + q(t)$. Differentiating this expression with respect to t, x and twice with respect to x and substituting it in the PDE, we obtain

$$x^2 \cdot (p'(t) - p^2(t) + 2ap(t) + 1) + q'(t) + p(t) \cdot \sigma^2 = 0.$$

This equation must hold for **all** x. This is only possible if both

$$p'(t) - p^2(t) + 2ap(t) + 1 = 0 \text{ and } q'(t) + p(t) \cdot \sigma^2 = 0.$$

From the constraint $0 = V(T, x) = p(T) \cdot x^2 + q(T)$, which must also hold for all x, it further follows that $p(T) = 0$ and $q(T) = 0$.

Once we have calculated the function p from the first equation (which is an ordinary quadratic differential equation + constraint for p), we can immediately determine q from the second equation. We then get $q(t) = -\sigma^2 \cdot \int_t^T p(s)ds$ (because, among other things, $q(T) = 0$).

What remains to be done is to solve the differential equation $p'(t) - p^2(t) + 2ap(t) + 1 = 0$ on $[0, T]$ with the constraint $p(T) = 0$. A differential equation of this form is called a Riccati equation. We can solve the equation using Mathematica and get $p(t) = \frac{(A+a)\cdot\left(1-e^{2A(t-T)}\right)}{1+e^{2A(t-T)}\cdot\frac{A+a}{A-a}}$, where we have set A for $A = \sqrt{a^2-1}$.

The optimal control process u is given by $-\frac{V_x}{2}$, and because $V(t,x) := p(t) \cdot x^2 + q(t)$, we get $\boldsymbol{u} = -\frac{V_x}{2} = -p(t) \cdot x = \boldsymbol{-x \cdot \frac{(A+a)\cdot\left(1-e^{2A(t-T)}\right)}{1+e^{2A(t-T)}\cdot\frac{A+a}{A-a}}}$.

The optimally controlled process $X^{(u)}$ therefore has the dynamics

$$dX^{(u)}(t) = X^{(u)}(t)\left(a - \frac{(A+a)\cdot\left(1-e^{2A(t-T)}\right)}{1+e^{2A(t-T)}\cdot\frac{A+a}{A-a}}\right)dt + \sigma \cdot dW(t)$$

and

$$X^{(u)}(0) = 0.$$

Thus the process is intervened in to the extent that the trend factor a is continuously reduced by the value $p(t) = \frac{(A+a)\cdot\left(1-e^{2A(t-T)}\right)}{1+e^{2A(t-T)}\cdot\frac{A+a}{A-a}}$.

In Fig. 2.43 we see the form of $p(t)$ in the time range $[0, T]$ for $T = 1$ and for a couple of different values of a.

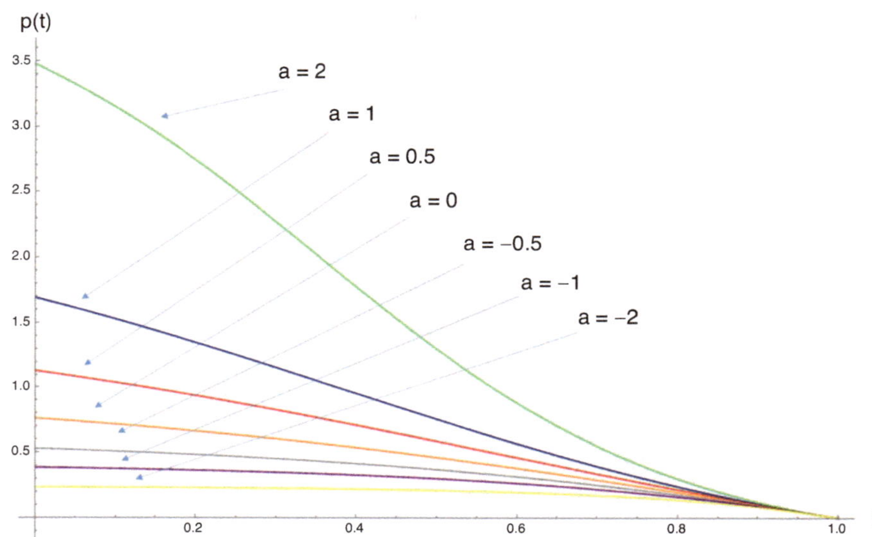

Fig. 2.43 Trend reduction function $p(t)$ in the linear regulator for different values of a

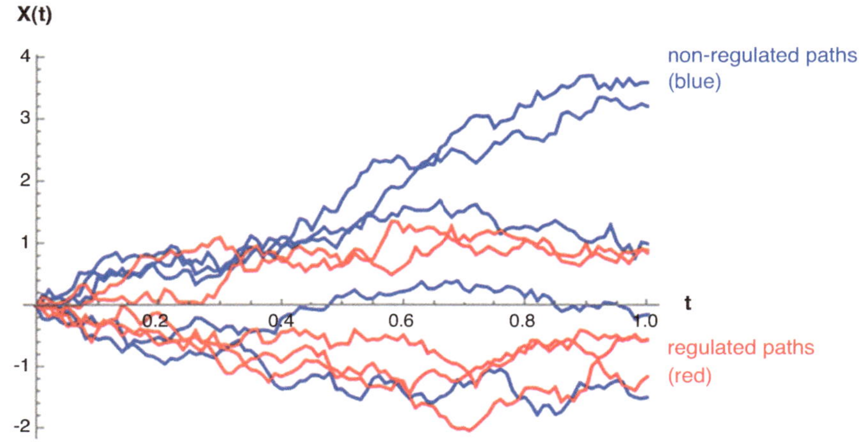

Fig. 2.44 Unregulated and optimally regulated paths in the linear regulator ($a = T = 1$)

In Fig. 2.44, we see five unregulated paths (blue) and five optimally regulated paths (red) for the parameter $a = 1$ and for $T = 1$.

An actual reduction of the trend parameter a is noticeable in all cases (even where a is already negative). The strength of that reduction decreases over time and approaches zero. The larger the a is, the more it is reduced.

2.17 Solving the Optimal Consumption Investment Problem

We now return to the optimal consumption investment problem addressed in Sect. 2.14. The task was:

Determine

$$\max_{c,u} E\left(\int_0^T e^{-\delta \cdot t} \cdot c(t)^\gamma \, dt + \Phi(X_{c,u}(T))\right),$$

taking the maximum over all allowable consumption strategies c and all trading strategies u. Furthermore, it is required that $X_{c,u} \geq 0$ is always satisfied.

A variant of this task was: Determine

$$\max_{c,u} E\left(\int_0^\tau e^{-\delta \cdot t} \cdot c(t)^\gamma \, dt\right).$$

Here, the random variable τ is defined as "the first point in time t at which $X_{c,u}(t)$ is equal to 0 (or $\tau = T$ if $X_{c,u}$ remains positive at all times)". Thus, in this version of the problem, $\Phi \equiv 0$, which means that it is of no relevance if capital is still available at time T.

In both cases, the value process X has the dynamics

$$dX(t) = X(t) \cdot (u(t) \cdot \alpha + (1 - u(t)) \cdot r) \cdot dt - c(t) \cdot dt + X(t) \cdot u(t) \cdot \beta \cdot dW(t).$$

We will only solve the second version of the problem here. However, for this we also need a variant of the theorem stated in the previous section on the HJB equation. More specifically, we need a version that is applicable to this type of problem with stopping time as a bound in the integral. Such a variant can be derived in an analogous way, and so we will simply state this variant here without proof:

Theorem 2.2 (HJB Equation, Variant) *If, for all $t \in [0, T]$, the stochastic control problem*

$$J(t, x, \boldsymbol{u}) := E_{t,x}\left(\int_t^\tau F(s, X^{(\boldsymbol{u})}(s), \boldsymbol{u}(s, X^{(\boldsymbol{u})}(s)))ds + \Phi(\tau, X^{(\boldsymbol{u})}(\tau))\right) \to \max$$

with (for simplicity, a one-dimensional process $X^{(\boldsymbol{u})}$)

$$dX^{(\boldsymbol{u})}(s) = \mu(s, X^{(\boldsymbol{u})}(s), \boldsymbol{u})ds + \sigma(s, X^{(\boldsymbol{u})}(s), \boldsymbol{u})d\boldsymbol{W}(s) \;\; \textit{and} \;\; X^{(\boldsymbol{u})}(s) = x,$$

where $\tau := \min\left\{T, \inf\left\{t \mid t \geq 0, (t, X^{(\boldsymbol{u})}(t) \notin D\right\}\right\}$, where $D = \bigcup_{t\in[0,T)}(t \times A_t)$ with intervals $A_t := [a_t, b_t] \subset \mathbb{R}$ for all t (the intervals A_t may also be unbounded), and with $\boldsymbol{u}(t, x) \in U_t$ for all $t \in [0, T)$ and $x \in \mathbb{R}$ has a solution $\hat{\boldsymbol{u}}$ and if $V(t, x) := J(t, x, \hat{\hat{\boldsymbol{u}}}) \in C^{1,2}$, then it follows that:

$$0 = V_t(t, x) + \sup_{\boldsymbol{u}}\left[F(t, x, \boldsymbol{u}) + \mu(t, x, \boldsymbol{u}) \cdot V_x(t, x) + \frac{\sigma(t, x, \boldsymbol{u})^2}{2} \cdot V_{xx}(t, x)\right],$$

where the following constraint must be satisfied for V:
$V(t, x) = \Phi(t, x)$ for all (t, x) with $x = a_t$ or $x = b_t$.

For all $(t, x) \in [0, T] \times \mathbb{R}$, the supremum is attained for $\boldsymbol{u} = \hat{\hat{\boldsymbol{u}}}(t, x)$.

If we are to determine the minimum instead of the maximum of $J(t, x, \boldsymbol{u})$, then the same equation applies but with an infimum instead of the supremum.

We now use this theorem to solve the variant of the optimal consumption investment problem:

In this setting, we have:

$\Phi = 0$,
$F(s, x, u) =: F(s) := e^{-\delta \cdot s} \cdot c(s)^\gamma$,
$A_t = [0, \infty)$ for all t.

Furthermore (in shorthand)

$$\mu = x \cdot (u \cdot \alpha + (1-u) \cdot r) - c$$
$$\sigma = x \cdot u \cdot \beta$$

The associated HJB equation can then be represented (in shorthand) as

$$0 = V_t + \sup_{u,c} \left(e^{-\delta \cdot t} \cdot c^{\gamma} + V_x \cdot (x \cdot (u \cdot \alpha + (1-u) \cdot r) - c) \right.$$
$$\left. + \frac{(x \cdot u \cdot \beta)^2}{2} \cdot V_{xx} \right).$$

We determine the values for c and u for which the supremum is attained in the above equation by differentiating the function to be maximized with respect to c and with respect to u and setting both derivatives to zero. We get the equations:

$$\frac{d}{du} = V_x \cdot (x \cdot (\alpha - r) + V_{xx} \cdot u \cdot (x \cdot \beta)^2 = 0$$
$$\frac{d}{dc} = \gamma \cdot e^{-\delta \cdot t} \cdot c^{\gamma - 1} - V_x = 0$$

From this, we immediately get the solutions

$$u = \frac{-(\alpha - r) \cdot V_x}{\beta^2 \cdot x \cdot V_{xx}} \text{ and } c = \left(\frac{V_x}{\gamma \cdot e^{-\delta \cdot t}} \right)^{\frac{1}{\gamma - 1}}.$$

(Here, of course, we would have to follow up with a discussion as to whether we actually get a maximum of the expression under consideration for these values u and c. However, we will simply assume this in the following.)

What follows now is a longish, unpleasant computation, which we will not carry out in detail here. Instead we will only describe the steps, as follows:

- Substituting the expressions for u and c in the HJB equation yields a partial differential equation PDE for V.
- For V we choose an approach of the form $V(t, x) := e^{-\delta \cdot t} \cdot x^{\gamma} \cdot h(t)$ where a function $h(t)$ is to be determined. Hence the constraint $V(t, 0) = 0$ is satisfied for all t. Because of the condition $V(T, x) = 0$, it is necessary that h satisfies the constraint $h(T) = 0$.
- Substituting this approach for V in the PDE yields the following ordinary differential equation for h:

$$h'(t) + A \cdot h(t) + B \cdot h(t)^{\frac{\gamma}{\gamma - 1}} = 0 \text{ with } h(T) = 0,$$

where

$$A = \frac{\gamma \cdot (\alpha - r)^2}{2\beta^2 \cdot (1-\gamma)} + r \cdot \gamma - \delta \text{ and } B = 1 - \gamma.$$

The equation for h is a so-called Bernoulli equation and can be solved explicitly. We will come back to this later.

If we substitute the expression $V(t,x) := e^{-\delta \cdot t} \cdot x^{\gamma} \cdot h(t)$ in the expressions for u and for c, i.e. in $u = \frac{-(\alpha-r)\cdot V_x}{\beta^2 x \cdot V_{xx}}$ and $c = \left(\frac{V_x}{\gamma \cdot e^{\delta \cdot t}}\right)^{\frac{1}{\gamma-1}}$, then, because $V_x = \gamma \cdot e^{-\delta \cdot t} \cdot x^{\gamma-1} \cdot h(t)$ and $V_{xx} = \gamma^2 \cdot e^{-\delta \cdot t} \cdot x^{\gamma-2} \cdot h(t)$, we get the following results for u and for c:

$$\boldsymbol{u = \frac{\alpha - r}{\beta^2 \cdot (1-\gamma)}} \text{ and } \boldsymbol{c = c(t,x) = x \cdot h(t)^{\frac{1}{\gamma-1}}}.$$

Before we address the specific form of h and conclude this example, we can already make an extremely interesting observation:

The optimal control variable u has a constant value that is independent of both t and x!

u denotes the fraction of the currently available wealth $X(t)$ that is to be invested in the portfolio S. This fraction has a constant value (often referred to as the "Merton ratio")!

In contrast, the rate of consumption c depends on both time and the amount of current wealth. In particular, there is a linear dependence on the amount of wealth!

We now turn our attention to the function h. For simplicity, we limit ourselves to a gamma value of $\gamma = \frac{1}{2}$ here.

Thus, the constants A and B in Bernoulli's equation for h become

$$A = \frac{(\alpha - r)^2}{2\beta^2} + \frac{r}{2} - \delta \text{ and } B = \frac{1}{2}.$$

The equation for h has the form

$$h'(t) + A \cdot h(t) + \frac{1}{2} \cdot h(t)^{-1} = 0 \text{ with } h(T) = 0.$$

As is easily verified, this equation for $A \neq 0$ has the solution

$$h(t) = \sqrt{\frac{e^{2A(T-t)} - 1}{2A}}.$$

For $A = 0$, we get $h(t) = \sqrt{T - t}$.

So, $h(0) = \sqrt{\frac{e^{2AT}-1}{2A}}$, which always yields a positive value and $h(T) = 0$.

The derivative $h'(t) = -\sqrt{\frac{A}{2}} \cdot \frac{e^{2A(T-t)}}{\sqrt{e^{2A(T-t)}-1}}$ is always negative.

The portion of wealth $X(t)$ that is consumed at time t (or the rate of consumption at time t), i.e. $c(t, x) = x \cdot h(t)^{-\frac{1}{2}}$, is thus monotonically increasing and tends towards ∞ at time T.

This is logical since we set $\Phi \equiv 0$ and wealth that is still available for investment at time T is no longer of any use and should therefore be completely consumed.

In what range can we expect the value $A = \frac{(\alpha-r)^2}{2\beta^2} + \frac{r}{2} - \delta$ to be, approximately? Interestingly, the first summand is exactly half the squared Sharpe ratio of the portfolio S. Typical portfolios have a Sharpe ratio between 0 and 1.

The δ value is often chosen in line with an inflation rate, so, generally in the range of 0 to $2r$. For the constant A, we can therefore expect values in the range of about -0.2 to 0.6.

In Fig. 2.45 we see the plot of the function h for some realistic values of A.

Finally, we note that the value function $V(0, x)$ that we are looking for has precisely the value $V(0, x) = x^{\gamma} \cdot h(0) = x^{\gamma} \cdot \sqrt{\frac{2^{2AT}-1}{2A}}$.

And in conclusion, just one final remark regarding the Merton ratio $u = \frac{\alpha-r}{\beta^2 \cdot (1-\gamma)}$. The question that arises here is under which circumstances the fraction of wealth that is to be invested in the risky asset S—as should mostly be the case in reality—is between 0 and 1.

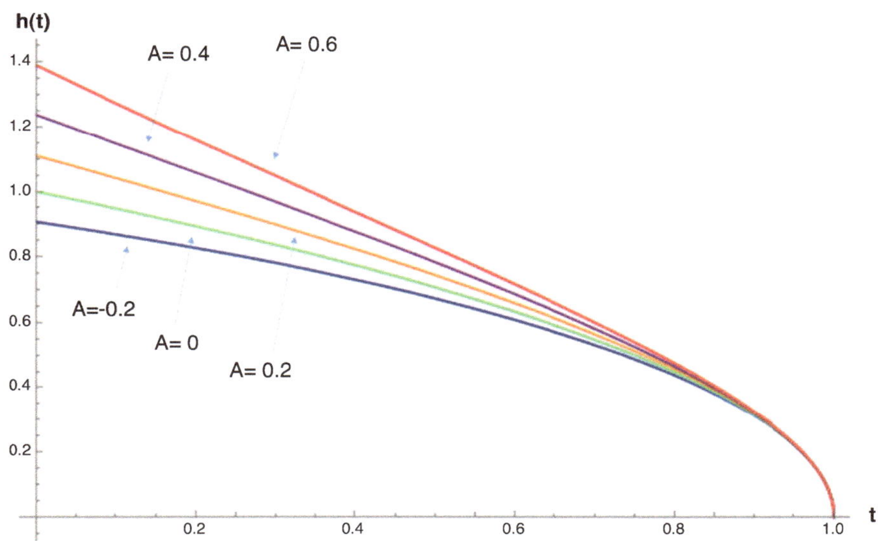

Fig. 2.45 The function $h(t)$ in the optimal investment consumption problem for different values of A

It is clear that the larger the trend factor α of S, the more is invested in S, and the larger the volatility factor β, the less is invested in S. And: The closer γ is to 1 (i.e. the more positive value is placed on consumption), the more is invested in S.

The condition $u \geq 0$ is satisfied if and only if $\alpha \geq r$. This is easily understood if we bear in mind that if the return to be expected from S was lower than the risk-free interest rate, then there would certainly be no need to invest in S.

Finally, the condition $u \leq 1$ is equivalent to $\alpha \leq r + \beta^2 \cdot (1 - \gamma)$. Specifically in the case of $\gamma = \frac{1}{2}$, we have the condition $\alpha \leq r + \frac{\beta^2}{2}$.

References

1. Wolfgang Breuer, Marc Gürtler, and Frank Schuhmacher. *Portfoliomanagement I*. Springer-Gabler Verlag, 2010.
2. Christoph Bruns and Frieder Meyer-Bullerdiek. *Professionelles Portfoliomanagement*. Schäffer-Pöschl Stuttgart, 2015.
3. Bernt Oksendal. *Stochastic Differential Equations*. Springer, 2010.
4. Tomas Björk. *Arbitrage Theory in Continuous Time*. Oxford Finance Series, 2009.

3 Case Studies

Keywords

Case studies: financial software based on machine learning · Churning · Pricing of interest swaps · Callable range accrual swaps · Analysis of put-write strategies · EUR CHF stop-loss order fiasco · Portfolio selection based on sustainability parameters · Basket derivatives · Liquidity risk

In this chapter we present a selection of examples of projects carried out by the author (and partly by members of his research group) in his capacity as a court-appointed expert and privately hired expert appraiser, as well as examples from his work as an analyst and consultant for various asset management, financial software, and fund management companies. A few select cases from his own research work are also included.

In choosing these examples, we ensured that they align with the knowledge that readers have acquired in the course of this book. Therefore, in some cases, we discuss and demonstrate only the fundamental aspects of the respective case while noting that further analysis using more sophisticated techniques would be called for.

As mentioned, all of the examples originate in actual project work that the author has undertaken. However, whenever necessary (especially in court cases), we have anonymized the examples and changed the numerical values.

Virtually all of the examples have been limited to brief excerpts or partial aspects of much broader analytical work. Often, the analyses were subsequently exposed to counter-opinions, which then had to be responded to. As is easily understood, it has not been possible to integrate all of these aspects into the following explanations.

It is important to note for almost all of the following examples that there is no such thing as *the* right approach to the problems described in these case studies. The chosen model, the way that parameters are estimated, and the method of calibration will always be individual decisions that impact the final outcome and could have

G. Larcher, *The Art of Quantitative Finance Vol. 3*, Springer Texts in Business and Economics, https://doi.org/10.1007/978-3-031-23867-3_3

been handled differently from the outset. The line of argumentation in any given controversial issue will also always depend on the analysing expert's position.

Therefore, the approaches presented in the following cases should not be seen as the only correct ones but rather as examples of possible ways to approach a particular problem.

And again: In some cases, we deliberately used simplified methods to ensure the presentation is readily understood by readers who are only familiar with the techniques discussed in this book.

On the other hand, the texts in each of the examples have largely been kept such that—after a brief introduction to the issue at hand and the specificities of each project—the actual discussion retains the form of an expert witness opinion, a project report, etc.

It is also important to note the following simplification: In the interest of ensuring a presentation that is clear and easy to follow, we should not have to concentrate on too many details of little relevance to the fundamental methodology and the tendency of the final result.

For example: In an expert witness analysis, certain relevant time periods must be measured precisely. If, for instance, a product is bought on 1 June of a year and expires on 31 August of the same year, an expert witness opinion would require that the exact number of trading days within this period be counted and used in the calculations. In our discussions we will not dwell on this, however, and will simply treat this period as a quarter of a year, hence as 13 weeks or $\frac{256}{4} = 64$ trading days. Yet in "real life," such an inaccuracy (even if it is completely irrelevant to the validity of the results, especially in view of other imponderables such as, above all, parameter estimation) would immediately be picked up and challenged by opposing expert witnesses or lawyers. Similarly, we will not discuss and argue down to the smallest detail which interest rates are to be used for discounting future cash flows and whether these rates necessarily provide the "correct" discount factors.

For better readability in the following, certain values (such as interest rates) are sometimes rounded to two (or fewer) decimals, for example, instead of using four decimals. In real expert witness opinions, however, the full number of decimals is always used, even if it has no bearing on the result, simply so as not to lay oneself open to any attack, however trivial.

You can safely assume, however, that all of these aspects were of course handled with the highest precision in the original expert opinions and project papers.

In terms of technical difficulty, the case studies presented in the following are of very different levels. Some of them involve complex financial mathematical considerations, while others require only basic technical knowledge (although in such cases—where the theoretical foundations may sometimes be somewhat lacking—a surprising idea or at least a skilful line of argumentation is occasionally needed to convince a potential user of the chosen approach).

The following case studies are purposely not organized by theme nor by difficulty, thus reflecting in a way the variety of the work in the field of quantitative finance.

3.1 Case Study I: The Fynup Ratio

Our first example is about a project placed by a company that had developed an advisory software for investment advisors. This software was based on extensive historical and fundamental data from a universe of more than 33,000 investment funds.

The developers of this program wished to develop a tool that would provide an indication of the quality of the future performance of each of these funds based on the available purely technical historical performance data. It was clear to all parties—both clients and contractors—that it is not possible to make truly valid forecasts about future fund performance from historical data.

The task was therefore to create a—science-based!—tool that can extract more information about future performance data from historical data than would be possible using a purely intuitive, naive approach. We then set about this task in the manner explained below.

3.1.1 The Task

When we started this project in early 2019, for each fund from a universe of 33,030 funds, we were provided with the annual returns from the respective fund's year of inception up to and including 2018. This universe had already been divided by sector (industries and regions), resulting in a total of 25 sectors.

The sector names and types were deliberately not disclosed to the developers of the forecast model. Forecasts were to be based solely on technical historical data. Knowledge of the sector description could have led to unintentional interpretations, which in turn could have biased the further course of action. By the same token, no other fund specifications were included in the analysis.

The funds had been in existence for the following number of years, respectively (compare Table 3.1).

Based on that, and purely informally at first, we were given the following task, in somewhat vague terms:

> Based solely on this data, create a forecast model that indicates a more or less positive future performance of the respective fund (individually or in comparison with the other funds within this fund universe).

The question was therefore *not* (as might be thought on a cursory reading): *Which of the funds from the fund universe performed best in the past?* Instead, the aim was to get an indication of future performance.

Table 3.1 Number of funds for each year since inception

Years since inception	Number of funds	Years since inception	Number of funds
1	4672	31	22
2	3583	32	15
3	1768	33	12
4	2061	34	16
5	2668	35	13
6	2124	36	7
7	2032	37	2
8	1530	38	3
9	1321	39	3
10	1678	40	2
11	1653	41	4
12	1495	42	0
13	924	43	1
14	729	44	0
15	662	45	0
16	653	46	0
17	776	47	3
18	711	48	0
19	568	49	7
20	435	50	0
21	237	51	5
22	203	52	1
23	153	53	0
24	153	54	0
25	87	55	0
26	93	56	0
27	98	57	1
28	84	58	1
29	73	59	1
30	34		

3.1.2 Definition of Terms and General Remarks

First of all, the rather vague definition of the task needed to be restated in more precise terms. In particular, the following questions needed to be addressed:

(a) What exactly should we define as the target variable of a "positive fund performance"?
(b) What is the future time horizon that the indication should (or can realistically) refer to?
(c) What can realistically be expected from an indication that is based purely on technical data?

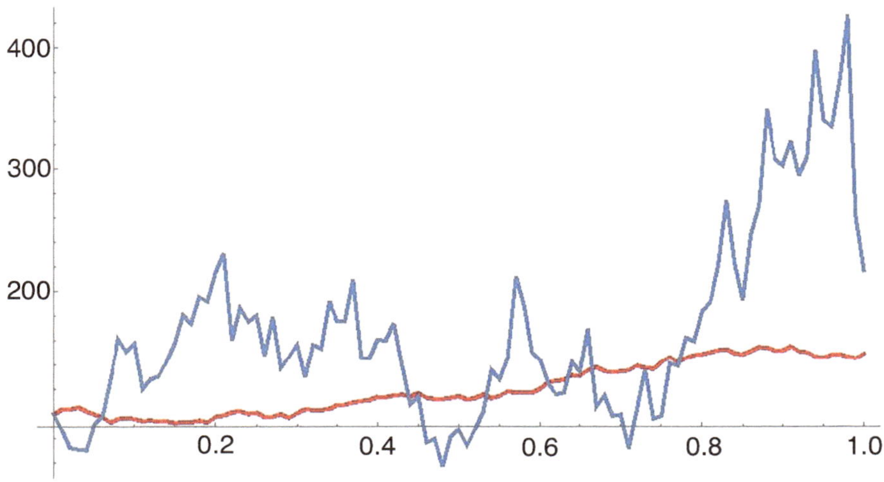

Fig. 3.1 Highly volatile performance (blue) versus steady low-volatility performance (red)

Ad (a)
The fynup ratio seeks to optimize two different target variables. To this end, we focus on two different types of investors. Rather risk-averse investors aim for a performance that is noticeably positive and highly consistently so. Risk-tolerant investors accept greater fluctuations in performance in the hope of significant upside potential (despite a potentially higher downside risk).

Thus, a risk-averse investor is more likely to be interested in a product promising a trend as shown in red in Fig. 3.1 than in a product with a typical trend as shown by the blue graph.

A more risk-tolerant investor (who may also want to be more actively involved in trading the respective product and, for example, using "good moments" for buying or selling) is likely to see more—and more intense—opportunities when investing in the "blue product" shown in Fig. 3.1.

Accordingly, we define two target variables in the following, namely, the respective fund's "expected return" for the indication period (which is yet to be defined) (for the risk-tolerant investor) and its "expected (modified) Sharpe ratio" for the indication period (for the risk-averse investor).

The Sharpe ratio (in its original form) is the ratio of a financial asset's excess return over the current risk-free interest rate to that asset's volatility (standard deviation of returns). We provide explicit information below on how this Sharpe ratio is modified in the present application case, as well as on the individual parameters and their determination.

Ad (b)
Of course, it would be desirable to have a reliable indication of an asset's future performance as far into the future as possible. However, this is impeded by certain

irrefutable facts. Here are just two arguments to illustrate why a long-term indication of "above-average quality" is not possible:

- Long-term indications would require empirical values that go far back into the past ("To what extent and based on what kind of approaches were long-term indications possible and successful in the past?").
  Yet we do not have sufficient historical data material for this. See the above table for our fund universe: Out of the 33,030 funds, only 1334 had been in existence for more than 20 years. This means that there was not enough data material for truly long-term analyses.
- We base our analyses solely on existing technical historical performance data. Thus, neither changes in a fund's management structure nor massive changes to a fund's investment philosophy can be taken into account in our analyses. A certain consistency in a fund's management and investment philosophy—and hence, a greater significance of purely technical data—can be assumed to hold to a greater degree over shorter periods of time than over longer ones.

Ad (c)
Even after conducting our analyses, we will not be able to answer this question explicitly in absolute numbers. We will only assess it in relation to other "intuitive," "average" indication approaches and show that the "optimized" indication approaches resulting from our analyses have historically been stable and significantly better, by a certain percentage, than average indication approaches.

Anyone operating in the financial market knows that there is no definitive way to make accurate predictions about the future performance of risky financial products. However, it will be possible to define systematic indication methods that, at least in the past, would have provided steadily better forecasts than average intuitive approaches based on the same model assumptions. The aim of this analysis and the software developed in parallel is to identify and make available the "best" forecasting methods for this purpose, based purely on technical data.

3.1.3 Technical Preparations

Before starting the actual analyses, it is necessary to prepare the existing technical data. From the available annual returns, we can compute various other technical data (on an annual basis), such as 1-year, 2-year, 3-year, ... all-time performances of the individual funds; 2-year, 3-year, ... all-time volatilities of the individual funds; 2-year, 3-year, ... all-time Sharpe ratios of the individual funds; maximum drawdowns of the individual funds; Sortino ratios of the individual funds; price moves in specific critical time periods; and much more. The annual returns of the funds thus contain all the information we need to work with historical technical data.

As described above, we will try to identify those funds for which—based on the available technical data—a high expected return (in the case of risk-tolerant investors) or a high expected (modified) Sharpe ratio can be predicted for the next 5 years.

In the following, we will limit our exposition to the explicit procedure for determining the "expected (modified) Sharpe ratio". The procedure for the "expected return" is completely analogous.

In the following, we are going to use a Sharpe ratio that has been modified as follows: The conventional Sharpe ratio of a financial asset for a period of n years is given by $\frac{r-r_d}{\sigma}$, where r denotes the asset's return per annum over n years, r_d denotes the current risk-free interest rate (per annum), and σ denotes the per-annum volatility (standard deviation of the returns) that has been calculated from the annual returns. In the analyses carried out specifically with the performance data up to 2018, the risk-free interest rate r_d was set at 0.

One problem associated with the use of the Sharpe ratio is this:
As soon as a financial asset has essentially constant returns (e.g. as is approximately the case with bonds), that is, when the asset's volatility is close to 0, the Sharpe ratio will have—practically irrespective of any positive excess return $r - r_d$—a very high value (the denominator being very close to 0). So the significance of the Sharpe ratio is greatly reduced in this case. For example, an asset A with a return of 1% at a volatility of 0.1% has the exact same Sharpe ratio of 10 as an asset B with a return of 100% at a volatility of 10%. Figure 3.2 shows two typical trends of two assets with such parameter choices (A in red, B in blue).

In our modified Sharpe ratio, the denominator (volatility) is bounded below by the value of 3%, and the modified Sharpe ratio is $\frac{r-r_d}{\max(\sigma,3)}$.

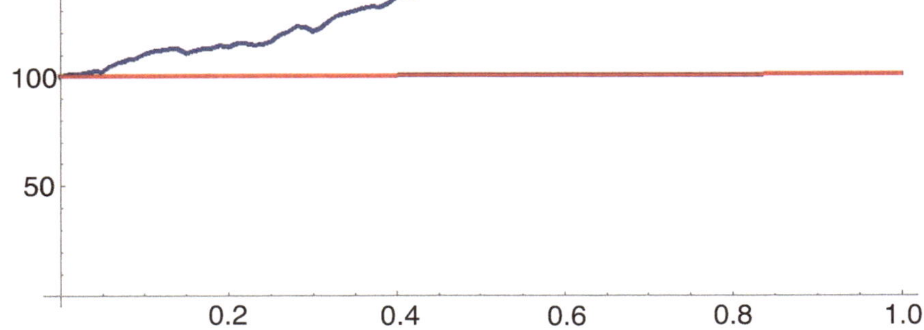

Fig. 3.2 Two assets with approximately equal Sharpe ratios

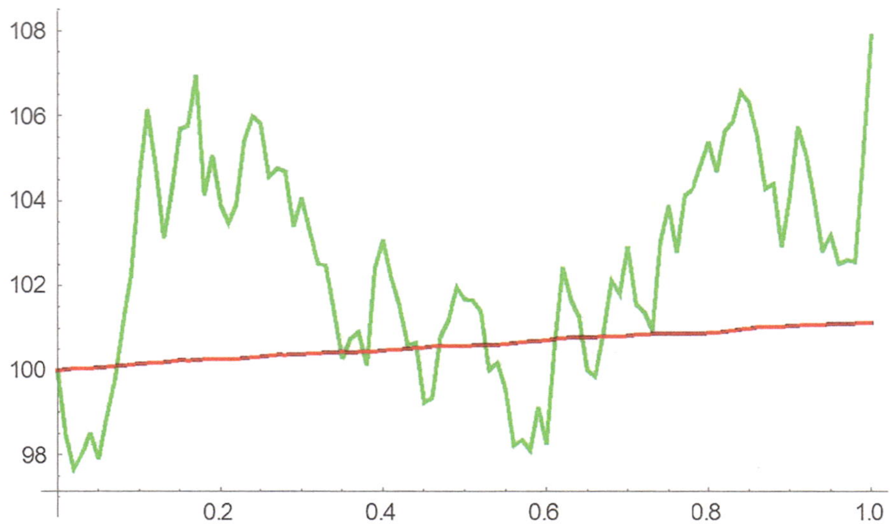

Fig. 3.3 Two assets with approximately equal modified Sharpe ratios

According to this definition, asset A has a modified Sharpe ratio of $\frac{1-0}{\max(0.1,3)} = \frac{1}{3}$, while asset B would still have a modified Sharpe ratio of 10. Based on this modified Sharpe ratio, A would now be comparable with a financial asset C with volatility 10% and return 3.3% (see Fig. 3.3 with asset C shown in green).

So first, we compute all of the above-mentioned variables (multi-year performances, multi-year volatilities, modified multi-year Sharpe ratios, maximum drawdowns, ...) for all assets and can then draw on them for all further analyses.

3.1.4 One Possible Intuitive Approach

The question we now have to ask is:
"Which of these performance parameters from a fund's past performance contributes how much to a positive performance of that fund in the future?"

We know of course that even with all these technical data we will never be able to fully determine the future performance of a fund. On the other hand, it is intuitively obvious that certain historical technical data will be more relevant than others for estimating a fund's future performance.

For example, viewed on average across all funds, the per-annum return of a fund over the past 3 years will, in all likelihood, be a stronger indication of future performance and opportunities than, say, the volatility in the fund's performance from 1997 to 2000.

So, while existing technical data may provide only little potential for future performance forecasts, we can try to make the best use of the material we have at our disposal and at least achieve demonstrable improvements in average intuitive approaches.

What could such an "intuitive", "average" approach look like, in principle?

Intuitive Approach 1 For example, one could work only with returns and use the mean of the returns for each fund. The forecast could then read as follows:
"The higher the mean of the returns, the better the forecast for a high modified Sharpe ratio for the next 5 years".

A somewhat more subtle approach, also using returns only, would consist in taking a weighted average across returns, with returns of the most recent years being given more weight than returns of years further back in time. An example of such a weighting would be:

Intuitive Approach 2 If we denote by r_{-1} the return of last year, by r_{-2} the return of the year before last, and so on, and by r_{-n} the return of the first year in which the fund was launched, then we could take the mean approximately in the following form:

$$\frac{1}{2} \cdot r_{-1} + \frac{1}{2^2} \cdot r_{-2} + \frac{1}{2^3} \cdot r_{-3} + \ldots + \frac{1}{2^i} \cdot r_{-i} + \ldots + \frac{1}{2^{n-1}} \cdot r_{-(n-1)} + \frac{1}{2^{n-1}} \cdot r_{-n}$$

With this weighting (the weight $\frac{1}{2^{n-1}}$ for the return r_{-n} having been chosen deliberately so), the sum of the weights is again equal to 1 (ensuring better comparability). And the corresponding forecast could then read:
"The higher this weighted mean of the returns, the better the forecast for a high modified Sharpe ratio for the next 5 years".

In another approach, one could work with some of the performance metrics provided above (multi-year performances, multi-year volatilities, modified multi-year Sharpe ratios, maximum drawdowns, ...). For example, one could proceed as follows:

Intuitive Approach 3 One could use the performance data of the last 3, 5, and 10 years, the all-time Sharpe ratio, and the maximum drawdown. We denote these five technical parameters by $K_1, K_2, \ldots, K_5$ and will again take a weighted average of these parameters as our "forecast measure". With one caveat however: The maximum drawdown is a negative measure (the larger the maximum drawdown, the poorer the fund performance is, as opposed to measures like return, multi-year performance, or multi-year Sharpe ratio). For this reason, we should not use weightings of the form $a_1 \cdot K_1 + a_2 \cdot K_2 + a_3 \cdot K_3 + a_4 \cdot K_4 + a_5 \cdot K_5$ with positive weights $a_1, a_2, \ldots, a_5$, but rather weightings of, for example, the form $a_1 \cdot K_1 + a_2 \cdot K_2 + a_3 \cdot K_3 + a_4 \cdot K_4 - a_5 \cdot K_5$ with positive weights $a_1, a_2, \ldots, a_5$, the sum of which should again be equal to 1. If we do this, a large maximum drawdown will then actually be measured as a negative parameter. The corresponding forecast could then read:

"The higher this weighted mean of the five technical parameters, the better the forecast for a high modified Sharpe ratio for the next 5 years".

The last approach above immediately begs two questions: What is the best way to select the technical parameters, and which weights $a_1, a_2, \ldots, a_5$ are the best choice?

To answer at least the first question, wouldn't the most obvious way be to simply use all of the available technical parameters?

The answer is "no"!

Why not? Because there are dependencies and redundancies between many of those parameters. The use of too many interdependent parameters is unnecessary and weakens the validity of the method.

Obvious dependencies and redundancies are, for example, the simultaneous consideration of *3-year return, 3-year volatility, and 3-year Sharpe ratio* (the last one being essentially the quotient of the first and second parameters), the simultaneous consideration of *7-year performance, 8-year performance*, and *9-year performance* (the *8-year performance* value provides little to no significant additional information if the *7-year and 9-year performance* is already included in the analysis).

So there is actually a conceivable variety of more or less plausible intuitive approaches to determining a "forecast measure". But "how good" are these approaches, how can the quality of these approaches "be measured", and most importantly, which of these approaches is "the best" one?

3.1.5 The Principle of the Fynup-Ratio Approach: Determination of Weights – Schematic

In the following we are going to present the first version of the fynup ratio. Improved variants are under development.

The fundamental principle of the fynup ratio approach is to have the system learn, by itself, which weights are the best ones to generate a forecast measure from historical trends. Additional information that accumulates over time should be processed continuously and lead to a constant improvement of the method.

Schematically, the procedure is roughly as follows (the detailed description follows in the subsequent sections): Based on extensive testing, a combination of seven metrics is identified as the most efficient choice (in terms of forecasting relevance on the one hand and in terms of independence and low redundancy on the other hand).

The combinations of seven metrics, on which the following optimizations are based, are:

$K_1 =$ 5-year return, $K_2 =$ 10-year return, $K_3 =$ all-time return
$K_4 =$ 5-year Sharpe ratio, $K_5 =$ 10-year Sharpe ratio, $K_6 =$ all-time Sharpe ratio
$K_7 =$ (actual) fund life to date

These are all positive ratios, so we can limit the following discussion to positive weights $a_1, a_2, \ldots, a_7$ in each case. (In more recently developed variants, the maximum drawdown is also included in the analyses.)

An immediate objection raised here might be as follows: Some of these values may not be available for many funds that have not been in existence for long. This valid objection will be addressed later.

In any case, these ratios are adequately normalized before being used any further, so that they can attain values between 0 (worst value) and 100 (best value) (details will follow later).

We now start **Phase 1**, that is, **determination of optimal weights**:

We begin with a schematic description of the procedure for determining the weights (the technically exact explanation follows)!

In the ideal (in no way realistic) case, our target variable, the "modified Sharpe ratio in the next 5 years (mSR)", would be a (linear) function of the variables $K_1, K_2, \ldots, K_7$ from the previous years, i.e. for **each** fund F, we would then have

$$mSR(F) = a_1 \cdot K_1(F) + a_2 \cdot K_2(F) + a_3 \cdot K_3(F) + \ldots + a_6 \cdot K_6(F) + a_7 \cdot K_7(F)$$

with suitable (positive!) fixed weights $a_1, a_2, \ldots, a_7$. (Here, for a given fund F from our universe, we denote by $mSR(F)$ the " modified Sharpe ratio" and by $K_i(F)$ the value of the parameter K_i for this fund.)

As noted, this is not realistic, of course! But one could try to determine the weights $a_1, a_2, \ldots, a_7$ such that this equality is approximately satisfied for as many funds as possible.

If we denote by $F_1, F_2, \ldots, F_N$ all the funds in our fund universe, then the vector

$$\begin{pmatrix} mSR(F_1) \\ \vdots \\ mSR(F_n) \end{pmatrix}$$

in the n-dimensional space should be close to the vector

$$\begin{pmatrix} a_1 \cdot K_1(F_1) + a_2 \cdot K_2(F_1) + a_3 \cdot K_3(F_1) + \ldots + a_6 \cdot K_6(F_1) + a_7 \cdot K_7(F_1) \\ \vdots \\ a_1 \cdot K_1(F_N) + a_2 \cdot K_2(F_N) + a_3 \cdot K_3(F_N) + \ldots + a_6 \cdot K_6(F_N) + a_7 \cdot K_7(F_N) \end{pmatrix}$$

We measure the distance (AB) of these vectors with the square of the usual Euclidean distance of two vectors, i.e. with

$$AB = \sum_{i=1}^{N} (mSR(F_i) - (a_1 \cdot K_1(F_i) + a_2 \cdot K_2(F_i) + \ldots +$$

$$+ a_6 \cdot K_6(F_i) + a_7 \cdot K_7(F_i)))^2 .$$

And now we try to determine the weights $a_1, a_2, \ldots, a_7$ such that this distance AB becomes minimal. We confine ourselves to positive weights $a_1, a_2, \ldots, a_7$.

This is the basic, schematically explained approach. Only now will we go into the technical details, since there are still some questions that need to be clarified or where we have to proceed in a slightly different way.

3.1.6 The Principle of the Fynup-Ratio Approach: Determination of Weights – Technical Details

Technical Detail 1 The first essential question regarding the details in determining the weights is the following:
From what point in time are the data $mSR(F)$ and $K_1(F), K_2(F), K_3(F), \ldots, K_6(F), K_7(F)$ taken to obtain the weights $a_1, a_2, \ldots, a_7$?

The most recent (closest) point in time that we could choose (in spring 2019) was the beginning of 2014. Why? Because in spring 2019, we can use the annual fund returns for 2014, 2015, 2016, 2017, and 2018, to compute the "modified Sharpe ratio for the next 5 years 2014–2018" from the perspective of 2014.

The values $K_1(F), K_2(F), K_3(F), \ldots, K_6(F), K_7(F)$ refer to the historical data from the perspective of the beginning of 2014, i.e. based on the annual returns for the years 2013, 2012, ... until the year of inception for each fund.

To determine the weights $a_1, a_2, \ldots, a_7$ in this way, however, it is necessary that these values $mSR(F)$ and $K_1(F), K_2(F), K_3(F), \ldots, K_6(F), K_7(F)$ for the funds which are used for determining the weights do indeed exist from the perspective of the year 2014! This means that only funds that had already been in existence for at least 10 years in 2014, i.e. that were already traded at the beginning of 2004 at the latest, can be used to determine the weights.

As we can see from the earlier table listing the funds with years since inception, there were $M_{2004} = 5139$ such funds. If we denote these funds by $F_1, F_2, \ldots, F_{M_{2004}}$ and use only these funds to determine the weights, then we actually have to minimize the distance

$$AB_{2004} := \sum_{i=1}^{M_{2004}} (mSR(F_i) - (a_1 \cdot K_1(F_i) + a_2 \cdot K_2(F_i) + \\ + \ldots + a_6 \cdot K_6(F_i) + a_7 \cdot K_7(F_i)))^2$$

Thus, we determine positive values $a_1, a_2, \ldots, a_7$ such that the value AB_{2004} becomes minimal (or at least approximately minimal) and obtain a set of "good" weights $a_1, a_2, \ldots, a_7$.

Technical Detail 2 However, in determining good weights $a_1, a_2, \ldots, a_7$, we do not want to limit the test period solely to the time range 2004–2018. It might be the case, for instance, that this time range provides an atypical set of weights. Therefore, we perform the same procedure from the perspective of the beginning of 2013, as well as the beginning of 2012, and of 2011, 2010, 2009, and 2008.

However, the further back the period, the fewer funds we have from which to determine the optimal weights. For the period from the perspective of the beginning of 2013, there were still 4477 funds; from the perspective of 2012, there were 3824; from the perspective of 2011, there were 3048; from the perspective of 2010, there were 2337; from the perspective of 2009, there were 1769; and from the perspective of 2008, there were 1334 funds.

Thus, going further back into the past would hardly have yielded representative results any more.

For each of these periods, we obtain a set of optimal weights $a_1, a_2, \ldots, a_7$ that we will work with in the following.

In fact, for the first two periods (the ones furthest in the past), we use the 10 best weight sets (i.e. the 10 weight sets that yield the best values in the Monte Carlo simulation for approximating the optimal weights, which we will explain in more detail later); for each of the next four periods, we use the 15 best weight sets; and for the most recent period, we use the 20 best weight sets. So in total, we determine 100 weight sets for the subsequent valuation process.

Using this approach, however, it has turned out that the weight sets that provided good values for one period almost consistently provided very good values (i.e. very small values for AB) for the other time periods too.

Technical Detail 3 Note that in the above description of minimizing

$$AB_{2004} := \sum_{i=1}^{M} (mSR(F_i) - (a_1 \cdot K_1(F_i) + a_2 \cdot K_2(F_i) + \\ + \ldots + a_6 \cdot K_6(F_i) + a_7 \cdot K_7(F_i)))^2$$

(and also in minimizing all other AB values for the other time periods), we do not work with the absolute values of $mSR(F_i)$ or $K_l(F_i)$; instead, to ensure comparability, we normalize the values first, as follows:

For each of the relevant funds F (launched no later than 2004), we first determine the value $mSR(F)$ (i.e. the modified Sharpe ratio for 2014–2018), denoting by min the smallest and by max the largest of these values. Then each fund F is assigned the value $\widetilde{mSR(F)} := 100 \cdot \frac{mSR(F)-\min}{\max - \min}$. The fund with the smallest modified Sharpe ratio thus gets the value 0, and the fund with the largest Sharpe ratio gets the value 100.

For each fund F, the values $K_l(F)$ are then also normalized in this way to values $\widetilde{K_l(F)}$ between 0 and 100.

The final goal is to minimize the variable

$$\widetilde{AB} := \sum_{k=1}^{M} \Big(m\widetilde{SR(F_k)} - (a_1 \cdot \widetilde{K_1(F_k)} + a_2 \cdot \widetilde{K_2(F_k)} + \\ + \ldots + a_6 \cdot \widetilde{K_6(F_k)} + a_7 \cdot \widetilde{K_7(F_k)}) \Big)^2$$

for each of the relevant periods. We can limit ourselves here to weights $a_1, a_2, \ldots, a_7$ that are positive and whose sum is 1.

3.1.7 Optimization Using a Monte Carlo Method

So, the task to determine the "best weights" $a_1, a_2, \ldots, a_7$ is:
For each of the relevant time periods, find positive values $a_1, a_2, \ldots, a_7$ with $a_1 + a_2 + \ldots + a_7 = 1$ such that

$$\widetilde{AB} := \sum_{k=1}^{M} \Big(m\widetilde{SR(F_k)} - (a_1 \cdot \widetilde{K_1(F_k)} + a_2 \cdot \widetilde{K_2(F_k)} + \\ + \ldots + a_6 \cdot \widetilde{K_6(F_k)} + a_7 \cdot \widetilde{K_7(F_k)}) \Big)^2$$

becomes minimal.

This problem could be solved by means of quadratic optimization, but we choose a (more flexible) numerical approach with Monte Carlo simulation.

To this end, an independent sample of 1,000,000 equally distributed 7-tuples $x_1, x_2, \ldots, x_7$ of random numbers in the range between 0 and 1 is generated in each optimization process.

Using $y_i := \frac{x_i}{x_1 + x_2 + \ldots + x_7}$, we then obtain 1,000,000 random scenarios $y_1, y_2, \ldots, y_7$ of positive random values y_i with $y_1 + y_2 + \ldots + y_7 = 1$.

For each of these 1,000,000 scenarios $y_1, y_2, \ldots, y_7$, we compute the value

$$\widetilde{AB} := \sum_{k=1}^{M} \Big(m\widetilde{SR(F_k)} - (y_1 \cdot \widetilde{K_1(F_k)} + y_2 \cdot \widetilde{K_2(F_k)} + \\ + \ldots + y_6 \cdot \widetilde{K_6(F_k)} + y_7 \cdot \widetilde{K_7(F_k)}) \Big)^2$$

In the (completely unrealistic) ideal case, this "forecast error" would be equal to 0 for a choice of weights $y_1, y_2, \ldots, y_7$.

For better comparison, we determine the average value D over all these forecast errors and consider the individual forecast errors $\widetilde{AB}$ in relation to the average error D, i.e. we consider $100 \cdot \frac{\widetilde{AB}}{D}$ (the respective forecast error as a percentage of the

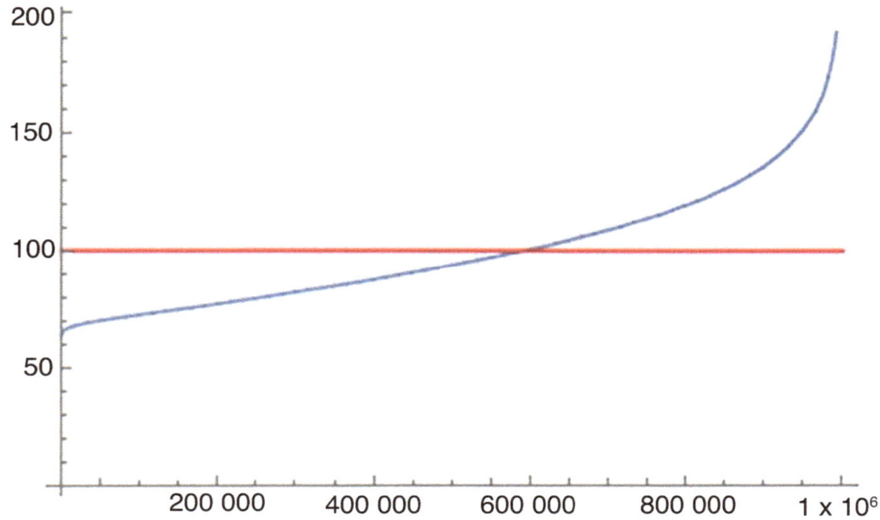

Fig. 3.4 Distribution of forecast errors

average error). The average error D is thus given the reference value 100. If we enter the resulting values into a frequency diagram, we typically obtain a graph of the following form:

We can see from the graph that the smallest forecast error is approximately 65% of the average forecast error. (The average error is represented by the red line in Fig. 3.4.) The largest forecast error is approximately 200% of the average forecast error. Slightly more than half of the weight selections result in a below-average forecast error.

For the subsequent fund valuation, we choose the sets of weight selections (10, 15, or 20 depending on the time period) for which we obtain the smallest forecast errors.

3.1.8 Adaptation of Weight Selections

When optimizing the weights as described above, the following becomes apparent: The optimal weight sets show a practically consistently strong weighting of the shortest-term parameters, i.e. of the historical 5-year return and the historical 5-year Sharpe ratio.

The longer-term parameters, as well as the parameter *fund life to date* show only a weak weighting.

From a purely technical point of view, this means that for an indication of the future performance of a fund, it is primarily the performance in the last few years before the indication date that is relevant and that fund performance data from years

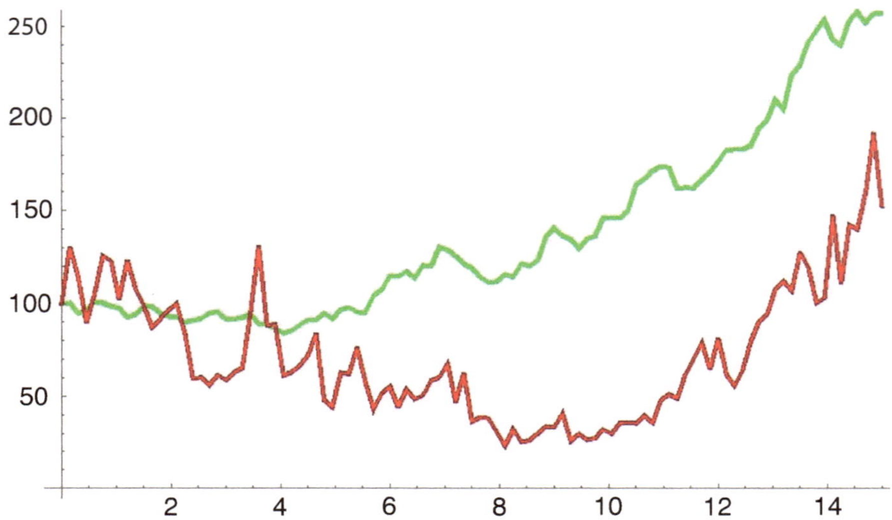

Fig. 3.5 Two examples of fund performance (better original indication for the red fund)

further in the past have only little influence on the indication. (This is also quite logical, from a rational and fundamental point of view: A fund's performance in the past often has little influence on future performance simply because changes in the fund's management and/or investment philosophy are more likely to have taken place.)

This fact occasionally leads to—at first glance—irritating results in the subsequent valuation of funds (or the subsequent computation of indications). For example, a fund with a very good performance over the past 5 years, but a very much below-average performance in the years before that, may get a much better indication than a fund with a consistently positive performance over many years, but with a slightly less positive performance in the last 5 years than the former fund.

For example, with this original weight construction, the red fund in Fig. 3.5 would get a clearly more positive indication than the green fund, since the red fund performed much better over the last 5 years (10–15) than the green fund did. The fact that the green fund had a much better positive performance in years 0–10 plays less of a role here.

In an effort to give more positive weight to longer-term stable developments in fund performance (especially during the financial crisis of 2007–2009), the process for determining optimal weights was adapted in the following way:
Only the scenarios $y_1, y_2, \ldots, y_7$ where y_7 (i.e. the weight of the entire fund life) had a value of at least 0.1 and where y_6 (the weight of the all-time Sharpe ratio) had a value of at least 0.15 were tested in the Monte Carlo simulation.

Now, restricting the weights in this way could, of course, result in significantly larger forecast errors than when choosing the weights without limitation. Yet this is not the case. Although the restriction brings about a slight loss in the forecast quality (thus slightly increased forecast errors), the indication results are practically as good as with unrestricted optimization. The main advantage of this approach is that long-term historical stability of funds now has a significantly more positive influence on the indication (while the indication quality weakens only marginally). In many cases, this also leads to intuitively more meaningful results.

Moreover, as we will discuss below, this adaptation of the weight determination approach is also necessary to be able to give, at least informally, an indication for funds with a (much) shorter history that is at least approximately consistent with the indication for funds with a longer history.

The process described in Chap. 3.1.7 for optimization using Monte Carlo simulation therefore has to be adapted in such a way that only scenarios that satisfy the conditions $y_7 > 0.1$ and $y_6 > 0.15$ are considered.

3.1.9 Indication for Funds with at Least 10 Years of History

Using the method described above, we have now generated 100 sets of weight vectors $(y_1, y_2, \ldots, y_7)$, each with positive entries y_i, with $y_7 > 0.1$ and $y_6 > 0.15$, and with $y_1 + y_2 + \ldots + y_7 = 1$.

For the funds in our fund universe that have been in existence for at least 10 years, we now want to create an indication for their performance in terms of the target variable (modified Sharpe ratio in the next 5 years) from the current perspective (which is early 2019).

This is done by determining the parameters $K_1(F), K_2(F), \ldots, K_7(F)$ (from the perspective of early 2019) for all the funds F to be valued. Then in the next step (as described above), the normalized values $\widetilde{K_l(F)}$ are calculated from those parameters.

For each of the 100 weight vectors $(y_1, y_2, \ldots, y_7)$, we then determine the forecast value $y_1 \cdot \widetilde{K_1(F_k)} + y_2 \cdot \widetilde{K_2(F_k)} + \ldots + y_6 \cdot \widetilde{K_6(F_k)} + y_7 \cdot \widetilde{K_7(F_k)}$. This is a value between 0 and 100. We obtain one such value for each of the 100 weight vectors and then finally take the mean over all of these values.

Actually, we do not work only with funds that have a history of at least 10 years here. For purposes of practical applications, it is desirable to have at least an **informal indication** for all the funds in the fund universe, including **also the funds with shorter histories**. Of course, this begs the question as to how the above procedure is to be carried out if no 10-year returns and 10-year Sharpe ratios or not even the corresponding 5-year values can be calculated for lack of sufficient data.

3.1.10 Informal Indication for Funds with Less Than 10 Years of History

In order to be able to obtain an indication (which would be informal only) for funds with a history of less than 10 years, all funds are assigned an artificial history of at least 15 years. This also applies to funds with a history between 10 and 14 years.

The procedure is as follows: As noted at the beginning of this chapter, the 33,030 funds are divided into 25 sectors. For each sector and for each of the last 15 years (2004–2018), we determine the average return of all the funds from that sector in that year. For each fund launched later than 2004, the return data for the years 2004 to its first full year of trading is supplemented by the average return of its sector in that year.

With this artificially extended history, it is then possible to determine all values $K_1(F), K_2(F), \ldots, K_6(F)$ for **all** funds F. The value $K_7(F)$, however, is still the **actual** life to date for each fund F.

Thus, since the weights y_7 always have a value of at least 0.1, the funds' actual life to date remains of significance in the valuation process.

Extending the return history in this way has an additional advantage: It means that the all-time parameters (K_3 and K_6) will not interfere with the 10-year parameters, as would otherwise be the case for funds with a history of around 10 years.

The procedure for valuing the funds in the fund universe is now as described in Sect. 3.1.9, except that (as noted in Sect. 3.1.10) we work with all the funds.

3.1.11 The Result and the Percentile Representation

The procedure described in detail above is updated at the beginning of each new year, as soon as the previous year's returns are available. Thus, one always gets an indication that is valid for one year at a time.

In addition, an update is also made during the year if necessary (especially when new funds are added to the fund universe). It is important to note that, due to the way that the indication is generated, any update that is required, for example, because new funds have been added to the fund universe, will affect not only those new funds but also all the funds in that universe.

Now, with the described procedure, we obtain a valuation that assigns an indication between 0 and 100 to all the funds in our fund universe. Only a fund whose score among all the funds in that universe is the best on each of the seven metrics $K_1, K_2, \ldots, K_7$ would be assigned an indication of 100, and only a fund whose score among all the funds in that universe is the worst on each of the seven metrics $K_1, K_2, \ldots, K_7$ would be assigned an indication of 0.

Based on the valuations in early 2019, the indications were distributed as follows (Fig. 3.6):

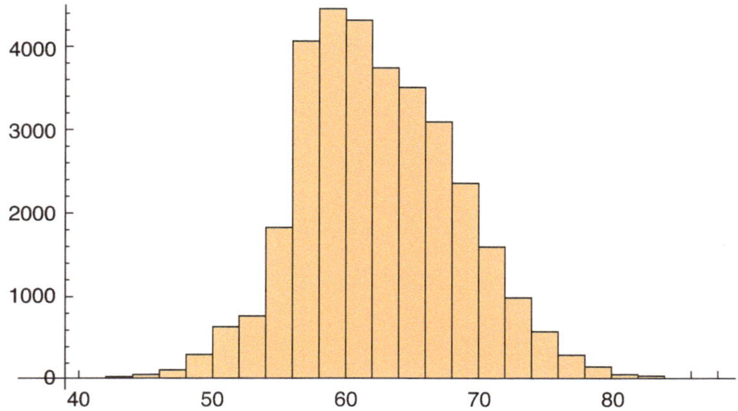

Fig. 3.6 Distribution of indications (2019 valuations)

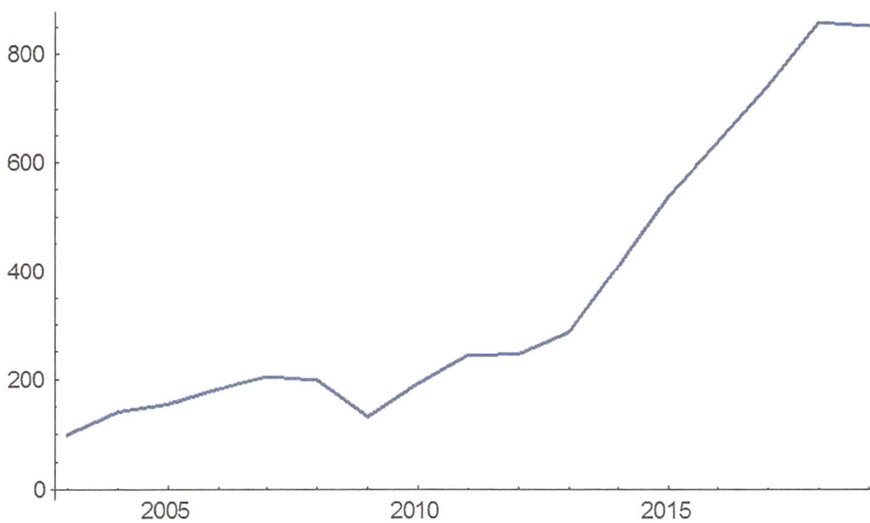

Fig. 3.7 Performance (in percent) for the fund with the best indication as determined in 2019

The indications range from 28.54 to 81.85. The majority of the valuations are in the range from around 55 to around 70.

The best indication in this valuation process was assigned to a fund whose first available annual return dated back to 2003, with annual percentage returns of 41.21, 9.57, 17.76, 12.48, −3.23, -33.26, 45.36, 26.96, 0.75, 16.14, 43.34, 31.68, 18.41, 16.34, 15.50, and −0.73 from 2003 to 2018, a per-annum return since inception of 14.33%, a per-annum return over the last 10 years of 20.47%, and a per-annum return over the last 5 years of 25.36%. The performance of this fund over the years, expressed as a percentage, can be seen in Fig. 3.7.

The indication determined with the fynup ratio approach always refers to a fund's position relative to the entire fund universe. It is not an absolute figure independent of the fund universe it belongs to!

In terms of informative value, the ranking percentile of the fund within the surrounding fund universe is therefore more relevant than the indication. This is why, in our valuations, we always give the percentile in addition to the absolute indication. For example, a percentile of 20% for a fund F means as follows: 20% of the funds in our fund universe have a higher indication value than this fund F, and 80% of the funds in our fund universe have a lower indication value than this fund F.

3.1.12 Informative Value of the Indication Results – Conclusions

It is clear that **an exact forecast of future performances** (or future successes of fund managers) cannot be derived purely from analysing quantitative historical data.

What the **fynup-ratio approach** attempts to achieve is to use **information** that is actually **contained in the existing technical data in the best way possible** (based on certain premises)!

The premises (target parameters, linear approach based on selected ratios, specification of lower bounds for certain weights, artificial extension of too short price histories, ...) are debatable. However, if the premises are accepted as a working hypothesis after careful debate, then the system itself will (by learning from the analysis of historical data) determine the optimal parameters.

Thus, what is debatable is the underlying modelling, not the optimization techniques based on it. In fact, the best choice of weights for calculating indications is indeed found on the basis of the chosen modelling.

The development of the fynup ratio is not complete, but rather a dynamic process. The version described here reflects the status at the end of 2019.

The resulting quality statement: *With the optimized fynup-ratio weights that we found based on the chosen modelling it is possible to achieve an improvement of about 35% in the indication compared to an average indication method (based on the chosen modelling)* is a relative statement compared to other indication approaches. It does not allow any conclusion as to whether or not a reliable indication can indeed be obtained from it.

The fynup ratio is therefore not a universal measure for valuing a fund and cannot provide a reliable forecast for a fund's future performance.

But: **The fynup ratio is a science-based approach to using existing technical data in an optimal way, based on a specific modelling. As such, it is another important piece of the mosaic in an investment decision-making process.**

Table 3.2 Call options with various underlyings and maturities

Underlying	Number of contracts	Remaining life in days	Strike	Price of underlying when traded	Price of option per unit	Fees per contract
AA	30	183	80	48.97	0.16	100
BB	4	115	40	29.72	0.44	100
CC	2	152	36	27.92	0.47	100
DD	19	141	50	33.04	0.53	100
EE	67	264	30	21.84	0.55	100
FF	21	135	50	33.00	0.60	100
GG	19	293	60	42.25	0.75	100
HH	80	259	27	25.21	0.85	100
II	17	146	45	43.82	1.20	100
JJ	40	217	14	12.94	1.25	100

3.2 Case Study II: Expert Witness Opinion on Churning

The following example is from a court case in which a financial service provider was accused of what is referred to as "churning".

3.2.1 Specifics of the Case

The US Commodity Futures Trading Commission (CFTC) defines churning as follows:
"Churning:[1]*Excessive trading of a discretionary account by a person with control over the account for the purpose of generating commissions while disregarding the interests of the customer".*

It is not the objective of this chapter to present that course case in its entirety or discuss it in detail. What we want to do here is focus on a specific question that was put to the author in his capacity as an expert witness in the court proceedings.

The financial services company, which we will refer to as XY in the following, had purchased a large number of options (approximately 200,000 contracts in a total of approximately 10,000 transactions), primarily call options, on US equities for customers over a period of several years (from approximately 2003 to 2008).

Most of these options had a remaining life of 3 months to 1 year and a strike price far above the current price of the underlying asset. This means that they were mostly far out of the money and thus relatively cheap. From among the large number of transactions, we pick out the following ten for our discussion in this chapter. All options are call options, the currencies are all stated in USD, and the contract sizes are 100 units each (see Table 3.2):

[1] https://www.cftc.gov/LearnAndProtect/EducationCenter/CFTCGlossary/glossary_c.html.

Some of these options were held until expiration, others were closed before expiration. Even after just a brief analysis, we notice the following:
In most cases, the commission charges are (sometimes significantly) higher than the capital used for buying the options!

This looks suspiciously like a case of churning. Did XY deliberately buy cheap options for their customers without a visible and justified trading strategy, so as to be able to buy a large number of contracts with their customers' investment capital and thus generate a very high commission income for XY?

To illustrate: In the situation described above (see Table 3.2), options were purchased for a total price of 21,967 USD. For this, customers had to pay a total of 29,900 USD in commission charges.

A ruling by the German Federal Court of Justice in another case stated that the following, among others, may be indicative of churning:

- Commission-to-equity ratio (CER) in excess of 17% monthly as a " material or strong indication"
- Short-term transactions that make no sense economically
- Lack of a recognizable trading strategy

"Commission-to-equity ratios are determined by dividing the total commission charges incurred in a month by the average daily equity in the account. The average daily equity in the account is determined by summing the total equity in the account for each day on which a transaction takes place during the month, and then dividing that sum by the number of days on which a transaction takes place during the month".

Using a simplified form of CER for the options shown in the table above, namely, taking as the total equity the value of the options when they were purchased, the CER in this case would be $CER \approx 100 \cdot \frac{29{,}900}{21{,}967}\% = 136.11\%$.

The first of the above-listed indications of possible churning is thus met in an impressive (or rather alarming) way.

To shed some light on the other two possible indications mentioned, the author as an expert witness in this case was asked the following question, among others:
"What were the profit probabilities for the customers in these options transactions?"

We are not going to discuss in the following how much sense this question makes or how useful the answer is for drawing material conclusions as to the two other indications of churning. We just take this question as is and will answer it below without any further reference to the court case.

In the case at hand, we had to process a total of around 4000 transactions. From those, we will take a closer look at the ten transactions that we listed as examples in the table above.

And: For simplicity, we will assume that these **ten transactions** were **executed for one client** and **at approximately the same time**, i.e. **("at time 0")**.

Before we start our analysis, it is imperative that we clarify the question—which was actually posed exactly as stated above.

The question was either "deliberately" or most likely "unintentionally" phrased in a way that is definitely imprecise and probably also incorrect (in terms of what the court actually wanted to know):

Presumably, the court's intention was to get an answer to the following question: *"What was the **customer's expected profit/loss** from these transactions?"*

For purposes of reaching its conclusions, it would likely have been more useful for the court to get information on profit **expectations** than on profit **probabilities** (whatever that is supposed to mean, which still remains to be clarified.)

One might argue, of course, that an expert witness should provide answers to both questions in their report—recognizing the fact that answering a different question would probably be more helpful. However, caution is advised: If an expert's answer to a question that the court did not ask in this form brings something to light that might incriminate one of the parties, then the attorneys of that incriminated party might construe this as reason to claim expert witness bias.

In the present case, the expert was not yet present in court at the hearing in which the question was posed and ruled on; otherwise he would of course have reacted to its phrasing **at the hearing** itself.

Here, of course, we also want to address the question in the more useful form, which, incidentally, is also much easier to answer. And to do that, we rephrase the question as follows:

*"What was the **customer's expected profit/loss** from these transactions?"*

Or more precisely:

*"What was the customer's expected profit/loss from these transactions **from the perspective of time 0**?"*

The (short) **answer** to **this** question would be:

All of these options are exchange-traded products. And we are dealing with a large number of options on a wide variety of stocks. We can (and must) therefore essentially assume an efficient market. This means: We can essentially assume that, on average, the traded options are priced at fair value. The discounted payoff to be expected from an option is therefore essentially equal to the price of the option.

From the perspective of time 0, the future expected payoff corresponds exactly to the current price of the option. Thus, if one were to pay only the price of the option (without commissions), the expected profit would be—on average—equal to 0.

However, since from the customer's perspective, the transaction needs to be valued inclusive of commissions, **the customer's expected profit is, on average, equal to the amount of the negative commissions!**

Thus, the customer's expected profit from the perspective of time 0 and relative to time 0 (assuming the EMH) is essentially equal to the negative commission charges, i.e. equal to −29,900 USD.

The customer's total investment is 21,967 + 29,900 = 51,867 USD. Therefore, an expected loss of 29,900 USD represents an **expected loss of 57.65%**.

One might be tempted here to raise the following objection:

"Due to the considerable (cost-intensive) analytical work done by highly specialized derivatives traders and the selection of options to be traded on the basis of these

analyses, the profit probabilities and thus the expected profit could have changed in the customer's favour so that in this case the EMH can no longer be assumed".

Without contemplating further whether it is at all possible to "beat the market through analyses", the above objection was immediately refuted in this case: There were no activities within Company XY that would have indicated an intensive market analysis. XY's activities were almost exclusively geared towards customer acquisition.

That would have been the simple—and low-cost—answer to the question that the court in all likelihood had actually intended to ask.

3.2.2 The Question in a More Precise Wording

Let us now return to the question as it was actually put to us at the time, namely: *"What were the profit probabilities for the customers in these options transactions?"*

Let us first rephrase the question in a more precise way: We assume that the question is what the probability is for the customer to make a profit from each of the transactions, despite having to pay a commission in each case. Thus:
*"What is **for each transaction** the **probability** that the **payoff achieved is greater than the commission plus option price**?"*

Possible alternative interpretations of the question—which we will also answer in the expert report—would be:
"*For comparison: What is **for each individual transaction** the **probability** that the **payoff achieved is greater than the option price (without commissions)**?*"
and:
*"What is the **probability for the customer** to achieve a **total payoff from all of the transactions** that is **greater than the sum of the option prices (with and without commissions)**?"*

For all of these three versions of the question, we need to clarify one more essential point. Namely: *"Exactly what kind of probability is meant here?"*

Here are several possible interpretations, each with and without commissions, and for the individual positions and the portfolio as a whole:

(a) Probability if held to expiration?
(b) Probability of a profit opportunity to occur at some point before expiration (probability as to whether a profit can be realized by closing positions at some point before expiration)?
(c) The previous point more precisely: When would such positions be closed? After all, closing positions **immediately on achieving** a positive result (just slightly above 0) would not result in any profit—on the contrary, it would in fact preclude any positive profit! So only a rule of the following form would make sense here: *"Positions are to be closed as soon as a profit of x% can be achieved from the trade".*
(d) The probability in the case of another specifically defined trading strategy that Company XY executes for the customer?

Variant (d) can be excluded in the present case: Company XY was not able to present any concept of any trading strategy (or exit strategy) specifically agreed with their customer.

We have therefore addressed **variants (a)** and **(c)** only (variant (b) being included in variant (c)). Only the approach **with commissions** made sense here, i.e. proceeding from the situation in which the customer actually found themselves in each case (of course, an analogous approach can be taken for the case without commissions).

And we will look at **both individual positions and the portfolio as a whole**, as shown in the table above.

3.2.3 Modelling the Underlying Assets

We will discuss individual positions first.

First of all, we obviously need a suitable and recognized model for the underlying asset's price path.

We note here that in practice, the Wiener model, i.e. the geometric Brownian motion, is still the most widely used and proven model for simulating stock price paths, despite certain well-known shortcomings ("too little allowance for the fat-tail phenomenon", "assumption of independent returns").

Thus, for the price path $AA(t)$ of the stock AA, for example, we assume a path of the form

$$AA(t) = AA(0) \cdot e^{\mu t + \sigma \cdot W(t)}$$

with a standard Brownian motion $W(t)$.

Which parameters μ and σ are we going to use here?

As stated above, the most reasonable assumption is that of an EMH and thus of a "risk-neutral world".

This means: The discounted price path $e^{-rt} \cdot AA(t)$ forms a martingale. However, as we know from Volume I Section 4.18, this is the case if and only if $\mu = r - \frac{\sigma^2}{2}$. The risk-free interest rate r for the respective point in time can be obtained from financial information systems. We used the 13-week Treasury Bill Yields on a monthly basis for our calculations. The tabulated options were all traded within 1 month. For this trading month, the interest rate $r = 1.74\%$ was used.

Furthermore, the EMH assumption implies that each option is essentially fairly priced. However, in the assumed Wiener model for the underlying asset, the fair price of an option is given by the Black-Scholes formula. Thus, the volatility σ to be used is the implied volatility, which can be calculated from the given option price. We already described in detail how to do this in Volume II Section 5.7.

Based on this, we have explicitly given the model to be used. The following Table 3.3 lists the required model parameters for each underlying stock.

Table 3.3 Model parameters for each underlying stock

Underlying	Price of underlying when traded	Implied volatility σ	$r - \frac{\sigma^2}{2}$
AA	48.97	0.3468	−0.0427
BB	29.72	0.4186	−0.0702
CC	27.92	0.3359	−0.0390
DD	33.04	0.4868	−0.1011
EE	21.84	0.3289	−0.0367
FF	33.00	0.5160	−0.1158
GG	42.25	0.2983	−0.0271
HH	25.21	0.1508	0.0060
II	43.82	0.1279	0.0092
JJ	12.94	0.3898	−0.0586

Later, upon determining profit probabilities for the entire portfolio, we will also need correlations between the returns of the individual underlying stocks. Of course, they could be estimated from historical data of the individual stocks or, as shown in the single-factor model in Chap. 2, from the respective correlations of the individual stocks to the SPX (almost all of the underlying stocks were included in the SPX). In the present case, this would have required far too much effort (and thus cost) and would not have been justifiable in view of the relevance of the results. For simplicity, a correlation value of 0.3 was therefore assumed for any two stocks. This roughly corresponds to the average correlation between two stocks in the SPX.

3.2.4 Profit Probabilities for the Individual Positions When Held to Expiration

First, we now calculate, for each individual position, the probability that holding the options until they expire will yield a profit.

From the perspective of time 0 in the case without commissions, we need to calculate the probabilities Pr_1 that at time T, the price of the underlying stock is greater than $K + C$, where K denotes the strike price of the option and C the price of the call option. And from the perspective of time 0 in the case with commissions, we need to calculate the probabilities Pr_2 that at time T, the price of the underlying stock is greater than $K + C + Sp$, where Sp denotes the amount of the commissions (per option). As we already demonstrated in Volume I Section 4.12, these two probabilities can be easily determined from the formula $S(t) = S(0) \cdot e^{\left(r-\frac{\sigma^2}{2}\right)\cdot t+\sigma \cdot W(t)}$.

We have:

$$Pr_1 = 1 - \mathcal{N}\left(\frac{\log\left(\frac{K+C}{S(0)}\right) - \left(r - \frac{\sigma^2}{2}\right)T}{\sigma\sqrt{T}}\right),$$

$$Pr_2 = 1 - \mathcal{N}\left(\frac{\log\left(\frac{K+C+S_p}{S(0)}\right) - \left(r - \frac{\sigma^2}{2}\right)T}{\sigma\sqrt{T}}\right).$$

Substituting the respective parameters for the individual positions in these formulas and evaluating them with Mathematica, for example, we get the results as tabulated in Table 3.4:

The average profit probability, if commissions are not taken into account, is therefore approximately 12.48%. With commissions the average profit probability is 9.52%. The difference is therefore approximately 2.96%.

Note that this difference *must not* be interpreted to mean that the customer can expect a result that's only about 2.96% worse due to commissions! (Even if this is self-evident to people constantly dealing with such topics, it is necessary to explicitly state this fact in court proceedings, where people who are mostly only very peripherally concerned with the matter have to read such an expert opinion and take it into account in their decisions)!

The **effect is actually much more dramatic**. Let us illustrate this with a relevant trading example:

In the example of the underlying stock EE, the company bought contracts of a call option with strike price 30 and 264 days to expiration. The purchase price per contract was 55 USD. The commissions per contract were 100 USD. The price of the underlying stock (EE) was 21.84 at the time of the transaction.

Figure 3.8 shows the profit functions of this trade once without commissions (blue) and once with commissions (red). On the y-axis, the profit made from this

Table 3.4 Extension of Table 3.3 with the additionally computed parameters Pr_1 and Pr_2

Underlying	Price of underlying when traded	Implied volatility σ	$r - \frac{\sigma^2}{2}$	Pr_1	Pr_2
AA	48.97	0.3468	−0.0427	1.81	1.60
BB	29.72	0.4186	−0.0702	8.00	6.57
CC	27.92	0.3359	−0.0390	9.55	7.60
DD	33.04	0.4868	−0.1011	6.26	5.50
EE	21.84	0.3289	−0.0367	9.77	7.93
FF	33.00	0.5160	−0.1158	6.70	5.93
GG	42.25	0.2983	−0.0271	7.49	6.66
HH	25.21	0.1508	0.0060	22.65	15.20
II	43.82	0.1279	0.0092	27.15	19.13
JJ	12.94	0.3898	−0.0586	25.39	19.11

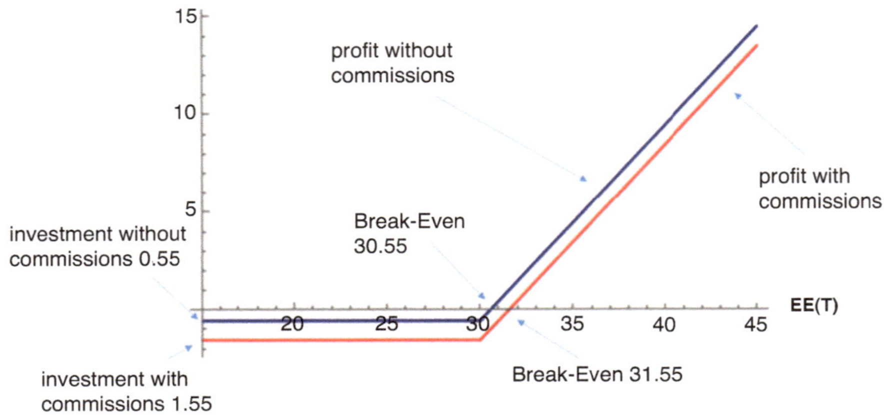

Fig. 3.8 Comparison of profit functions in **absolute** numbers

trade (per unit of the option) is shown as a function of the underlying asset's price $EE(T)$ at expiration (x-axis).

The "break-even point" 30.55 marks the threshold that the underlying stock needs to reach to yield a profit in the case without commissions.

The "break-even point" 31.55 marks the threshold that the underlying stock needs to reach to yield a profit in the case with commissions.

Of course, the profit in the case without commissions is higher in absolute terms and occurs earlier than in the case with commissions.

But: What is relevant here is *not* the amount of such profit (the difference in profits) in **absolute** numbers, but the profit in **relative** terms, that is, in relation to the invested capital! And that is where we see a striking difference, clearly shown in the table below. The table lists the profit as a percentage of the invested capital for different values of the underlying stock at expiration. This is calculated for the case without and with commissions.

As noted above: This clearly illustrates the essential difference! **The probability as to whether or not a profit is made** in the case with or without commissions is less relevant; what matters is **the profit as a percentage of the investment** in each case.

Here's an Example If the underlying stock had risen to 35 by expiration, the customer would have realized a profit of 809.01% (in the case *without* commissions) and a profit of 222.58% (in the case *with* commissions; see Table 3.5).

Table 3.5 Possible profits obtained by a customer

Price of underlying at expiration	Profit in absolute numbers *without* fees	Profit in % of invested capital *without* fees	Profit in absolute numbers *with* fees	Profit in % of invested capital *with* fees
25	0	−100%	0	−100
28	0	−100%	0	−100
30	0	−100%	0	−100
31	0.45	**+81.82%**	0	**−100**
32	1.45	**+263.64%**	0.45	**+29.03%**
35	4.45	**+809.01%**	3.45	**+222.58%**
40	9.45	**+1718.18%**	8.45	**+545.16%**

3.2.5 Profit Probabilities for the Individual Positions in the Case of Profit-Taking Before Expiration

To determine the profit probabilities under the assumption of early profit-taking if a certain profit target of $x\%$ is reached, we use Monte Carlo simulation.

And we proceed as follows: Based on the parameters determined in the Wiener model in Sect. 3.2.3, we simulate a large number (10,000 scenarios each) of possible price realizations on a daily basis (each day representing a new simulation step) for each underlying stock. In parallel, we calculate the call option's current value for each individual trading day and each resulting path using the Black-Scholes formula.

In the case **without** commissions:

As soon as the call option's current value exceeds the option's initial price by $x\%$, we assume to have closed the option for a profit of $x\%$. If this never happens until the option expires, we keep the option until expiration and then realize either a profit (of less than $x\%$) or a loss.

We then calculate the percentage of simulation scenarios where a profit occurs and take this percentage as an approximation for the probability of making a profit.

In the case **with** commissions (which is the only relevant case in the present situation):

As soon as the call option's current value exceeds the option's initial price plus accrued commissions (i.e. the entire investment) by $x\%$ (of the entire investment), we assume to have closed the option for a profit of $x\%$. If this never happens until the option expires, we keep the option until expiration and then realize either a profit (of less than $x\%$ of the entire investment) or a loss. We then calculate the percentage of simulation scenarios where a profit occurs and take this percentage as an approximation for the probability of making a profit.

If we let the profit target x become larger and larger in the above procedure, then the probabilities that we just obtained will approximate the probabilities calculated in the previous section. As an example, we run the simulation process again using the option on the underlying stock EE, yet this time tabulating the results for all ten considered option types.

In the upper part of Fig. 3.9, we see two path realizations for the price of the underlying stock EE in the Wiener model with the parameters given or calculated for EE. The red horizontal line in the upper part shows the price that EE should have **at expiration** (i.e. at least strike price + option price + commissions = 30 + 0.55 + 1 = 31.55) to make a profit from the payoff at expiration. In both example paths, no profit would have been made at expiration, since both paths end well below the strike price of 30. This means that a total loss would have occurred in both cases.

In the lower part of the chart, we see the option price moving in parallel with the above price paths. The option price obviously starts at 0.55 in both cases. The light blue path of the option price relates to the dark blue path of the underlying stock in the upper part of the picture, and the light green path of the option price relates to the dark green path of the underlying stock in the upper part of the picture.

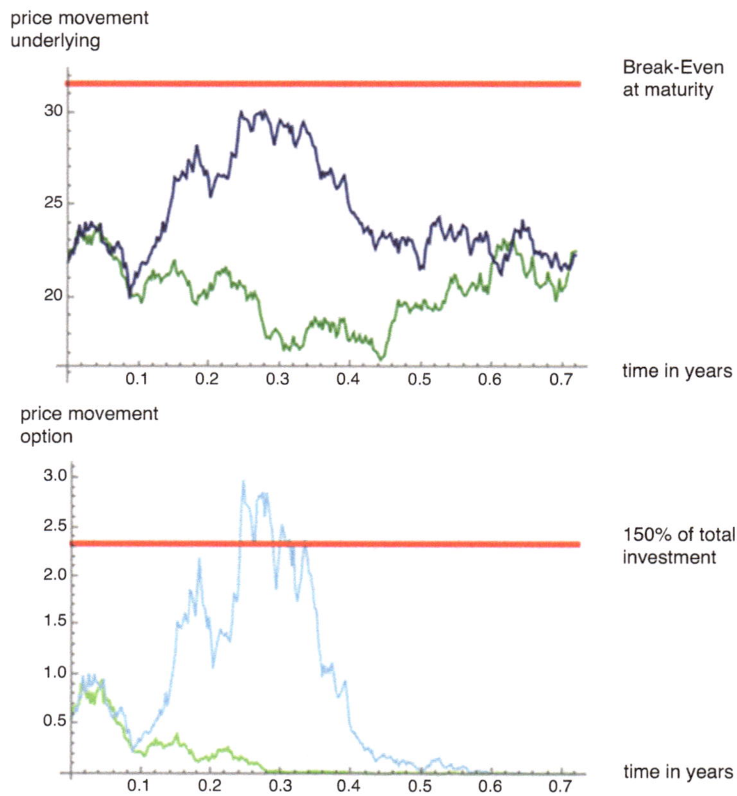

Fig. 3.9 Option price movement in relation to the simulated price paths of the underlying stock and potential profit-taking at 50% profit

Table 3.6 Profit-taking probabilities at 10, 50 or 100% with commissions

Underlying	Profit probability at expiration	Profit probability if closed at 10%	Profit probability if closed at 50%	Profit probability if closed at 100%
AA	1.60	8.90	8.20	5.90
BB	6.57	23.70	16.70	12.90
CC	7.60	25.40	20.30	14.70
DD	5.50	26.20	20.90	12.40
EE	7.93	29.10	22.10	17.80
FF	5.93	29.00	19.40	14.30
GG	6.66	32.90	26.00	18.00
HH	15.20	30.70	21.10	18.60
II	19.13	37.10	25.10	22.70
JJ	19.11	42.00	31.00	23.60

The red horizontal line in the lower part of the picture is at $1.5 \times (0.55 + 1) = 2.33$. If the option price exceeds this value before expiration and the option is then closed immediately, then a profit of 50% is made on the investment of 1.55. In the case of the blue price path, closing and 50% profit-taking would have occurred approximately after a quarter of a year.

The following Table 3.6 lists the profit-taking probabilities, calculated using Monte Carlo simulation, at 10, 50, or 100% in the case with commissions.

3.2.6 Profit-Taking Probabilities for the Option Portfolio

To calculate the probabilities of making a profit at expiration or of making a profit through early profit-taking for the entire portfolio, we basically proceed in the same way as above; only this time, we simulate the ten price paths of the ten underlying stocks in parallel according to the respective correlation (remember: we assumed a correlation of 0.3 between each pair of return paths).

The price paths are generated using the corresponding Cholesky decomposition of the correlation matrix, as explained in Volume I Section 4.8.

When it comes to computing the **profit-making probability at expiration**, the resulting ten payoffs at the (different!) expiration dates $T_1, \ldots, T_{10}$ are added up for each scenario. If this sum is higher than the sum of the initially paid option prices plus commissions, then a profit was made in this scenario, otherwise a loss. We run 10,000 simulations again and divide the number of scenarios that generate a profit by 10,000. This gives us an estimate of the profit-making probability.

When it comes to computing the **profit-making probability under the assumption of potential profit-taking of x%**, the ten option prices that result at **any** point in time are added up for each scenario. If at any point in time, this sum is more than $\left(1 + \frac{x}{100}\right)$ times the sum of the initially paid option prices plus commissions, then a profit-taking (and therefore a profit) has occurred in this scenario; otherwise

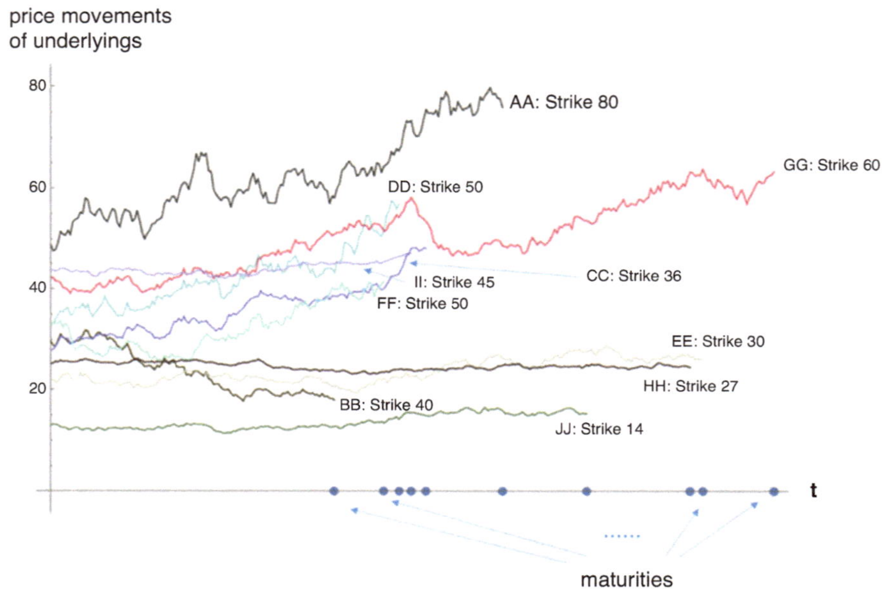

Fig. 3.10 Possible scenario of the correlated paths of the ten underlying stocks

we would have to wait for the expiration date to see whether or not a profit would occur then. We run 10,000 simulations again and divide the number of scenarios that generate a profit by 10,000. This gives us an estimate of the profit-making probability.

Of course, we could also analyse here the procedure for the case that each individual option is closed as soon as a profit of $x\%$ can be achieved from it.

In the following, we limit our analyses to the probability of making a profit at expiration (i.e. the first of the two cases described above).

In Fig. 3.10 we see a possible scenario of the correlated paths of the ten underlying stocks until the respective options expire.

In this scenario, the ten underlying stocks have the following values at expiration and thus the following payoffs:

So in this scenario, the total payoff would have been 30,537 USD, at an investment in options of 21,967 USD, yet a total initial payment (including commissions) of $21{,}967 + 29{,}900 = 51{,}867$ USD (compare Table 3.7).

Taking the commissions into account, this scenario would therefore have resulted in a loss.

If we now run this calculation 10,000 times, we obtain an approximation for the portfolio's profit-making probability of 8.35% in a (representative) simulation.

Table 3.7 Payoffs of the ten underlying stocks

Underlying	Price at expiration	Strike	Payoff per unit	Number of contracts	Total payoffs	Option price	Total option prices	Fees
AA	75.83	80	0	30	0	0.16	480	3.000
BB	17.95	40	0	4	0	0.44	176	400
CC	48.12	36	12.12	2	2.423	0.47	94	200
DD	56.92	50	6.92	19	13.150	0.53	1.007	1.900
EE	26.20	30	0	67	0	0.55	3.685	6.700
FF	41.52	50	0	21	0	0.60	1.260	2.100
GG	63.23	60	3.23	19	6143	0.75	1.425	1.900
HH	24.47	27	0	80	0	0.85	6.800	8.000
II	47.21	45	2.21	17	3752	1.20	2.040	1.700
JJ	15.27	14	1.27	40	5069	1.25	5.000	4.000
Total					30,537		21,967	29,900

3.3 Case Study III: Pricing an Interest Rate Swap and Analysing Its Suitability for Optimizing a Loan Portfolio

This case study deals with one particular aspect of a much larger issue in a court case that had caused quite a stir and public interest.

It is important to emphasize here, more so than in any of the other case studies discussed in this chapter, that the problem we present here is only one small aspect of a much larger investigation, that we describe only one of various approaches, and that all the statements, findings, and argumentations made in this context have been subject to sharp rebuttals, objections, and controversial discussions by lawyers and expert witnesses with counter-opinions. The following should be seen in this light.

3.3.1 Presentation of the Case, the Products, and the Issues

A large municipality in the euro area, we will refer to it as "Investor *A*", had taken out a large CHF loan in 2005 with a bank *C* with a variable interest rate and maturity in 2017.

In 2007, Investor *A* entered into a swap arrangement (referred to as "the swap in the following") with another Bank *B* in which Bank *B* essentially (see below) committed to taking over the variable loan interest due from Investor *A*, yet subject to a side agreement linked to the EUR CHF exchange rate.

The intention of this transaction—and this is material to the following—was, in the wording of that contract, *to optimize the loan portfolio of Investor A*.

However, unfortunately for Investor *A*, the EUR CHF exchange rate took such a fatal course over the following years that the arrangement led to a fiasco for *A*. Investor *A* subsequently refused to fulfil its obligations under the side agreement any further and was promptly and understandably sued by Bank *B*.

In the ensuing legal proceedings, there were two issues in particular (among many others) that we will address in this chapter. Before we go into these issues, we want to first describe how the products in this case were structured:

The Loan
Borrower: Investor *A*
Loan period: 1 October 2005 to 30 September 2017
Loan amount: 200,000,000 CHF
Loan interest: CHF 6-month LIBOR + 0.05%
Payments: Semi-annually on 31 March and 30 September of each year

The movement of the CHF 6-month LIBOR during the loan period is shown in Fig. 3.11. The red part of the curve represents the LIBOR movement from when the loan was taken out, on 1 October 2005, to when the swap arrangement was entered into, on 1 February 2007; the blue part of the curve represents the movement from when the swap was entered into on 1 February 2007 to when the loan became due for repayment on 30 September 2017. Upon opening the loan, the CHF 6-month LIBOR was 0.88%.

On 1 February 2007, the CHF 6-month LIBOR was 2.32%. Not surprisingly at that point, Investor *A* began to think about possible ways to hedge against strong hikes in the CHF interest rate and thus growing borrowing costs. And so, on 1 February 2007, the following swap arrangement was entered into with Bank *B*.

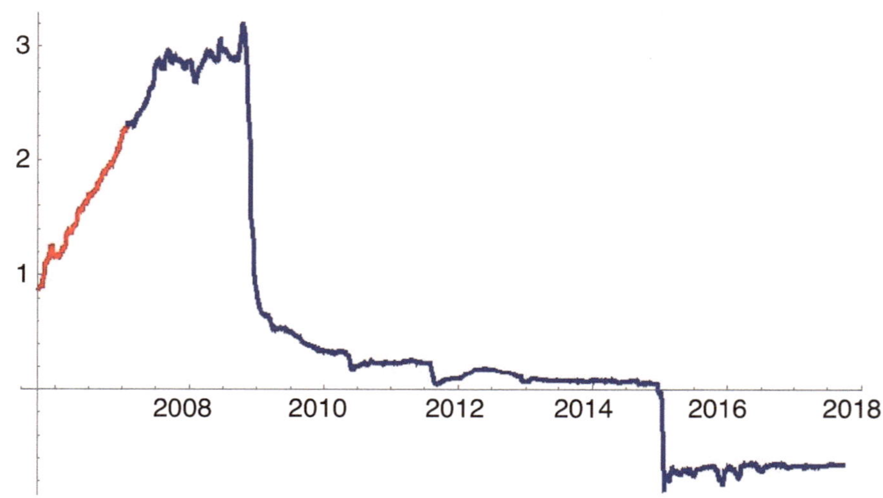

Fig. 3.11 Evolution of the CHF 6-month LIBOR over the loan period (start of loan to start of swap in red, start of swap to maturity of loan in blue)

The Swap
Loan period: From 1 February 2007 to 31 January 2017
Notional principal amount: 200,000,000 CHF

Bank B pays to Investor A:
Six-month CHF LIBOR as determined on the preceding payment date. Fixed rate for the first payment date of 2.18% p.a. (Note that the bank does not use the loan interest rates that would actually accrue as per the respective **current** LIBOR value but instead uses the interest rate that was valid on the last payment date.)
Payments: Semi-annually on 31 March and 30 September of each year

Investor A pays to Bank B:
0.07%
In addition, Investor A pays to Bank B:
If the EUR CHF exchange rate is below 1.54 on a payment date, the interest rate $\frac{1.54-\text{EUR CHF}}{\text{EUR CHF}} \cdot 100\%$ of the notional principal amount.
Payments: Semi-annually on 31 March and 30 September of each year

For example:
If the EUR CHF exchange rate is 1.60 on a payment date, then A pays to B 0.07% per annum of 200 million CHF. If the EUR CHF exchange rate is 1.40 on a payment date, then A pays to B $0.07\% + \frac{1.54-1.40}{1.40} \cdot 100\% = 0.07\% + 10\% = 10.07\%$ per annum of 200 million CHF.

When the swap agreement was entered into on 1 February 2007, the EUR CHF exchange rate was 1.62.

In Fig. 3.12, where we show the amount of payments from A to B as a function of the EUR CHF exchange rate, we see what dire consequences this agreement can have for Investor A.

Figure 3.13 shows how the EUR CHF exchange rate moved during the swap period, while Fig. 3.14 shows which payments were to be made by Investor A under this swap.

The effects of the massively strengthening Swiss franc were fatal for A. The following Table 3.8 lists the payments that were due on each of the payment dates under the swap.

The total amount of payments was 810,846,210 CHF (i.e. more than four times the original loan amount of 200 million CHF).

When Investor A stopped making payments in the course of 2011 and the bank subsequently instituted a lawsuit, the investor's line of defence was based on two claims:

1. The swap entered into on 1 February 2007 (at a price of 0) was already highly disadvantageous for the investor. At that time, the swap already had a strongly negative value. The bank had therefore taken massive advantage of the investor at the time, and the transaction should therefore be reversed.
2. In 2007, the bank had sold the swap to the investor with the wording that the intent was "to optimize a loan portfolio". This wording had been massively misleading,

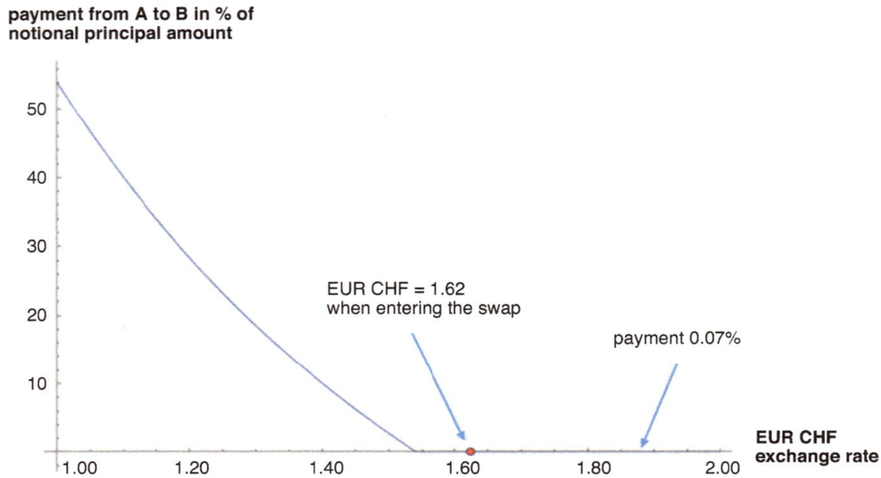

Fig. 3.12 Payments in percent per annum from *A* to *B* as a function of the EUR CHF exchange rate

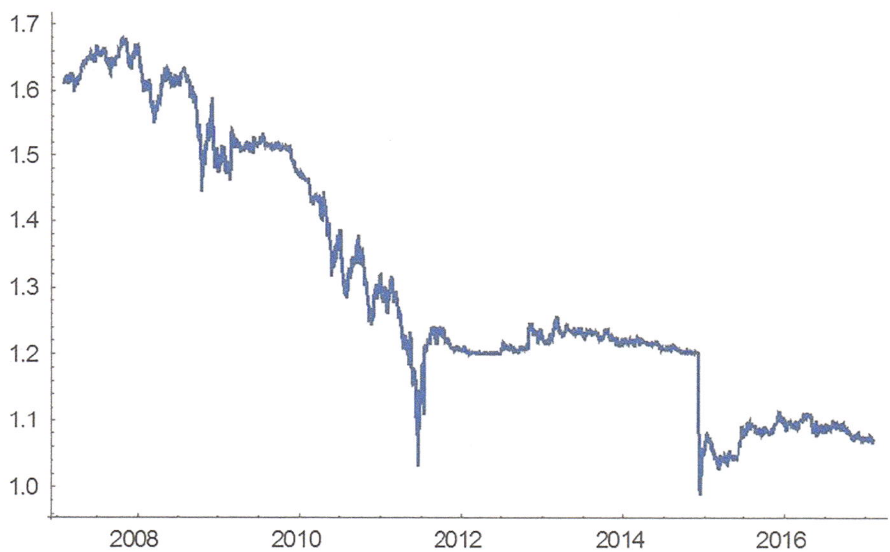

Fig. 3.13 Movement of the EUR CHF exchange rate during the swap period from 1 February 2007 to 31 January 2017

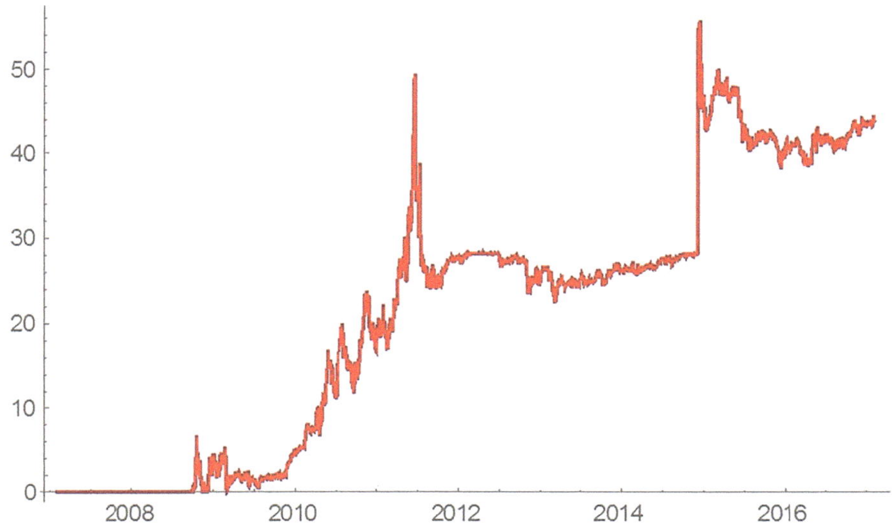

Fig. 3.14 Payments by *A* to *B* as a per-annum percentage of the notional amount

Table 3.8 Payments due on each payment date of the swap

Date	Payment
7 March	140,000
7 Sept.	140,000
8 March	140,000
8 Sept.	140,000
9 March	4,128,400
9 Sept.	3,173,600
10 March	16,448,400
10 Sept.	30,062,000
11 March	36,838,000
11 Sept.	53,500,800
12 March	55,964,800
12 Sept.	54,979,200
13 March	53,133,200
13 Sept.	51,776,000
14 March	53,034,320
14 Sept.	55,466,200
15 March	95,074,400
15 Sept.	83,020,230
16 March	81,547,030
16 Sept.	82,139,630

so the investor argued, the bank had therefore given incorrect advice and the transaction should therefore be reversed.

In the following, we will discuss to what extent these claims by the investor hold up to objective scrutiny.

3.3.2 Fair Value of the Swap at Inception: Part 1 – Interest Rate Swap Component

The swap in this case is a conventional interest rate swap combined with a derivative (actually a bundle of options) on an exchange rate as the underlying asset.

The interest swap component of that swap (part 1) exchanges the fixed interest rate of 0.07% per annum (A to B) for the floating CHF 6-month LIBOR per annum (B to A) on the payment dates of 31 March and 30 September of each year from 2007 to 2016.

(So, on every semi-annual (!) payment date, half the amounts are paid.)

The option component of the swap (part 2) is a bundle of 20 European options expiring on 31 March and 30 September, respectively, in each of the years 2007 to 2016. On each of these expiration dates, a payment is made in an amount that depends on the EUR CHF exchange rate on that day. Thus the options are not path-dependent.

We **first determine the value of part 1 of the swap:**

The **fixed payments** of 0.07% each that Investor A has to pay in CHF at 20 points in time T_i in the future are to be discounted with the respective discount factor $d_i^{(\text{CHF})}$, which is calculated from $f_{0,T_i}{}^{(\text{CHF})}$, thus $d_i^{(\text{CHF})} = \frac{1}{\left(1+f_{0,T_i}{}^{(\text{CHF})}\right)^{T_i}}$. Here, $f_{0,T_i}{}^{(\text{CHF})}$ denotes the risk-free interest rate in CHF for the time period $[0, T_i]$.

We reiterate here the relevant argument: If at time 0 (now), I want to secure a payment of X CHF at time T, I need to take out a loan now of $\frac{x}{\left(1+f_{0,T}{}^{(\text{CHF})}\right)^{T}}$ CHF and maturity in T. So, the cost of a guaranteed payment of X CHF at time T is $X \cdot d^{(\text{CHF})}$.

A **variable payment** that is paid out at T_i in the amount of the CHF 6-month LIBOR (denoted here by z_i), determined at T_{i-1} (z_i being the interest rate that applies at time T_{i-1} for semi-annual interest payments for the following half-year. A total amount of 1 CHF for this half-year thus yields an interest income of $\frac{z_i}{2}$ CHF.), has a value of $\frac{1}{2} \cdot d_i^{(\text{CHF})} \cdot f_{T_{i-1},T_i}{}^{(\text{CHF})}$ from the perspective of time 0. Here, $f_{T_{i-1},T_i}{}^{(\text{CHF})}$ denotes the forward interest rate (of the second type) from the perspective of time 0.

While this should already be clear from the definition of these forward interest rates and also from the discussion in Volume II Section 3.8, we present the relevant argument again here for the sake of completeness or as a refresher:

To secure a payment of $\frac{z_i}{2}$ CHF at time T_i at time 0, the first thing we can do is agree on the interest rate $f_{T_{i-1},T_i}{}^{(\text{CHF})}$ for a loan taken out for the period $[T_{i-1}, T_i]$.

Then we invest the amount of $\frac{1}{2} \cdot d_i^{(\mathrm{CHF})} \cdot f_{T_{i-1},T_i}{}^{(\mathrm{CHF})}$ CHF until time T_i at an interest rate of $f_{0,T_i}{}^{(\mathrm{CHF})}$. By time T_i, this amount will have attained the value $\frac{1}{2} \cdot f_{T_{i-1},T_i}{}^{(\mathrm{CHF})}$ CHF.

At T_{i-1}, we take out a loan in the amount of 1 CHF at the agreed semi-annual interest rate of $f_{T_{i-1},T_i}{}^{(\mathrm{CHF})}$, and we invest this amount at the then applicable interest rate of z_i for half a year until T_i. By T_i, this amount will have attained the value $1+\frac{z_i}{2}$ CHF.

However, at T_i we have to repay the loan, i.e. $1+\frac{1}{2} \cdot f_{T_{i-1},T_i}{}^{(\mathrm{CHF})}$. All in all, what is left is an amount of $\frac{z_i}{2}$. The cost spent on this at time 0 was $\frac{1}{2} \cdot d_i^{(\mathrm{CHF})} \cdot f_{T_{i-1},T_i}{}^{(\mathrm{CHF})}$.

Caveat Here it is important to note that the LIBOR paid at time T_i is in fact the value determined at T_{i-1}; otherwise such a "simple" model-independent type of valuation would not be possible, and we would have to use model assumptions about the LIBOR movement instead!

Thus, to determine the fair value of part 1 of the swap, we only need the interest rates $f_{0,T_i}{}^{(\mathrm{CHF})}$ from 1 February 2007. From these we calculate both the $d_i^{(\mathrm{CHF})}$ and the $f_{T_{i-1},T_i}{}^{(\mathrm{CHF})}$, and the value of the swap's first part from Investor A's perspective then has the value

$$100.000.000 \cdot \sum_{i=1}^{20} d_i^{(\mathrm{CHF})} \cdot \left(f_{T_{i-1},T_i}{}^{(\mathrm{CHF})} - 0.0007 \right).$$

Using the values for the interest rates determined or interpolated for 1 February 2007, we obtain Table 3.9:

Evaluating the above sum with these values, we get the **fair value of part 1 of the swap** of 47,321,000 CHF from the perspective of 1 February 2007. Based on the EURCHF exchange rate of 1.62 on 1 February 2007, this results in a value of **29,210,494 EUR**.

Table 3.9 $f_{T_{i-1},T_i}{}^{(\mathrm{CHF})}$ and $d_i^{(\mathrm{CHF})}$ for the determination of the fair value of part 1 of the swap

i	1	2	3	4	5	6	7	8	9	10
$f_{T_{i-1},T_i}{}^{(\mathrm{CHF})}$	2.18	2.44	2.67	2.70	2.74	2.73	2.77	2.79	2.80	2.80
$d_i^{(\mathrm{CHF})}$	0.98	0.97	0.96	0.94	0.93	0.92	0.90	0.89	0.88	0.87

i	11	12	13	14	15	16	17	18	19	20
$f_{T_{i-1},T_i}{}^{(\mathrm{CHF})}$	2.82	2.89	2.92	2.99	3.02	3.04	3.07	3.11	3.13	3.15
$d_i^{(\mathrm{CHF})}$	0.85	0.84	0.83	0.82	0.80	0.79	0.78	0.77	0.75	0.74

3.3.3 Fair Value of the Swap at Inception: Part 2 – Option Component

To determine the fair value of part 2 of the swap—the option component—as of 1 February 2007, we proceed using Monte Carlo simulation as proposed in Volume II Section 2.2 and Section 2.16.

In doing so, we need to be mindful of the following: The payoffs of the swap component are in CHF, and the underlying asset is the price of one EUR in CHF. We therefore have to discount using CHF discount factors as well as define, in the risk-free exchange rate representation, the CHF interest rates as the "home currency interest rates" and the EUR interest rates as the "foreign currency interest rates".

For each individual maturity date T_i, we therefore model the movement of the EUR CHF exchange rate $S(t)$ using a risk-neutral Wiener model of the following form:

$$S(t) = S(0) \cdot e^{\left(r(c)_i - r(e)_i - \frac{\sigma_i^2}{2}\right) \cdot t + \sigma_i \cdot w_t}.$$

Here, $S(0) = 1.62$ and W_t is the standard Brownian motion.

We write $r(c)_i$ as short for the $f_{0,T_i}{}^{(\mathrm{CHF})}$ that we already used above, and $r(e)_i$ denotes the corresponding EUR interest rate $f_{0,T_i}{}^{(\mathrm{EUR})}$.

Again, we do not give the individual interest rates in detail here, and we are also not going to discuss the correct choice of the respective interest rates any further at this point. We only give the required difference $r(c)_i - r(e)_i$ (as a percentage in the Table 3.10; the calculations obviously require absolute numbers) for the different time periods.

By σ_i we denote the implied volatility of the EUR CHF exchange rate, calculated as of 1 February 2007 from EUR CHF options expiring (approximately) by T_i. The following chart (Fig. 3.15) from Bloomberg presents a breakdown of such implied EUR CHF volatilities (calculated from both bid and ask prices of various OTM, ATM, and ITM options with different expiries) from roughly the relevant date of 1 February 2007. Again, these values are given as percentages but must be used as absolute numbers in the calculations.

Of course, for a full (and, as in this case, highly delicate) expert opinion, it is necessary to discuss in depth which volatilities exactly are used here and for what reason. Again, this is a discussion we will not be going into here. We see that the volatilities that are relevant to us are almost exclusively in the range between 2.5 and

Table 3.10 Difference of $r(c)_i$ and $r(e)_i$ as percentage

i	1	2	3	4	5	6	7	8	9	10
$r(c)_i - r(e)_i$	−1.64	−1.64	−1.59	−1.55	−1.53	−1.51	−1.51	−1.50	−1.50	−1.49

i	11	12	13	14	15	16	17	18	19	20
$r(c)_i - r(e)_i$	−1.47	−1.47	−1.47	−1.47	−1.45	−1.44	−1.44	−1.44	−1.44	−1.43

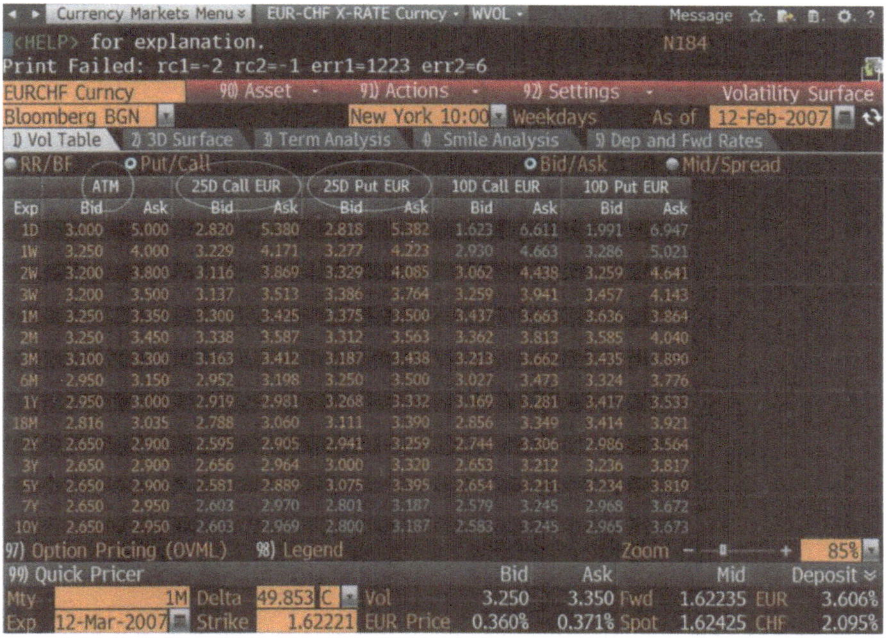

	ATM		25D Call EUR		25D Put EUR		10D Call EUR		10D Put EUR	
Exp	Bid	Ask	Bid	Ask	Bid	Ask	Bid	Ask	Bid	Ask
1D	3.000	5.000	2.820	5.380	2.818	5.382	1.623	6.611	1.991	6.947
1W	3.250	4.000	3.229	4.171	3.277	4.223	2.930	4.663	3.286	5.021
2W	3.200	3.800	3.116	3.869	3.329	4.085	3.062	4.438	3.259	4.641
3W	3.200	3.500	3.137	3.513	3.386	3.764	3.259	3.941	3.457	4.143
1M	3.250	3.350	3.300	3.425	3.375	3.500	3.437	3.663	3.636	3.864
2M	3.250	3.450	3.338	3.587	3.312	3.563	3.362	3.813	3.585	4.040
3M	3.100	3.300	3.163	3.412	3.187	3.438	3.213	3.662	3.435	3.890
6M	2.950	3.150	2.952	3.198	3.250	3.500	3.027	3.473	3.324	3.776
1Y	2.950	3.000	2.919	2.981	3.268	3.332	3.169	3.281	3.417	3.533
18M	2.816	3.035	2.788	3.060	3.111	3.390	2.856	3.349	3.414	3.921
2Y	2.650	2.900	2.595	2.905	2.941	3.259	2.744	3.306	2.986	3.564
3Y	2.650	2.900	2.656	2.964	3.000	3.320	2.653	3.212	3.236	3.817
5Y	2.650	2.900	2.581	2.889	3.075	3.395	2.654	3.211	3.234	3.819
7Y	2.650	2.950	2.603	2.970	2.801	3.187	2.579	3.245	2.968	3.672
10Y	2.650	2.950	2.603	2.969	2.800	3.187	2.583	3.245	2.965	3.673

Fig. 3.15 Implied EUR CHF volatilities as of 12 February 2007 (source: Bloomberg)

3.5%. For all our computations, we will therefore perform evaluations once with the value 2.5% for all σ_i and once with 3.5% for all σ_i.

For valuation purposes, we are now going to proceed in the exact same way as we did in Volume II Section 2.16 for pricing options using Monte Carlo. (We note here that these options could also be priced using an exact pricing formula for so-called quanto options. However, since we have not derived this formula so far, we will again resort to the Monte Carlo method.)

We price the 20 options individually (using the exact same scheme, of course) and add up the individual values to the swap component's total value.

For this purpose, we again simulate a large number of possible scenarios, i.e. paths (the step length for the simulations being 1 week each). For each path, we determine the resulting payoff for that path (= interest rate payable by Investor A at time T_i in the option component), discount it to time 0 using the discount factors $d_i^{(\mathrm{CHF})}$, and calculate the mean over these discounted payoffs. The result is the fair value of the options from the perspective of 1 February 2007, expressed in CHF. Dividing by 1.62 we get the value expressed in EUR.

Below we show only for the option with the longest period to expiration on 30 September 2016 and for volatility 3.5% (upper bound), a selection (Fig. 3.16) of 30 scenarios (paths) and the convergence behaviour (Fig. 3.17) towards the value of the option. In the following table, we list the values of the individual options as a percentage of the notional value.

Valuations of the individual options expiring on the ith payment date for two volatility values are given in the subsequent table (Table 3.11):

We now obtain the total valuation by adding up, i.e. 76.57% for $\sigma = 2.5\%$ and 93.65% for $\sigma = 3.5\%$.

If we apply the interest rates to the notional principal to determine the payment amounts, however, we need to divide these values by 2, since the results are per-annum interest rates, yet interest is paid semi-annually. The value of the total

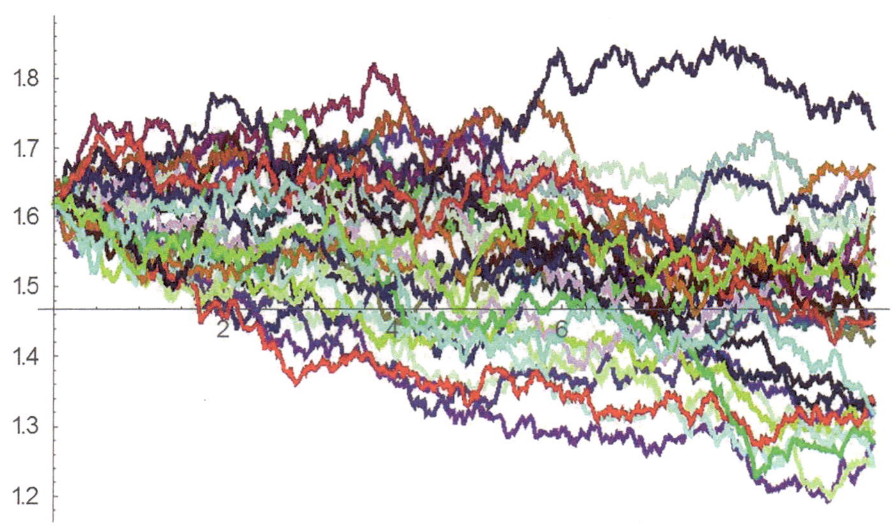

Fig. 3.16 30 scenarios (= risk-neutral paths) of EUR CHF from 2007 to 2017

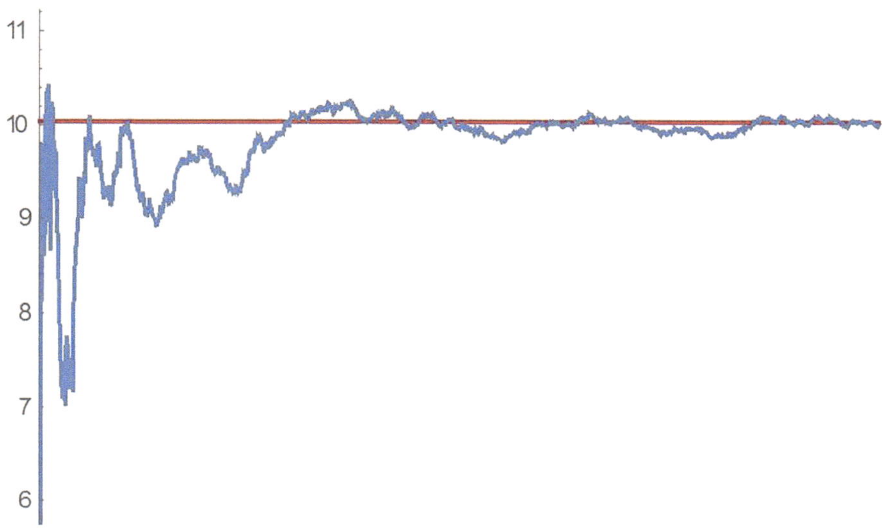

Fig. 3.17 Convergence behaviour in Monte Carlo simulation of the option with the longest expiration period and volatility 3.5 (approximate value 10.04)

Table 3.11 Valuations of the individual options expiring at the ith date for different σ

i	1	2	3	4	5	6	7	8	9	10
Value of option for $\sigma = 2.5\%$	0.07	0.08	0.26	0.43	0.79	1.30	1.89	2.08	2.58	3.13
Value of option for $\sigma = 3.5\%$	0.69	0.16	0.53	0.98	1.49	1.94	2.59	2.76	3.62	4.33

i	11	12	13	14	15	16	7	18	19	20
Value of option for $\sigma = 2.5\%$	3.73	4.51	4.86	5.43	6.10	6.62	7.26	7.73	8.38	9.34
Value of option for $\sigma = 3.5\%$	4.77	5.35	6.16	6.47	7.04	7.69	8.59	8.88	9.57	10.04

option component is therefore CHF 76,570,000 if we choose $\sigma = 2.5\%$ and CHF 93,650,000 if we choose $\sigma = 3.5\%$.

Converted to EUR at the rate of 1.62, we get:

The value of the total option component is approximately 47,265,000 EUR if we choose $\sigma = 2.5\%$ and 57,809,000 EUR if we choose $\sigma = 3.5\%$.

From Investor A's perspective, these are negative values of course.

Thus, for the entire swap, taking into account the value of part 1 of approximately 29,210,000 EUR, we obtain an upper threshold and a lower threshold for the entire swap from the perspective of Investor A as follows:

29,210,000–57,809,000 < swap value from the perspective of Investor A < 29,210,000–47,265,000,
i.e. **−28,599,000 EUR < swap value when entered into < −18,055,000 EUR**.

So in fact, for purposes of Investor A, this swap had a clearly negative value when it was entered into on 1 February 2007.

Of course, a real expert opinion would require additional discussions of concomitant circumstances (liquidity of the product, types of parameters to use, ...). Yet the calculated outcome in this case is so clear that even such discussions would not change the fundamental conclusion.

3.3.4 Regarding the Intent of Achieving "Portfolio Optimization" by Using the Swap

Of course, any spontaneous reaction along the lines of *"Obviously this swap is anything but portfolio optimization, being that it ended in such a fiasco!"* would be overly hasty, since we would be looking at it a posteriori. Rather, we must ask: *"Did the bank present a sound and stringent rationale a priori that justified using this product for portfolio optimization?"*

Here we will simply explain—for reasons that we will go into later—our fundamental approach to this question and present a selection of findings. We will not go into the details of the calculations or simulations.

Preliminary Remark
If a financial product (especially a potentially high-risk one) is proposed by the bank to the customer as a portfolio optimization measure, then the associated ex ante analyses and simulations, taking into account the forecasts prevailing on the market at the time of the analyses, must show that the addition of the product to the portfolio does indeed lead to optimization in the vast majority of realistic scenarios.

For the parameters and models used below, we are going to use the following terminology:

- Parameter or model choices that seek to optimally reflect real-market conditions and forecasts will be referred to as **"realistic choices"**.
- Parameter or model choices that are more likely than the realistic choice to lead to a result where portfolio optimization is indeed present will be referred to as **"optimistic choices"** (from the perspective of the bank pushing such a product).
- Parameter or model choices that are less likely than the realistic choice to lead to a result where portfolio optimization is indeed present will be referred to as **"pessimistic choices"** (from the perspective of the bank pushing such a product).

Now, how should we define "portfolio optimization" for a given parameter choice? For this purpose, we recall the concept of the volatility/return diagram in the portfolio selection theory discussed in Chap. 2. "Portfolio improvement" is when volatility decreases significantly while returns remain essentially constant or increase or, at worst, decrease slightly or when returns increase significantly while volatility remains essentially constant or decreases or, at worst, increases slightly.

The relevant price movements in our case are those of the EUR CHF exchange rate and of the CHF 6-month LIBOR.

In the following, we will therefore select suitable models for simulating these two movements (paths), generate a large number of scenarios (100,000 each) for different parameter sets, and determine the payoffs (or the payments to be made by Investor A) once for the loan without swap and once for the loan with swap. In both cases, we will determine the average returns and standard deviations of the payments (as well as other technical ratios) and then compare the results for the two cases in terms of whether—in the light of the above—we can indeed speak of portfolio optimization.

The models we will use: For the EUR CHF exchange rate $S(t)$, we are going to use the Wiener model

$$S(t) = 1.62 \cdot e^{\left(\mu - \frac{\sigma^2}{2}\right) \cdot t + \sigma \cdot w_t},$$

where for the parameters μ and σ we will then make optimistic/realistic/pessimistic choices (as described above).

For the CHF 6-month LIBOR $L(t)$, we are going to use a mean-reverting Cox-Ross-Ingersoll model, which we introduced briefly in Volume II Section 3.3:

$$dL(t) = a \cdot (m - L(t)) \cdot dt + b \cdot \sqrt{L(t)} \cdot d\widetilde{W}_t$$

with constant $a > 0$, b and m and with fixed initial value $L(0)$.

The reason for choosing the mean-reverting Cox-Ross-Ingersoll (CIR) model was as follows: We anticipate (which won't be much of a surprise) that the result of the following analyses will be that it is not possible in the present case to speak of portfolio optimization. If we want this result, which speaks against Bank B, to hold up against inevitable criticism, we must ensure that the choice of the model does not provide attackable weaknesses.

The mean-reverting Ornstein-Uhlenbeck model (Vasicek model) and the mean-reverting CIR are the main models used for modelling interest rates. The CIR leads to higher fluctuations in the interest rate than the Vasicek model.

Higher fluctuations in the interest rate tend to lead to higher risk in the loan, and so using a swap makes more sense. In terms of portfolio optimization through the swap, one is therefore more likely to obtain positive results when using the CIR model than when using the Vasicek model. Thus, if even the CIR model leads to the conclusion that there is generally no portfolio optimization, then that conclusion is all the more valid for the Vasicek model. The choice of the CIR model is therefore not open to attack in this respect.

So far in this book, we have not looked into parameter calibration in the CIR model. Therefore, we will not be able to explicitly justify the specific parameter estimates here and will only present the procedure in principle and some select results.

Let us now look at the parameters used in both models:

The **parameters for the currency paths:**

As noted in the previous subsection, we can presuppose values between 2.5 and 3.5 for the **volatility** σ. A **realistic choice** would be the mean $\sigma = 3$.

High volatility is indicative of strong fluctuations in the exchange rate and thus a higher risk for the swap. From the bank's perspective, and with regard to potential portfolio optimization through the swap, the choice $\sigma = 3.5$ is therefore a **pessimistic choice** and $\sigma = 2.5$ an **optimistic choice**.

As regards the trend parameter μ: In the risk-free valuation (see previous section), this parameter is to be set as the difference $r(c) - r(e)$ of risk-free CHF interest rates $r(c)$ and risk-free EUR interest rates $r(e)$. This difference was—as we saw—essentially in the range of -1.50.

This implied ratio between CHF interest rates and EUR interest rates is of mathematical relevance, since it provides the risk-neutral measure on which the valuation of currency derivatives is based. It does not provide forecasts but must definitely be included in analyses.

The following Chart 3.18 shows the major banks' currency forecasts for the CHF in February 2007 with the corresponding statistics.

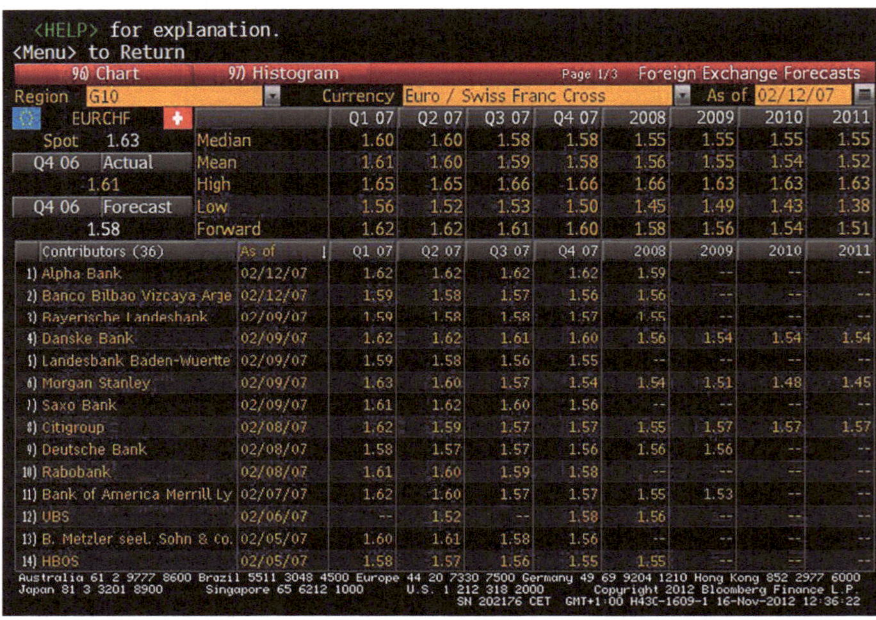

<HELP> for explanation.
<Menu> to Return

96) Chart 97) Histogram Page 1/3 Foreign Exchange Forecasts

Region G10 Currency Euro / Swiss Franc Cross As of 02/12/07

EURCHF		Q1 07	Q2 07	Q3 07	Q4 07	2008	2009	2010	2011
Spot 1.63	Median	1.60	1.60	1.58	1.58	1.55	1.55	1.55	1.55
Q4 06 Actual	Mean	1.61	1.60	1.59	1.58	1.56	1.55	1.54	1.52
1.61	High	1.65	1.65	1.66	1.66	1.66	1.63	1.63	1.63
Q4 06 Forecast	Low	1.56	1.52	1.53	1.50	1.45	1.49	1.43	1.38
1.58	Forward	1.62	1.62	1.61	1.60	1.58	1.56	1.54	1.51

Contributors (36)	As of	Q1 07	Q2 07	Q3 07	Q4 07	2008	2009	2010	2011
1) Alpha Bank	02/12/07	1.62	1.62	1.62	1.62	1.59	--	--	--
2) Banco Bilbao Vizcaya Arge	02/12/07	1.59	1.58	1.57	1.56	1.56	--	--	--
3) Bayerische Landesbank	02/09/07	1.59	1.58	1.58	1.57	1.55	--	--	--
4) Danske Bank	02/09/07	1.62	1.62	1.61	1.60	1.56	1.54	1.54	1.54
5) Landesbank Baden-Wuertte	02/09/07	1.59	1.58	1.56	1.55	--	--	--	--
6) Morgan Stanley	02/09/07	1.63	1.60	1.57	1.54	1.54	1.51	1.48	1.45
7) Saxo Bank	02/09/07	1.61	1.62	1.60	1.56	--	--	--	--
8) Citigroup	02/08/07	1.62	1.59	1.57	1.57	1.55	1.57	1.57	1.57
9) Deutsche Bank	02/08/07	1.58	1.57	1.57	1.56	1.56	1.56	--	--
10) Rabobank	02/08/07	1.61	1.60	1.59	1.58	--	--	--	--
11) Bank of America Merrill Ly	02/07/07	1.62	1.60	1.57	1.57	1.55	1.53	--	--
12) UBS	02/06/07	--	1.52	--	1.58	1.56	--	--	--
13) B. Metzler seel. Sohn & Co.	02/05/07	1.60	1.61	1.58	1.56	--	--	--	--
14) HBOS	02/05/07	1.58	1.57	1.56	1.55	1.55	--	--	--

Australia 61 2 9777 8600 Brazil 5511 3048 4500 Europe 44 20 7330 7500 Germany 49 69 9204 1210 Hong Kong 852 2977 6000
Japan 81 3 3201 8900 Singapore 65 6212 1000 U.S. 1 212 318 2000 Copyright 2012 Bloomberg Finance L.P.
SN 202176 CET GMT+1:00 H43C-1609-1 16-Nov-2012 12:36:22

Fig. 3.18 Currency forecasts (February 2007) (source: Bloomberg)

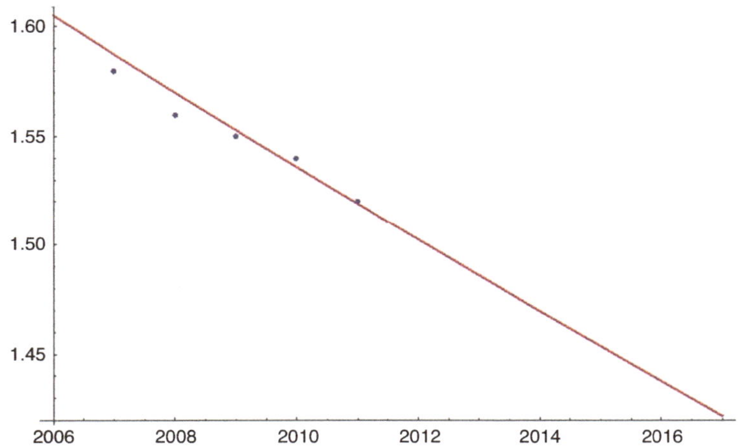

Fig. 3.19 Forecasted trends of the EUR CHF exchange rate from the perspective of February 2007

The mean value of these forecasts clearly shows a downward trend:

Spot: 1.63, 4th Quarter 2007: 1.58, 2008: 1.56, 2009: 1.55, 2010: 1.54, 2011: 1.52

To analyse Investor *A*'s portfolio with regard to "portfolio optimization" through a high-risk construction, it is imperative that these currency forecasts from February 2007 be included in that analysis. The negative trend shown in these forecasts is above the trend of the implied CHF to EUR interest rate ratio of -1.5%. Continuing the forecasts (black dots) to the relevant year 2017, we get approximately the following Picture 3.19:

The red line shows an exponentially decreasing approximation to these forecasts with a trend of -1.1% p.a. In our further analyses, we will therefore use the drift of $\boldsymbol{\mu = -1.1}$, which results from the banks' currency forecasts in February 2007, as the **realistic estimate** for the exchange rate trend.

A stronger negative trend leads to substantially higher risk for the swap construction. The choice $\boldsymbol{\mu = -1.5}$ (which follows implicitly from the interest rate ratio) is therefore to be regarded as the **pessimistic choice**.

Thus, a definitely **optimistic choice** would be $\boldsymbol{\mu = 0}$.

The **parameters for the interest rate paths:**

As we argued when choosing the CIR model, in the case of widely fluctuating interest rates, it will tentatively make more sense to use a swap to mitigate the risk of the loan. The same applies, of course, if interest rates rise sharply. Thus, a higher mean reversion parameter m and a higher diffusion parameter b are optimistic parameter choices in the CIR model. The calibration analyses, which we cannot go into any further here, have thus lead to:

Optimistic choices $\boldsymbol{m = 5\%}$ and $\boldsymbol{b = 0.39}$
Realistic choices $\boldsymbol{m = 3.44\%}$ and $\boldsymbol{b = 0.318}$

Pessimistic choices $m = 2\%$ and $b = 0.25$
which are the ones we will work with in the following.

Of interest for the following simulations is also the question as to a potential dependence between the two Brownian motions W_t and $\widetilde{W}_t$. However, such a dependence between the currency path and the interest rate path is already implicitly present since the choice of the drift μ is influenced, in a certain way, by the interest rate differentials. Additional modelling of a dependence between W_t and $\widetilde{W}_t$ will hardly bring about any additional insights nor change the results.

We used Mathematica again for our modelling and give below, exported directly from Mathematica, the simulation results for the case in which all parameters were chosen realistically. One hundred thousand scenarios were simulated, respectively.

The discounted payments **without the swap range** from
17,952,300 CHF
to
81,491,200 CHF.
The **average amount** of these discounted payments **without the swap** is
36.011.800 CHF
with a **standard deviation** of
6,644,670.
The discounted payments **with the swap range** from
1.0760620
to
423,662,000.
The **average amount** of these discounted payments **with the swap** is
47,585,200
with a **standard deviation** of
50,682,200.

Increase in return through swap: Factor 0.756784
Increase in risk through swap: Factor 7.62751

The following Plots 3.20 and 3.21 show the first 1000 of 100,000 simulations of possible currency and CHF-6m LIBOR paths, respectively. In addition to that, we display a histogram of payments with and without swap (see Fig. 3.22) and the associated convergence behaviour in Fig. 3.23.

So, as we can see from this thoroughly "realistic choice" of parameters, there is a very strong increase in risk (expressed by the ratio of standard deviations and by the much wider range of payments and hence the much higher VAR of payments), while the expected return decreases slightly. The increase in risk and the reduction in return are reflected by the two factors given above:

Increase in return through swap: Factor 0.756784
Increase in risk through swap: Factor 7.62751

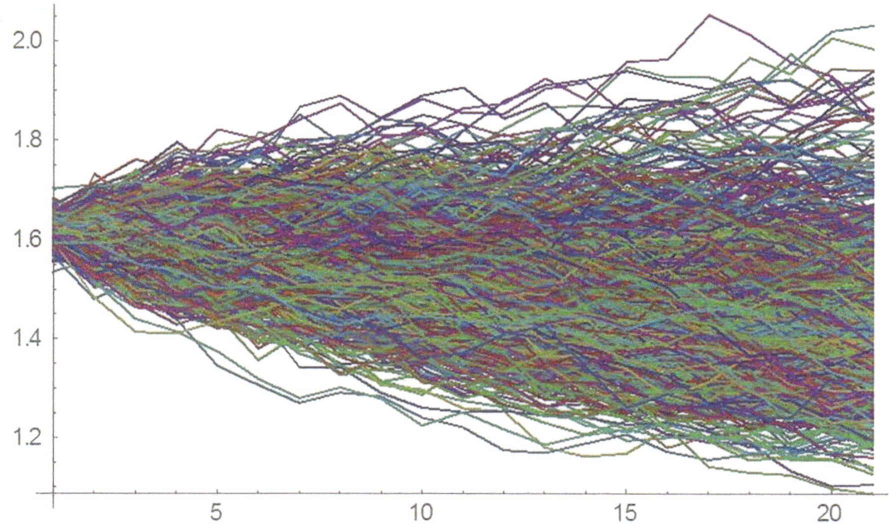

Fig. 3.20 First 1000 of 100,000 simulations of possible currency paths

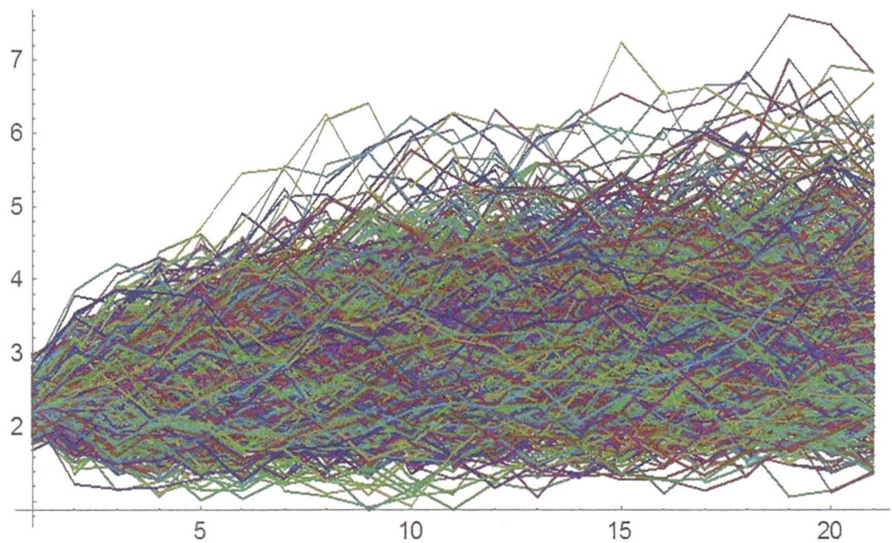

Fig. 3.21 First 1000 of 100,000 simulations of possible CHF-6m LIBOR paths

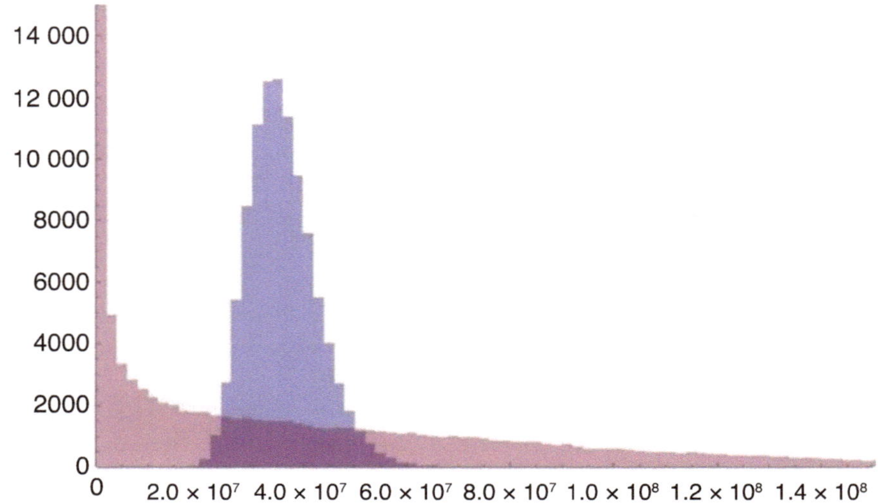

Fig. 3.22 Histograms of payments without swap (blue) compared to payments with swap (pink)

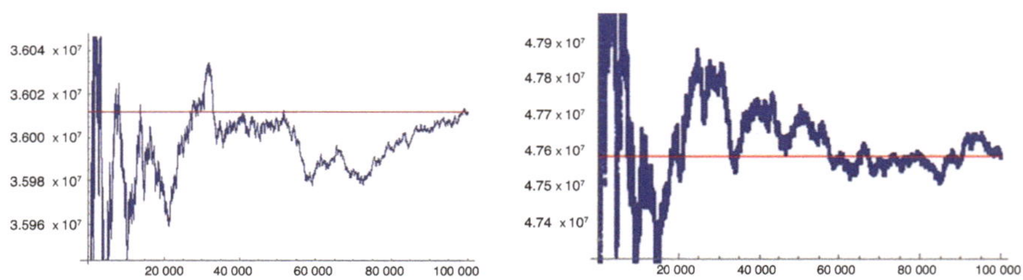

Fig. 3.23 Convergence behaviour when simulating average payment amounts without swap (left) and with swap (right)

If we run these simulations for all combinations of parameter choices and consolidate the return and risk increase factors, we get the following table (Table 3.12):

Only a few cases with very optimistic parameter estimates would allow the conclusion of potential portfolio optimization (highlighted in blue). For all other optimistic/realistic (and of course even more so for all pessimistic/realistic) parameter choices, there is no indication whatsoever of potential portfolio optimization; on the contrary, most parameter choices yield extremely adverse outcomes (highlighted in red).

Table 3.12 Result of the simulations for all possible parameter choices under the consideration of the return and risk increase factors

Currency volatility	Currency trend	Interest rate volatility	Interest mean reversion	Increase in return	Increase in risk
Realistic	Realistic	Realistic	Realistic	**0.76**	**7.6**
Realistic	Realistic	Realistic	Optimistic	**0.98**	**6.7**
Realistic	Realistic	Optimistic	Realistic	**0.75**	**6.3**
Realistic	Realistic	Realistic	Realistic	**2.63**	**3.8**
Optimistic	Realistic	Realistic	Realistic	**0.87**	**6.4**
Realistic	Realistic	Optimistic	Optimistic	**0.98**	**5.6**
Realistic	Optimistic	Realistic	Optimistic	**3.35**	**3.5**
Optimistic	Realistic	Realistic	Optimistic	**1.15**	**5.5**
Realistic	Optimistic	Optimistic	Realistic	**2.56**	**3.3**
Optimistic	Realistic	Optimistic	Realistic	**0.86**	**5.3**
Optimistic	Optimistic	Realistic	Realistic	**3.82**	**2.8**
Realistic	Optimistic	Optimistic	Optimistic	**2.55**	**3.2**
Optimistic	Realistic	Optimistic	Optimistic	**1.13**	**4.6**
Optimistic	Optimistic	Realistic	Optimistic	**5.03**	**2.5**
Optimistic	Optimistic	Optimistic	Realistic	**3.96**	**2.2**
Optimistic	Optimistic	Optimistic	Optimistic	**5.14**	**2.0**
Realistic	Pessimistic	Realistic	Realistic	**0.58**	**8.7**
Pessimistic	Realistic	Realistic	Realistic	**0.65**	**9.0**
Pessimistic	Pessimistic	Realistic	Realistic	**0.51**	**10.2**

3.4 Case Study IV: Valuation of Callable Range Accrual Swaps as Part of an Expert Opinion

As part of an expert opinion in a lawsuit, it was necessary to determine the value of the product described below at a specific point in time during its life:

3.4.1 The Product

The product was a so-called range accrual swap (RAS). This swap was entered into on ***2 May 2012*** *between a* ***Bank B*** *and an* Investor ***A***.

The **term** was 20 years **until 30 April 2032**.

The swap transaction's **notional principal amount** was **50,000,000 EUR**.

The swap's **underlying asset** was the **6-month Euribor**.

Payments were to be exchanged on 30 April 30 and 31 October of each year, beginning on 31 October 2012 and ending on 30 April 2032.

Investor *A* pays to Bank *B*:

On each payment date:

Half the value of the 6-month Euribor on the preceding payment date, applied to the principal amount.

(In particular, on 31 October 2012 (the first payment date), half the value of the 6-month Euribor on 30 April 2012.) In case of a negative 6-month Euribor, no payment will be exchanged either way.

Take, for example, the payment date of 30 April 2013:

The 6-month Euribor was 0.389% on the previous payment date on 31 October 2012. Therefore, on 30 April 2013, Investor *A* pays to Bank *B* an amount of $0.5 \times 0.00389 \times 50$ million $= 97{,}250$ EUR.

Bank *B* pays to Investor *A*:

On each payment date:

Half of 4% × "factor" × principal.

The "factor" is determined by the relative percentage of trading days in the previous period on which the 6-month Euribor was below a certain threshold. These thresholds are:

1% from 2 May 2012 to 30 April 2013
1.50% from 2 May 2013 to 30 April 2014
2% from 2 May 2014 to 30 April 2018
3% from 2 May 2018 to 30 April 2028
3.50% from 2 May 2028 to 30 April 2032

Take, for example, the payment date of 30 April 2013:

In the preceding period from 31 October 2012 to 30 April 2013, the 6-month Euribor (see Fig. 3.24) was continuously below 1%. Thus, the factor is equal to 1. Therefore, on 30 April 2013, Bank *B* pays to Investor *A* an amount of $0.5 \times 0.04 \times 1 \times 50$ million $= 1{,}000{,}000$ EUR.

Thus, the actual payment on 30 April 2013 was in the amount of $1{,}000{,}000 - 97{,}250 = 902{,}750$ EUR from Bank *B* to Investor *A*.

If, on the other hand, the 6-month Euribor in the previous period from 31 October 2012 to 30 April 2013 had been below the relevant threshold of 1% only on, say, 41 of the total 123 trading days, then the factor would have been $\frac{41}{123} = \frac{1}{3}$, and the payment from Bank *B* to Investor *A* would have been only $0.5 \times 0.04 \times \frac{1}{3} \times 50$ million $= 333{,}333$ EUR (the actual payment made would then have been 236,083).

However, this product has another essential feature that significantly impacts its value. **Bank B has a unilateral right of termination**:

Bank *B* may terminate (call back) the range accrual swap on any payment date (beginning on 31 October 2012). However, it still needs to make the payment that is due on the payment date on which the swap is terminated.

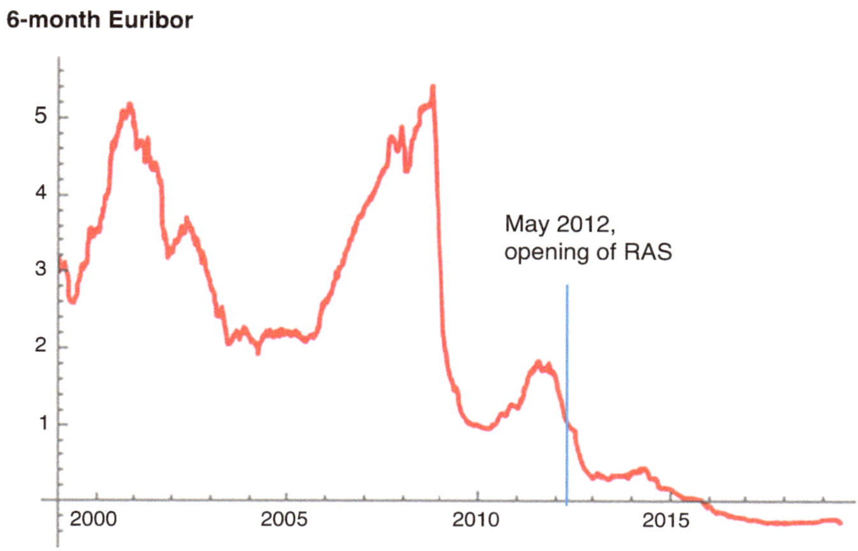

Fig. 3.24 Six-month Euribor in the period from January 1999 to June 2019

3.4.2 Motivation for Entering into the Swap and Further Background Information

Let us first take an informal look at the swap situation and gather some preliminary impressions. For this purpose, we temporarily disregard the bank's right of termination for the time being.

Investor A hopes that the 6-month Euribor will remain low over the next 20 years (from the perspective of May 2012) and will not increase strongly. If, ideally, the 6-month Euribor stays below the defined thresholds on every trading day during the swap's entire life (see illustration in Fig. 3.25), then the bank would always make payments of 4% per annum to the investor, while the investor would only have to pay the 6-month Euribor, which (being below the threshold values!) would then always be below the 4% mark.

Put simply, one could say: Investor A speculates that interest rates remain low or rise only slightly, while Bank B speculates on strongly increasing interest rates. Or: Investor A hedges against interest rates remaining low, and Bank B hedges against strongly increasing interest rates. The second case however requires that at least one of the two parties in the transaction has other investments whose risk is to be mitigated by the RAS.

What kind of investments could that be?

In fact, at the beginning of 2007, Investor A had taken out a 50-million EUR fixed interest loan due for repayment in 2032.The interest rate of this loan was 5%.

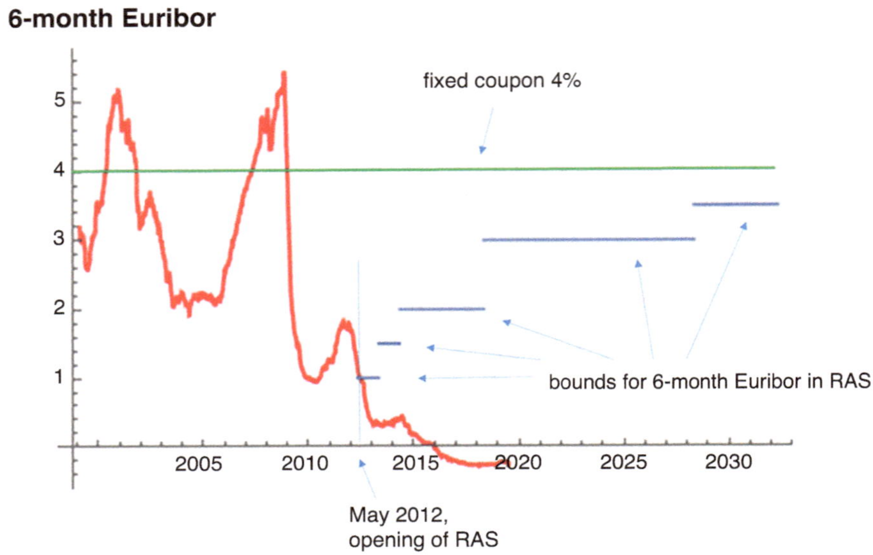

Fig. 3.25 Illustration of the 6-month Euribor thresholds used to determine payment amounts

Yet in the course of the financial crisis, market interest rates tumbled; at the beginning of 2012, the 6-month Euribor rate was around 1%, and the fixed interest rate of 5% for the loan was therefore very unfavourable from the investor's point of view. The investor's intention from the perspective of 2012 was therefore to improve this unfavourable situation to the greatest possible extent. This is what this range accrual swap was expected to do.

If the interest rates were to remain below the defined thresholds, then Bank *B* would take over the fixed loan interest for Investor *A* except for 1%. Investor *A* would thus only have to pay the 6-month Euribor plus 1% per annum in loan interest.

Yet while this arrangement might appear tempting at first glance, it has two major drawbacks:

Drawback #1:

If, for whatever reason, market interest rates were to rise again and stay consistently above 3%, for example, in the period from May 2018 to April 2028, then Bank *B* would pay nothing to Investor *A* during this period, while Investor *A* would have to pay both the fixed loan interest of 5% per annum and the then high 6-month Euribor interest on the 50-million EUR loan. This would be an extremely unfortunate situation for Investor *A*.

Drawback #2:

Even if interest rates remained constantly low, the RAS would not represent a permanent improvement of Investor *A*'s situation, since the bank can unilaterally terminate the RAS at any time. It is near to hand to assume that the bank would terminate this swap relatively soon if interest rates were to remain low or fall even further. This means that the investor can only secure a potentially favourable interest rate from one payment date to the next and never for any longer period beyond that.

Seen from today's perspective, knowing how the 6-month Euribor evolved from 2012 until today, we can safely say that Bank B would likely have terminated the RAS already in 2013 when the 6-month Euribor fell below 0.4% or at the latest in 2015 when the 6-month Euribor first dipped into negative territory.

In the context of a controversial review of the advisory activities provided by Bank B to Investor A, the author was commissioned by the courts in 2015 to review this RAS (which had been agreed at price 0, that is, without any further upfront payment between the two parties) to determine whether the conclusion of this swap could be classified as an approximately fair transaction. In particular, he was to determine whether Bank B, by concluding the transaction, had possibly taken advantage of Investor A.

This means: It was **not necessarily (!) required to determine an exact fair price** of the product. After all, a **lower threshold** for the fair price of the product from the viewpoint of Bank B could **already have sufficed** for drawing the requisite conclusions: If it turns out—for example!—that when the product was entered into, the fair price of the product from Bank B's vantage point was **at least** 1,000,000 EUR, and then one could already have concluded from this that the investor had been taken advantage of.

3.4.3 Modelling and Calibration of the Requisite Interest Rates in the Vasicek Model

As noted in Volume II Chapter 3, where we discussed (or rather: touched on) the basic principles of modelling interest rate curves and pricing interest rate derivatives, the valuation of interest rate derivatives of the present form is a thoroughly complex task.

In Volume II Chapter 3, we limited the discussion to valuations using spot rate models, specifically the Vasicek model and the Hull-White model (extended Vasicek) for spot rates, pointing out the need to use more complex models (such as the Cox-Ingersoll-Ross model) if we wanted truly reliable results.

In the original treatment of this task, the author used the Hull-White model. As we know, this model has the same basic structure as the Vasicek model, except that the parameters in the model can (partly) be chosen as time-dependent functions.

Here, we will take the somewhat simpler and more illustrative approach using spot rates, based on the Vasicek model and followed by Monte Carlo simulation (which will give us similar results as the ones obtained in the original approach using the Hull-White model). A particular challenge in the valuation will be the bank's right of termination.

We start by pricing the product without considering that termination right. In this section, we are going to model all of the required interest rate structures. For this purpose, we will also briefly recap the basic facts.

In the present RAS, the swap's underlying asset is the 6-month Euribor. To simulate the 6-month Euribor, we simulate a EUR interest spot rate $R(t)$ (i.e. a shortest-term interest rate). Specifically, we are going to simulate the "EONIA" (the Euro OverNight Index Average). We choose a Vasicek model for this purpose.

$dR(t) = a \cdot (b - R(t))dt + \sigma \cdot dW(t)$ with given initial value $R(0)$.

For certain technical reasons, we will not use the version $(b - a \cdot R(t))dt + \sigma \cdot dW(t)$ that we introduced in Volume II Section 3.13. The version that we are using here can also be written in the form $(ab - a \cdot R(t))dt + \sigma \cdot dW(t)$. Therefore, when using formulas from Volume II Section 3.13, we always need to replace the parameter b by $a \cdot b$.

Time $t = 0$ denotes 30 April 2012, and so, according to the historical interest rate database (see below), we have $R(0) = 0.00344$.

Caveat In the following, we will not simulate the percentage values of the interest rates but rather the interest rates in absolute figures!

The parameter σ in this model is estimated in exactly the same way as demonstrated in Section 7.5, based on the historical data of the EONIA index available on 20 April 2012 (which can be obtained, for example, from Bloomberg or also, for example, from https://www.ariva.de/eonia/historische_kurse?boerse_id=34).

Of course, the actual estimate for the parameter σ will depend on how far back we go in the historical rates used for calibration. In the specific case of the author's expert opinion, the estimate was $\sigma = 0.032558$.

The parameters a and b—as explained in Volume II Section 3.5—cannot and must not be determined on the basis of historical spot rate data, since to determine the value of derivatives, the discounted expected cash flow has to be determined on the basis of the "risk-neutral model" for the underlying asset. The parameters a and b must therefore be determined (in a roundabout way) from the prices of liquid derivatives on the underlying asset, as shown in Volume II Section 3.13.

In this particular case, we proceeded as follows:

For the euro spot rate R (EONIA) with parameters a, b and the above-determined parameter σ, the price of a bond over R at time t and maturing at time T is denoted by $p(t, T)$. If the spot rate R follows a Vasicek model, then the formula for $p(t, T)$ is

$$p(t, T) = e^{A(t,T) - B(t,T) \cdot R(t)},$$

where

$$B(t, T) = \frac{1}{a} \cdot \left(1 - e^{(-a \cdot (T-t))}\right)$$

and

$$A(t, T) = \frac{(B(t, T) - T + t) \cdot \left(a^2 \cdot b - \frac{\sigma^2}{2}\right)}{a^2} - \frac{\sigma^2 \cdot (B(t, T))^2}{4 \cdot a}.$$

As "liquid derivatives" on the underlying R, we choose in the case of the euro spot rate the 1-month Euribor $E01(t)$, the 6-month Euribor $E06(t)$, the 1-year swap rate $S1(t)$, the 2-year swap rate $S2(t)$, the 5-year swap rate $S5(t)$, the 10-year swap rate $S10(t)$, the 20-year swap rate $S20(t)$, and the 30-year swap rate $S30(t)$.

The fair prices of these derivatives can be expressed by the bond prices as follows:

$$E01(t) = -\frac{p\left(t, t + \frac{1}{12}\right) - 1}{\frac{1}{12} \cdot p\left(t, t + \frac{1}{12}\right)}$$

$$E06(t) = -\frac{p\left(t, t + \frac{1}{2}\right) - 1}{\frac{1}{2} \cdot p\left(t, t + \frac{1}{2}\right)}$$

$$S1(t) = \frac{1 - p(t, t + 1)}{\frac{1}{2} \cdot \left(p\left(t, t + \frac{1}{2}\right) + p(t, t + 1)\right)}$$

$$Sn(t) = \frac{1 - p(t, t + n)}{\sum_{i=1}^{2n} \frac{1}{2} \cdot p\left(t, t + \frac{i}{2}\right)} \qquad \text{for } n = 2, 5, 10, 20, 30.$$

(These formulas follow directly from our observations in Volume II Section 3.9 and, in the case of swap rates, from the fact that swap rates are defined on the basis of coupon bonds with semi-annual payments.)

This means that, at time 0 of the calibration process, the above derivatives are functions of the parameters a and b. The current values (as of 30 April 2012!) of Libor rates and swap rates are obtained from the market. Specifically, on 30 April 2012, we had the following values (in absolute figures!):

$E01 = 0.004$
$E06 = 0.00998$
$S1 = 0.00618$
$S2 = 0.00941$
$s5 = 0.01455$
$S10 = 0.02202$
$S20 = 0.02601$
$S30 = 0.02499$

So we have a total of eight equations to determine the unknowns a and b. Of course, this system cannot be solved exactly in this form.

As an example, we give the equation for the 6-month Euribor $E06$ here: We have:

$$0.00998 = \frac{1 - p\left(0, \frac{1}{2}\right)}{\frac{1}{2} \cdot p\left(0, \frac{1}{2}\right)} = \frac{2}{p\left(0, \frac{1}{2}\right)} - 2 = 2 \cdot \left(e^{-A\left(0,\frac{1}{2}\right)+B\left(0,\frac{1}{2}\right)\cdot R(0)} - 1\right) =$$

$$= 2 \cdot \left(e^{-\frac{\left(B\left(0,\frac{1}{2}\right)-\frac{1}{2}\right)\cdot\left(a^2\cdot b-\frac{\sigma^2}{2}\right)}{a^2}+\frac{\sigma^2\cdot\left(B\left(0,\frac{1}{2}\right)\right)^2}{4a}+B\left(0,\frac{1}{2}\right)\cdot R(0)} - 1\right)$$

and therefore

$$0.00998 = 2 \cdot \left(e^{-\frac{\left(B\left(0,\frac{1}{2}\right)-\frac{1}{2}\right)\cdot\left(a^2\cdot b-0.00053\right)}{a^2}+\frac{0.00106\cdot\left(B\left(0,\frac{1}{2}\right)\right)^2}{4a}+B\left(0,\frac{1}{2}\right)\cdot 0.01} - 1\right).$$

If we substitute for $B\left(0, \frac{1}{2}\right) = \frac{1}{a} \cdot \left(1 - e^{-\frac{a}{2}}\right)$, we get the equation

$$0.00998$$

$$= 2 \cdot \left(e^{-\frac{\left(\frac{1}{a}\cdot\left(1-e^{-\frac{a}{2}}\right)-\frac{1}{2}\right)\cdot\left(a^2\cdot b-0.00053\right)}{a^2}+\frac{0.00106\cdot\frac{1}{a^2}\cdot\left(1-e^{-\frac{a}{2}}\right)^2}{4a}+\frac{1}{a}\cdot\left(1-e^{-\frac{a}{2}}\right)\cdot 0.01} - 1\right).$$

The parameters a and b are now estimated approximately from the eight equations using a least-squares optimization method. However, it has been found advantageous to first restrict the definition domain for parameter a in order to prevent simulation results in which the volatility structure of the various interest rate types occurring in the swaps deviates strongly from realistic risk-neutral price movements.

Let us illustrate this observation with the following graphs (Fig. 3.26). They show some **typical** simulation paths over 20 years for $R(t)$ in the above Vasicek model with parameters $R(0) = 0.00344$ and $\sigma = 0.032558$, as well as $b = 0.03$ and for different values of a.

For values a smaller than 0 and for values larger than 1.5, we get very unrealistic paths. Negative values of a result in mean aversion (i.e. trending away from the long-term mean), while values of a greater than 1.5 result in a much too strong drift towards the long-term trend, causing prices to remain close to the long-term mean. These results are fairly independent of the value of the parameter b.

When it comes to finding the optimal parameters a and b, we can therefore limit our computations to a range for a within the bounds $0 < a < 1.5$.

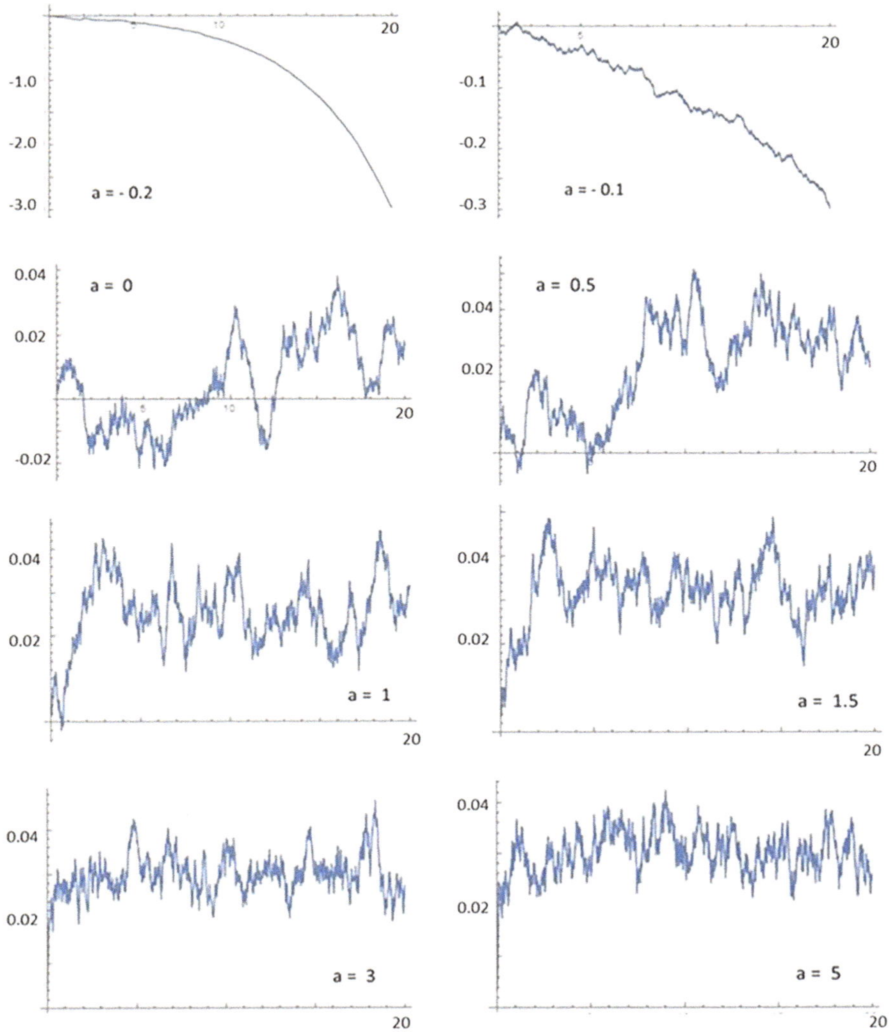

Fig. 3.26 Typical simulation paths for $R(t)$ over 20 years in the mean-reverting Ornstein-Uhlenbeck model $dR(t) = a \cdot (b - R(t))dt + \sigma \cdot dW(t)$ for fixed parameters b, σ,s and $R(0)$ and for different values of a

A least-squares optimization that we performed within these bounds yielded the following good risk-neutral estimates:

$$a = 0.283 \text{ and } b = 0.01.$$

Together with the parameters $R(0) = 0.00344$ and $\sigma = 0.032558$, we have now obtained a model for the EONIA overnight rate $R(t)$ that we can work with:

$$\boldsymbol{dR(t) = 0.283 \cdot (0.01 - R(t))dt + 0.032558 \cdot dW(t) \text{ and } R(0) = 0.00344}.$$

And so, for the price trend of our underlying 6-month Euribor $E06(t)$, we get an explicit representation of the form

$$\boldsymbol{E06(t)} = \frac{1 - p\left(t, t + \frac{1}{2}\right)}{\frac{1}{2} \cdot p\left(t, t + \frac{1}{2}\right)} = \frac{2}{p\left(t, t + \frac{1}{2}\right)} - 2 =$$
$$= \boldsymbol{2 \cdot \left(e^{-\left(A\left(t, t+\frac{1}{2}\right) + B\left(t, t+\frac{1}{2}\right) \cdot R(t)\right)} - 1\right)}$$

where

$$A\left(t, t + \frac{1}{2}\right) = \frac{\left(B\left(t, t + \frac{1}{2}\right) - \frac{1}{2}\right) \cdot \left(a^2 \cdot b - \frac{\sigma^2}{2}\right)}{a^2} - \frac{\sigma^2 \cdot \left(B\left(t, t + \frac{1}{2}\right)\right)^2}{4a}$$

and

$$B\left(t, t + \frac{1}{2}\right) = \frac{1}{a} \cdot \left(1 - e^{-a \cdot \frac{1}{2}}\right).$$

Based on these representations for $R(t)$ and $E06(t)$, we will now run path simulations for $R(t)$ and $E06(t)$ specifically for the 20-year period from 2 May 2012 to 30 April 2032. We limit the simulations to step lengths of 1 week, that is, we choose $dt = \frac{1}{52}$. For the entire period, we therefore need a total of $20 \times 52 = 1040$ simulation steps. Note that actually, for valuation of the swap, we would have to simulate daily rates in order to determine the number of trading days on which (in the respective simulation) the 6-month Euribor is below the respective thresholds. However, daily simulations would come at the expense of the number of possible scenarios (and the number of required scenarios is quite high, as we will see later when taking into account the bank's right to termination).

Therefore, instead of counting the number of days (in a scenario) in the respective half-years between two payment dates on which the relevant threshold is exceeded and dividing that number by the total number of trading days, we count the number of weeks in which the relevant threshold is exceeded and divide that by 26 to determine the factor for the respective half-year.

To illustrate this, Fig. 3.27 shows a simulation of the short rate (in light green) and the corresponding 6-month Euribor (in dark green). In Fig. 3.28, we see 20 simulations of the short rate and the 20 corresponding 6-month Euribor rates. The colours of the EONIA short rates and of the 6-month Euribor rates do not match in

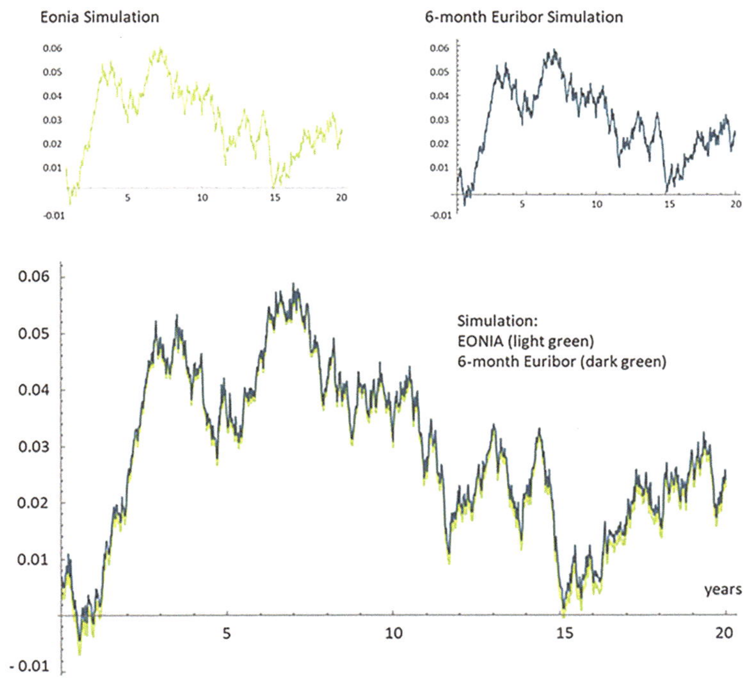

Fig. 3.27 A typical simulation path of the short rate with the risk-neutral parameters (light green) compared with the relevant 6-month Euribor (dark green)

this picture; our intent here is simply to illustrate the typical path structure of the two variables when using our simulation approach and our choice of parameters.

3.4.4 Valuation of the RAS Based on the Vasicek Model *without* Considering the Termination Right

To value the RAS without taking into account Bank B's right of termination, all that is left for us to do is determine the resulting cash flows for each path. The resulting payments for each simulation scenario are added up and divided by the number of scenarios. This gives us a Monte Carlo approximation for the fair value of the RAS (without termination right).

We simulate this total payment determination using one specific path of the 6-month Euribor. Figure 3.29 shows this simulation path as well as the bounds for determining the payments and the payment dates.

We see here that this price path would result in Investor A having to make payments in the periods from around 2026 to 2032 and possibly around 2022.

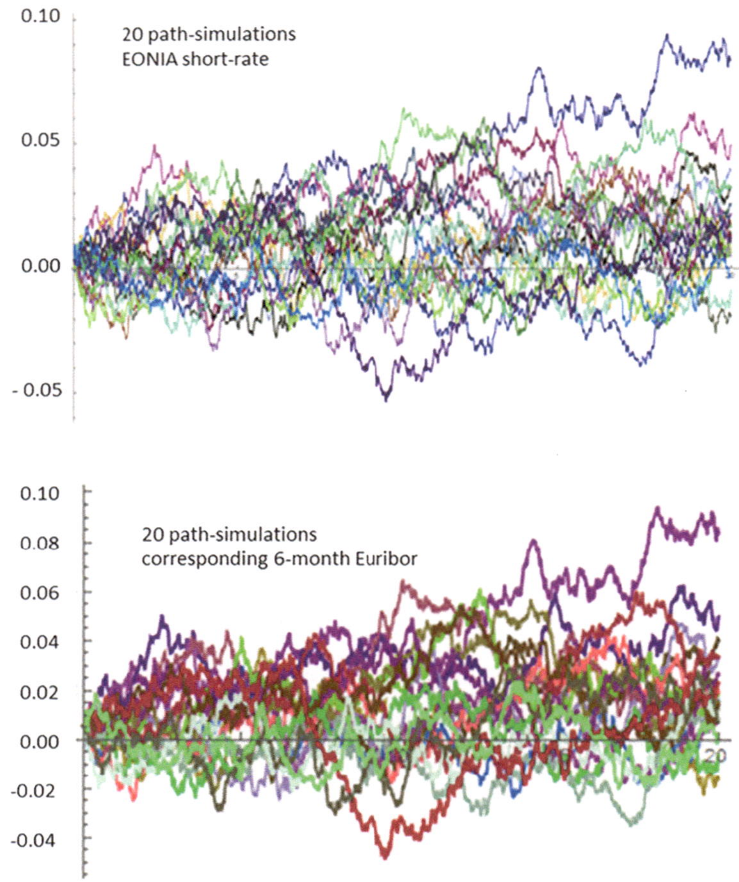

Fig. 3.28 Twenty typical simulation paths of the short rate with the risk-neutral parameters compared with the relevant 6-month Euribor

Specifically, the payments to be made by Investor A are:
−387636, −1125000, −1125000, −1125000, −1125000, −1125000, −1125000, −1125000, −1125000, −1125000, −1125000, −951924, −1120898, −580868, 38603, −1006841, −1081731, −931510, −299608, 798619, 135112, −200274, −1125000, −1125000, −1125000, −1125000, 491885, 1089243, 1203200, 1145903, 1317908, 1407489, 1434097, 1265866, 1313791, 1399135, 1238278, 1067592, 1392495, 2882928

The sum of payments in this scenario from the investor's perspective is −2,689,146 EUR.

Yet if we want to determine the fair value of this payment from the perspective of when the product is valued, we obviously have to discount the payments that will take place at dates later than 2 May 2012. To do this, we need the appropriate

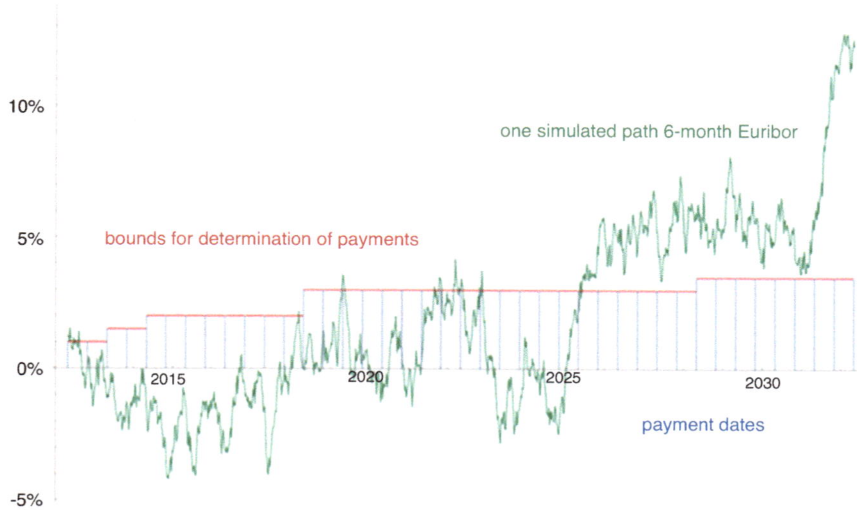

Fig. 3.29 Simulation path with payment dates and bounds for determining payment amounts

discount factors (we denote them by $d_{0.5}, d_1, d_{1.5}, \ldots, d_{20}$ from the perspective of 2 May 2012). And that in turn requires the interest rates $f_{0,0.5}, f_{0,1}, f_{0,1.5}, \ldots, f_{0,20}$ from the perspective of 2 May 2012. We interpolate them linearly from the Euribor rates given above and the swap rates on 2 May 2012, that is, from:

$E06 = 0.00998$
$S1 = 0.00618$
$S2 = 0.00941$
$S5 = 0.01455$
$S10 = 0.02202$
$S20 = 0.02601$

From this we then obtain the discount rates by means of $d_t = \frac{1}{(1+f_{0,t})^t}$. The discount rates that are then finally used are presented in the following list: 0.995047, 0.993858, 0.98842, 0.981442, 0.974787, 0.967359, 0.959176, 0.95026, 0.940633, 0.930321, 0.919894, 0.908918, 0.897416, 0.885412, 0.872931, 0.859998, 0.846639, 0.832882, 0.818752, 0.804278, 0.793937, 0.783578, 0.773202, 0.762816, 0.752422, 0.742026, 0.731632, 0.721242, 0.710862, 0.700495, 0.690146, 0.679817, 0.669513, 0.659237, 0.648992, 0.638784, 0.628613, 0.618485, 0.608402, 0.598368

Thus, from the perspective of 2 May 2012, the payments resulting from the above path have a discounted value of −6,942,004 EUR to Investor *A* (in this case, therefore, this is income for *A* in the discounted amount of almost 7 million euros).

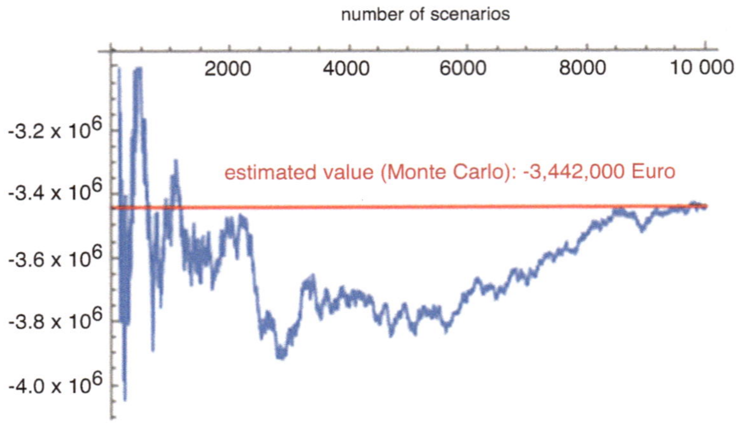

Fig. 3.30 Convergence behaviour when determining the fair value of the RAS without termination right using Monte Carlo method

The discounted value has a substantially higher negative value because the positive payments to be made by A do not occur until towards the end of the swap period and are therefore discounted to a much greater extent.

If we now run this simulation with 10,000 scenarios, we already obtain a relatively stable average value of $-3{,}442{,}000$ EUR of the discounted payments that Investor A has to make to Bank B. The convergence behaviour in this simulation is shown in Fig. 3.30.

From the perspective of Investor A, the product without termination right would therefore have a fair value on 2 May 2012 of approximately 3,442,000 EUR.

3.4.5 Valuation of the RAS Based on the Vasicek Model *with* Termination Right

Bank B has the right to terminate the RAS on any payment date starting 30 April 2012. In this case, the payment that is due at the moment of termination must still be made.

Let us start by looking at the first payment date.

On the first payment date, that is, on 31 October 2012, the bank has to make a payment (or receive a payment, as the case may be). On 31 October 2012, the bank can decide whether to continue the swap or terminate it.

If interest rates are relatively high on 31 October 2012, say, above the relevant 1% threshold for determining the payment amounts, then the bank will likely decide to continue the swap. Conversely, if interest rates are relatively low, then it will likely consider terminating it. If the bank decides on 31 October 2012 to allow the swap to continue, then it can next decide on 30 April 2013 whether or not to terminate.

The situation is again the same: If interest rates are relatively high on 30 April 2013, say, above the relevant 1.5% threshold for determining the payment amounts, then the bank will likely decide to continue the swap. Conversely, if interest rates are relatively low, then it will likely consider terminating it. The bank's strategy will therefore probably be something like this:

The bank will set itself a certain threshold L_i for each payment date Z_i and proceed as follows: As long as the value of the 6-month Euribor is above L_i on a payment date Z_i, the swap will continue. Once the value of the 6-month Euribor is below L_i on a payment date Z_i, the swap is terminated.

The question would then be which termination thresholds L_i would be the optimal choice from the bank's perspective.

The answer to this question requires deeper knowledge than what has been imparted so far, and we will therefore take a different approach.

We adopt a strategy that looks like it will produce relatively good results. For this purpose, we propose to perform some preliminary tests with the help of our Monte Carlo modelling in order to gain ideas for good strategies.

Specifically, let us take a closer look at the following strategy:

We consider only the payment dates $Z_1, Z_2, Z_3, Z_4, Z_5, Z_6$. *For these dates, we choose the termination thresholds* $L_1 = 0.005(0.5\%)$, $L_2 = L_3 = L_4 = L_5 = L_6 = 0.01(1\%)$. *If the swap has not been terminated by* Z_6, *it will not be terminated at any later point either.*

Now, if the bank implements the above termination strategy, we can easily calculate the approximate average discounted payment amount from the perspective of Investor A with the help of our simulation program, by simply modifying how the payments per path are determined and running a total of 10,000 scenarios.

We obtain the result illustrated in Fig. 3.31, which suggests an average payment amount to be made by the investor of approximately 1,320,000 EUR.

So if Bank B implements this particular strategy, the RAS has a fair value for the bank of about 1,320,000 EUR (and for Investor A of about −1,320,000 EUR).

In all likelihood, this strategy is far from the best strategy. If the bank implements an even better (or even optimal) strategy, this swap product has an even higher value for the bank and thus an even more negative value for Investor A.

This RAS was therefore (based on this modelling) a definitely disadvantageous deal for Investor A.

3.4.6 Test of the Risk-Free Model's Calibration Quality and Sensitivity Analysis of the Results

We now return to the determination of parameters a and b in the risk-free Vasicek model. The aim in determining a and b was that the values calculated from the spot rate and from the bond prices for the 1-month Euribor, 6-month Euribor, 1-year, 2-year, 5-year, 10-year, 20-year, and 30-year swap rates on 2 May 2012 should come as close as possible to the actual values on 2 May 2012.

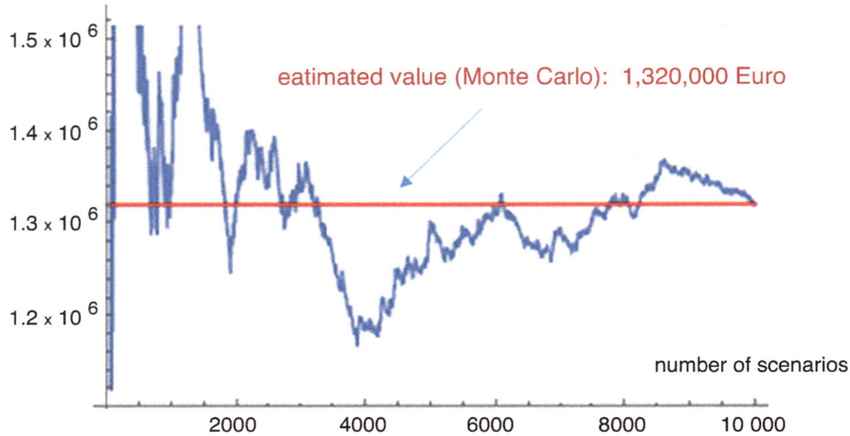

Fig. 3.31 Convergence behaviour when determining the fair value of the RAS with termination right according to the termination strategy described above, using Monte Carlo method

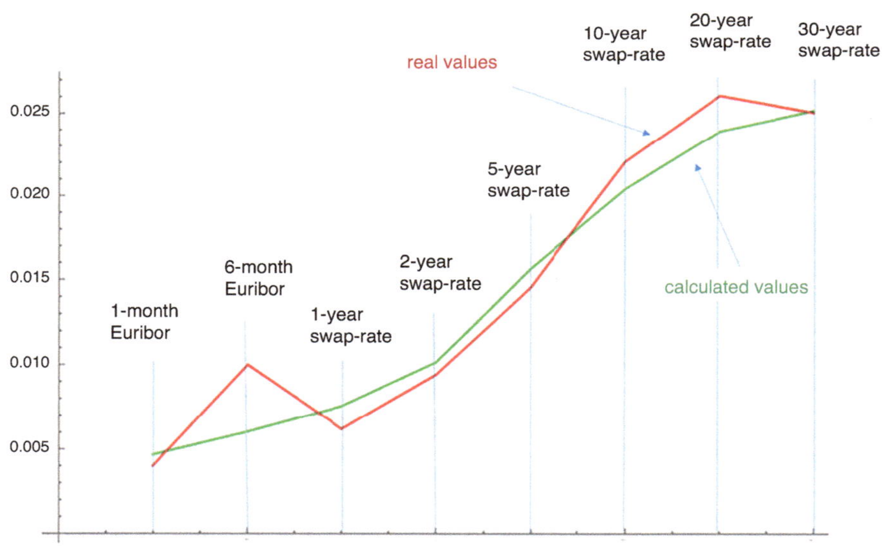

Fig. 3.32 Comparison of calculated and actual Euribor interest rates and swap rates with "optimal" parameters $a = 0.283$ and $b = 0.01$

Figure 3.32 shows the values calculated with the parameters that we chose as optimal, namely, $a = 0.283$ and $b = 0.01$ (red), compared to the actual values (green). The way they align is acceptable in principle, with a slightly larger deviation

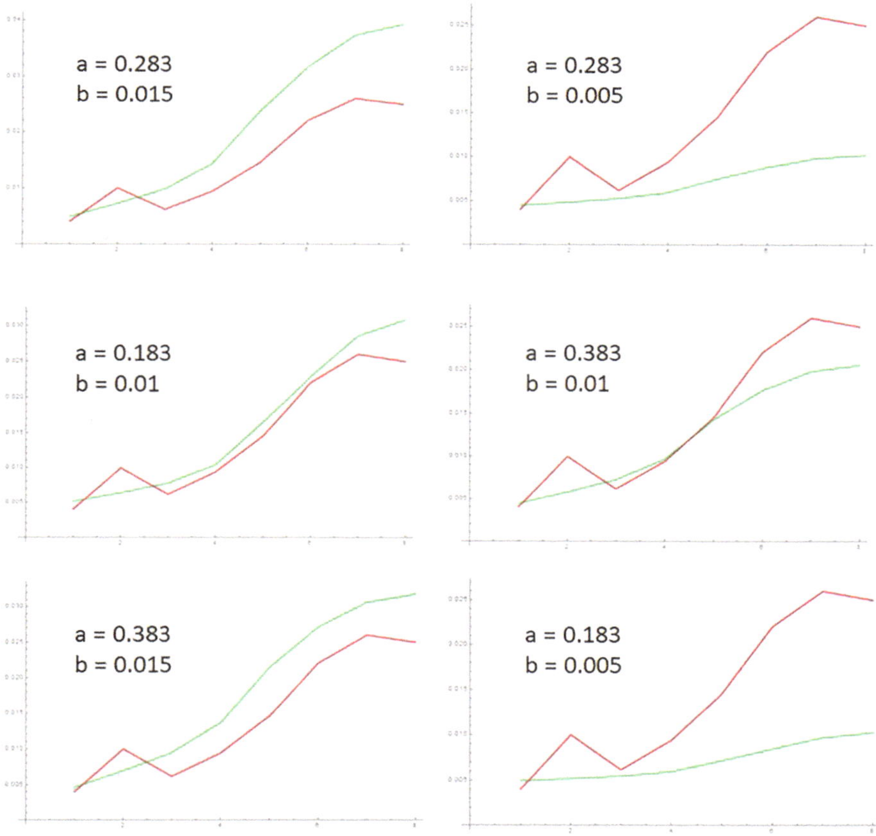

Fig. 3.33 Comparison of calculated (green) and actual (red) Euribor interest rates and swap rates with different parameters a and b

for the 6-month Euribor. As already noted, it is, of course, not possible to always choose the two parameters a and b such that all eight equations are exactly satisfied.

The sensitivity of this comparison between calculated and actual values to slight changes in the parameters a and b can be seen in Fig. 3.33. Here, the comparisons are made for slightly modified values of a and b.

For these (obviously less than perfect) parameters that we tested in Fig. 3.33, we also ran the Monte Carlo simulations to determine the fair values of the RAS without termination right, as well as with termination right and implementing the strategy described above, which gives us the following table (Table 3.13):

We see strongly fluctuating values for the swap without termination right. This fluctuation becomes significantly lower when the termination right is taken into account: In this case, all the values are clearly negative for all parameter choices.

Table 3.13 Results of the MC simulation for the determination of the fair values of the RAS

a	b	Value for A without termination right	Value for A with termination right and above strategy
0.283	0.01	3,442,000	−1,320,000
0.283	0.015	−1,033,000	−2,405,000
0.283	0.005	7,205,000	−633,000
0.183	0.01	−169,000	−2,592,000
0.383	0.01	4,431,000	−479,000
0.383	0.015	−765,000	−1,565,000
0.183	0.005	3,748,000	−1,851,000

This brief analysis is intended to show how difficult it is to obtain truly reliable estimates of the fair value of somewhat more complex interest rate derivatives. Still, the analysis for the case with the termination right clearly demonstrates that the product with termination right has a distinctly negative value for Investor *A* in every tested scenario.

3.5 Case Study V: Analysis of the Performance of Put-Write Strategies

In Volume II Section 1.16, we pointed out on various occasions that an underlying asset's implied volatility, that is, its volatility calculated from options, is often (or even systematically) significantly higher than the volatility that is then actually realized. Since calculations of implied volatility from options of different types and different strikes do not always—as we have seen (volatility smile, volatility skew)—yield the same implied volatility value, the extent of overestimation also depends substantially on what options are used for calculating the implied volatility.

However, such perceived or seeming inconsistencies between implied and realized volatilities could also simply be due to the use of inappropriate modelling in computing implied volatility (e.g. Black-Scholes model with its underestimation of heavy tails) and may therefore merely appear to be inconsistencies rather than being "real" inconsistencies.

The question as to whether these inconsistencies exist in reality can therefore not be answered by comparing realized and implied volatilities directly. Instead, we need to explore if suitable trading strategies (based above all on selling options whose implied volatility appears to be "too high" and are thus perceived to have a price that is "too high") can be used to systematically generate excess returns.

In the following, we limit the discussion to volatilities and options on the S&P500 index (SPX).

The questions that we are going to answer are:

Is it possible to systematically profit from a (real (?) or just alleged (?)) overestimation of the realized volatilities of the SPX, based on its implied volatilities, by consistently executing a suitable trading strategy? What would be the optimal trading strategy for this, and what risk-return ratios would such a strategy have? Overestimation of realized volatility is particularly evident in put options with a strike price in a certain out-of-the-money range.

The put-write index, which we looked at briefly in Volume II Section 1.17, already gave us an idea of the accuracy of an observed overestimation. And back then we already asked ourselves how we could make optimal use of such an overestimation. The put-write index refers to put options that are essentially at the money. However, such an overestimation is likely to have a stronger impact for put options in a certain out-of-the-money range.

Such observations are frequently mentioned in academic literature on finance. Just a few examples: In their 2009 article "The Cash-Secured PutWrite Strategy and Performance of Related Benchmark Indices" in the Journal of Alternative Investments, Ungar and Moran wrote [1]:

"Some of the more popular and successful strategies for increasing the risk-adjusted returns of portfolios have involved systematically selling index options. This is because historically, index options have often been richly priced, and option sellers rather than buyers have been rewarded. Since 1990 the average future volatility implied by the prices of the S&P500 Index options has generally been greater than the average subsequent actual volatility of the index".

In 1992, in an article titled "Stock Market Volatility and the Information Content of Stock Index Options" published in the Journal of Econometrics, Day and Lewis wrote [2]:

"Put options [note: meaning long positions in put options] have returns that are both economically and statistically negative [...]. The mean returns increase monotonically in the strike price, ranging from -14.56 percent for the most out-of-the-money options to -6.16 percent for those deeply in the money".,

and Santa-Clara and Saretto, in their article "Option Strategies: Good Deals and Margin Calls" in the 2009 Journal of Financial Markets, [3] analyse certain short put strategies, concluding:

"Selling 10% OTM put contracts earns 51% per month on average, with a Sharpe ratio of 0.306, and a Leland alpha of 30%".

In their article "modelling and Performance of Certain Put-Write Strategies" in the Journal of Alternative Investments, published by the author together with L. Del Chicca and M. Szölgyenyi in 2012 [4], a specific class of put-write strategies on the SPX for various strategy parameter choices was analysed thoroughly and under real trading conditions. The entire test results were available on more than 20,000 web pages. Such strategies did indeed show significant outperformance for most parameter choices. (However, it was pointed out in the article and demonstrated for various sample cases (e.g. in the context of the 2008 financial crisis) that operational risks can still arise and must be taken into account even when implementing these strategies "by the rules".)

For this book and this particular case study, we continued the analyses and findings of the 2012 paper for the period up to 2019, updating and extending the earlier findings. In addition, we have created a **software**—accessible on the website in the restricted area reserved for readers of this book—that can be used to test and track the detailed performance of these strategies as well as their risk-return parameters and their processes for any time periods and for any parameter choices. See https://app.lsqf.org/book/put-write-strategy.

For a collection of recent results, see also [5].

The motivation for us to conduct these analyses and develop the software was—apart from including them as a case study in this book—to update our research paper in the Journal of Alternative Investments from 2012 and secondly a commissioned work by an asset management company that implements such trading strategies for private customers on their trading accounts.

The author himself, in his work as an asset manager, also executed such put-write strategies for investors from 2002 to 2016. The invested sums were as high as up to 200 million euros during the boom period. Apart from the catastrophic year 2008 (which we will look at more closely in Sect. 3.5.6), these investments showed very promising results (in spite of sometimes very adverse regulations hampering the necessary flexibility in trading for customers).

In the following, we will not present comprehensive statistics on the performance of these SPX put-write strategies for various parameter sets (even just the space needed for that would be prohibitive). Instead, we are going to describe the strategies and their parameters as well as the realistic assumptions. In addition, we will give some selected performance results for selected parameter sets and encourage readers to read the original article by the authors in the Journal of Alternative Investments (JAI article) and in the recent article [5]. Above all, we encourage readers to run tests on their own, using the software provided on the website. That way readers can see for themselves what characteristics and potential these trading strategies have. Finally, we will again address some operational risks in executing such strategies.

3.5.1 Setup of the Tested Put-Write Strategies

So, in the following, we are going to analyse option strategies on the SPX that we intend to execute systematically over a longer period of time. These strategies will be clearly defined and will not contain any discretionary decision components. The performance of these strategies in past time periods can thus be tracked fairly reliably using historical option price data. Of course, the results of such "backtests" should always be treated with caution, and we will address applicable caveats in each case below. Yet we strive to formulate and use underlying assumptions that largely correspond to real trades. Therefore, our analyses cannot and should not be classified as merely theoretical tests "on paper".

The basic structure that all the strategies have in common, meaning that they differ only (with the exception of one variant, the so-called 2-M strategy) in the chosen parameters, is as follows:

We have an initial investment of I_0 at our disposal. For simplicity, we assume that investment has to be in US dollars (USD). All currencies are given in USD (even if not specifically noted).

Put options on the SPX with 1 month to expiration are traded on every third Friday of the month at the close of trading (thus, the relevant prices for our analyses here are the closing prices).

The currently available investment I_{akt} at each such trading date results from the performance of the investment I_0 at the start of trading and the trading profits or losses accumulated from the start of trading until now. (No money is added or withdrawn until expiration, and no further transactions are made. The entire available investment is always invested, to the greatest extent possible, in the strategy as described below.

As we will see, the available investment is always used only as margin and does not leave the investor's trading account. The investment can therefore earn additional interest during implementation of the strategy (if the market's interest rate environment allows relevant interest on credit balances). However, we will not pursue this additional possibility of enhancing the investment in detail here and will only analyse the performance of the pure option strategy. The JAI article discusses this interest issue in more detail.)

Specifically, a certain number A of combinations of short put contracts with the same strike price K_1 and of the same number of long put contracts with the same strike price K_2, where $K_2 < K_1$ are traded each month. Strike K_1 (and thus also K_2) is always below the respective current SPX price S_0.

The short positions with a higher strike K_1 have a higher contract price P_1 than the long positions with a lower strike K_2, whose price per contract we denote by P_2. So, for each short/long combination contract, we take a premium in the amount of $P_1 - P_2$, thus generating a total income of $A \cdot (P_1 - P_2)$ per transaction. Since the long position for each short position limits the maximum potential loss per combination contract to $100 \times (K_1 - K_2)$ dollars, the maximum required margin for each combination contract is $100 \times (K_1 - K_2)$ dollars. Because we want to make maximum use of the available investment amount I_{akt}, the number A of contracts traded each month is given by $A = \left[\frac{I_{akt}}{100 \cdot (K_1 - K_2)}\right]$ (here, $[x]$ denotes the largest integer less than or equal to x).

(In fact, due to the initial premium income of $A \cdot (P_1 - P_2)$ dollars, it would even be possible to trade $\left[\frac{I_{akt} + A \cdot (P_1 - P_2)}{100 \cdot (K_1 - K_2)}\right]$ contracts, but that would generally not really constitute any relevant improvement.)

If all the traded contracts were kept until expiration and if the SPX at expiration were above the short strike K_1, the options would expire worthless, and the investor would be left with the collected premium in the amount of $A \cdot (P_1 - P_2)$. If the SPX were below the long strike K_2 at expiration, then a loss in the amount of $100 \times (K_1 - K_2) \times \left[\frac{I_{act}}{100 \cdot (K_1 - K_2)}\right]$ dollars would have to be expected, i.e. almost the amount of I_{act}. In other words: The investor would basically sustain a total loss of the investment. An illustration of this is shown in Figs. 3.34 and 3.35.

However, in order to avoid high "existential" losses or, in the worst case, total losses and thus the end of the strategy, the strategies are equipped with (again, freely selectable) "exit strategies", which require all positions to be closed as soon as a

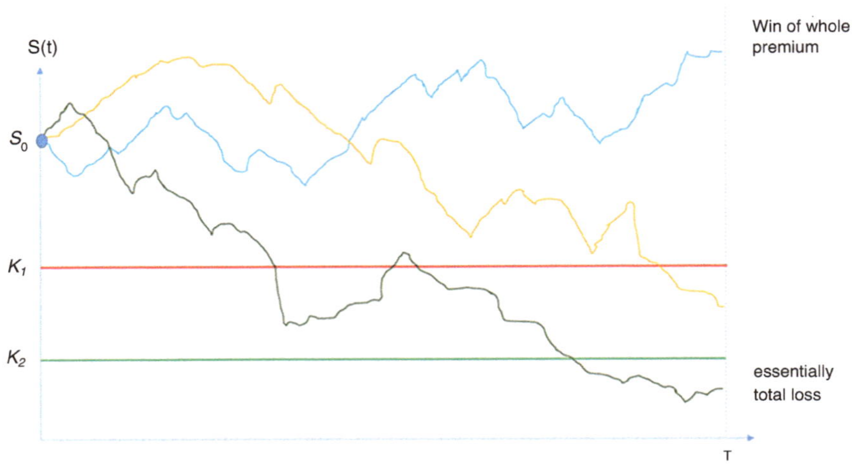

Fig. 3.34 Schematic diagram of put-write strategy when held to expiration

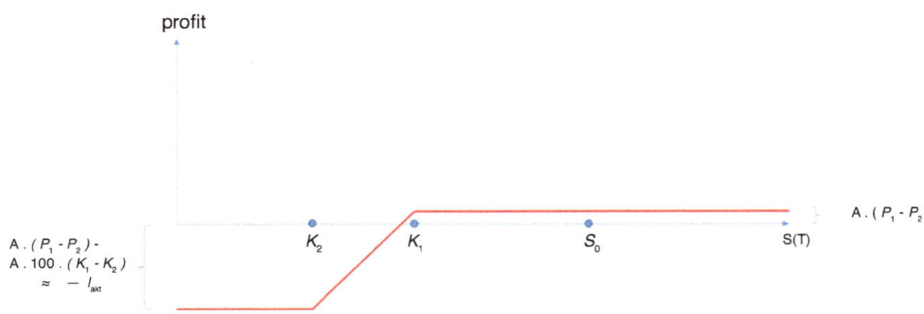

Fig. 3.35 Profit function for put-write strategy when held to expiration

certain predefined price decline of the SPX or a predefined maximum loss from the option positions occurs. In such a case, all positions are closed, and on the following third Friday of the month, the strategy is continuously unchanged with the then still existing investment. (A possible variant, which we will not pursue any further here, would be to immediately re-enter new positions and thus take advantage of increased implied volatilities, caused by the preceding SPX price decline, and thus of increased option premiums.) The strategy's variable parameters then are:

Choice of K_1
Choice of K_2
Choice of exit strategy

3.5.2 Discussion and Examples of Some Parameter Choices

For illustration, we choose an explicit real numerical example:

Trading date: 21 June 2019 (= third Friday in June)
Expiration date of the options to be traded is therefore **19 July 2019** (settlement price at opening of the exchange)
Available **investment** $\boldsymbol{I_{act}} = \mathbf{100,000}$ **USD**

The SPX closing price on 21 June 2019 was $S_0 = 2,950.46$.

The chosen strikes K_1 and K_2 could then be (note that here and in the following the strike price always means the largest available strike with a value less than or equal to the specified bound):

$\boldsymbol{K_1 = 95\%}$ **of** $\boldsymbol{S_0}$ (thus $K_1 \approx 0.95 \times 2.950.46 = 2.802.94$, hence $\boldsymbol{K_1 = 2.800}$)
$\boldsymbol{K_2 = K_1 - 75}$ (thus $\boldsymbol{K_2} = 2.800 - 75 = \mathbf{2.725}$)

The **exit strategy** could be: Close all positions as soon as the SPX falls below the value of K_1.

The number of option contracts traded is

$$\boldsymbol{A} = \left\lceil \frac{I_{akt}}{100 \cdot (K_1 - K_2)} \right\rceil = \left\lceil \frac{100.000}{7.500} \right\rceil = \mathbf{13\ contracts}.$$

The prices (closing prices of the bid//asks) of the put options to be traded with expiry on 19 July 2019 are taken from the Interactive Brokers' option price table on 21 June 2019, as shown in Fig. 3.36.

The quotes for the put option with strike 2800 are 10.30 // 10.60, and the quotes for the put option with strike 2725 are 5.50 // 5.80. In all likelihood, the sale can therefore be made at $P_1 = 10.40$ and the purchase at $P_2 = 5.70$. The transaction cost is 4 USD per combination contract, so 52 USD in total. Thus the **total premium collected** is $13 \times 100 \times (10.40 - 5.70) - 52 = \mathbf{6.058}$ **USD**.

If the SPX stays above the level of 2800 between 21 June 2019 and 19 July 2019 (i.e. if it does not fall by more than 5% during the trading month), the investor makes a **monthly profit of 6.06%**.

If, however, the SPX falls below the level of 2800 during the trading month, then all positions must be closed at the moment of it falling below this threshold. The expected loss in this case cannot be estimated in advance and depends on various factors, in particular on the implied volatility at the moment when the positions are closed and on the point in time at which the SPX falls below the threshold (rather early in the trading month or more towards the end of the trading month). Another factor of uncertainty is whether the trader succeeds in closing the positions immediately when the threshold is reached or whether a certain "reaction time" is to be expected until they are actually closed, maybe at a point where the SPX (in the

Strike	Bid x Ask
2725	5,50 x 5.80
2730	5.70 x 6.00
2735	5.90 x 6.20
2740	6.20 x 6.50
2745	6.50 x 6.70
2750	6.70 x 7.00
2755	7.00 x 7.30
2760	7.30 x 7.60
2765	7.60 x 7.90
2770	7.90 x 8.30
2775	8.30 x 8.60
2780	8.60 x 9.00
2785	9.00 x 9.30
2790	9.40 x 9.80
2795	9.80 x 10.20
2800	10.30 x 10.60

Fig. 3.36 Part of option chart on 21 June 2019 for SPX put options expiring on 19 July 2019

event of a potentially rapid decline) is already as low as, for example, 2790 points (operational risk!).

To get an approximate idea of the magnitude of the losses to be expected in such a case, see the graph in Fig. 3.37, which shows the difference $P_1(t) - P_2(t)$ of the Black-Scholes prices, on the assumption that the SPX price $S(t)$ has fallen to 2800 points, for the 28 days of the options' life and for different implied volatility values. Figure 3.38 is essentially the same graph but in this case showing the corresponding strategy losses (taking into account the initially collected premiums).

An alternative choice for the strikes K_1 and K_2 could be, for example:

$$\mathbf{K_1 = S_0 \times (100\% - 0.3 \times VIX\%)}$$

$$\mathbf{K_2 = K_1 - 75}$$

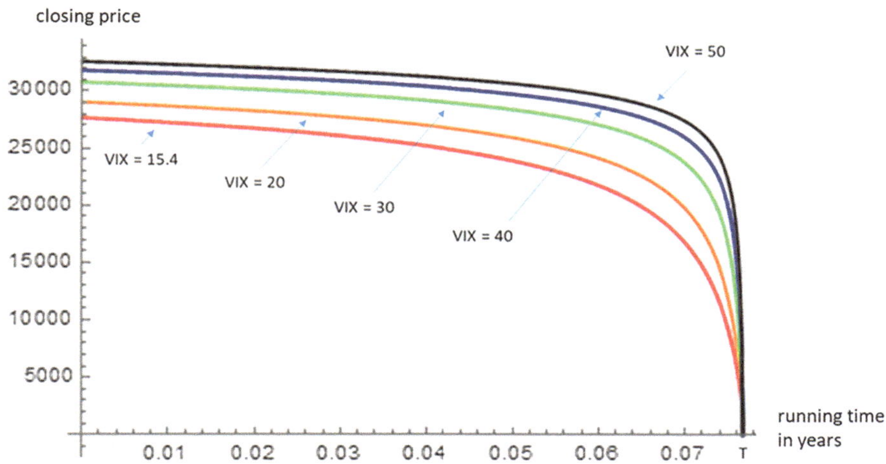

Fig. 3.37 Evolution of the price difference $P_1(t) - P_2(t)$ during the trading month for different implied volatility values and with the SPX at K_1

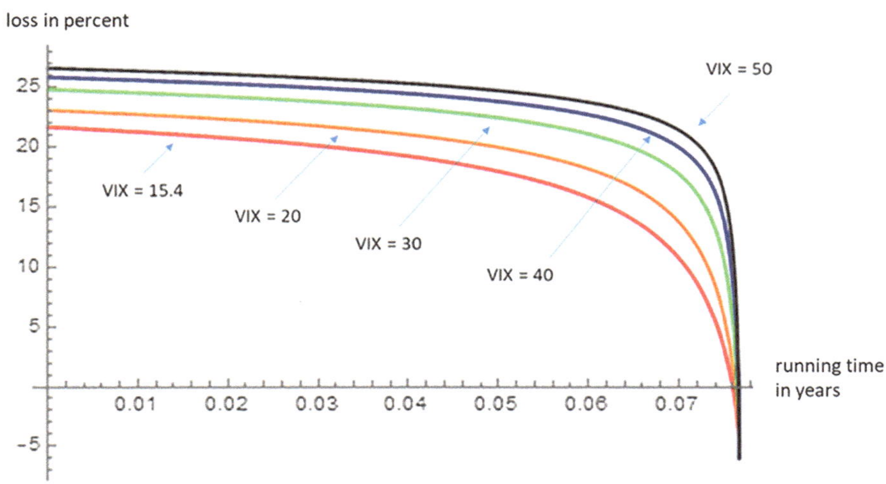

Fig. 3.38 Evolution of the strategy's loss as a percentage during the trading month for different implied volatility values and with the SPX at K_1

Here the distance ("safety distance") from K_1 to the current level of the SPX is chosen such that the greater the market's volatility is (as expressed by the VIX), the larger that distance has to be.

On 21 June 2019, the VIX (closing price) was 15.40 points. Thus, with this parameter choice, $K_1 \approx (1-0.3\times0.154)\times2950.46 = 0.9538\times2950.46 = 2814.15$, so $\boldsymbol{K_1 = 2810}$ and thus $\boldsymbol{K_2 = 2735}$.

An **alternative exit strategy** could be as follows: Close all positions as soon as the total portfolio value falls below 90% of the investment amount I_{akt} (i.e. as soon as the loss exceeds 10% of the investment amount). With this type of exit strategy, the loss to be expected in the event of a required close-out can be estimated relatively clearly. It should essentially be around 10% (in our numerical example).

Yet again, there is a caveat: The above is true provided that it is possible for the trader to react sufficiently fast to the close-out signal and actually make a loss on closing that is in the 10% target range. To account for this uncertainty in the backtest results, a freely selectable "buffer" was integrated in the test program on our website. If this buffer is preset to 2%, for example, then this means:

Positions will be closed as soon as a theoretical loss (based on market values) of 10% is reached. Yet we assume that the actual loss after closing (due to reaction times) is $10\% + 2\% = 12\%$.

Now, how do our backtests recognize that a 10% loss would theoretically have been reached due to market prices? (In real real-time trading, it is easy to track the open portfolio's value in real time and react when the 10% loss barrier is exceeded.) However, to be able to identify such loss exceedance events when performing backtests, we need historical tick data of the option prices. But due to the incredible abundance of options, working with tick data is not possible. So, in our tests and in our analysis program, we work with daily option price data where the opening, high, low, and last prices as well as the last bid//ask prices are available. Positions are then opened based on these last bid//ask prices. Due to our real trading experience in SPX options via Interactive Brokers, we thereby assume (in our analyses) that in each case, an actual buy is possible at 10 cents above the respective mean of bid and ask, and an actual sell is possible 10 cents below the respective mean of bid and ask.

But let us get back to our original question: How do our backtests recognize, based on historical daily option prices, that a 10% loss would theoretically have been reached due to market prices?

Within 1 day we can assume that the difference $P_1(t) - P_2(t)$ is monotonically decreasing in $S(t)$, i.e. in the price of the underlying asset. The smaller that $S(t)$ is, the larger $P_1(t) - P_2(t)$ and thus the higher the closing price of the options becomes.

Similarly, $P_1(t)$ and $P_2(t)$ are each monotonically decreasing in $S(t)$, that is, they increase as $S(t)$ decreases. From this we can conclude that the highest $P_1(t) - P_2(t)$ (and hence the maximum loss in the strategy) occurs when $P_1(t)$ resp. $P_2(t)$ are at their highest.

So, the difference $P_1(t) - P_2(t)$ is calculated from the **option price intraday highs**, and only if this difference induces a loss of 10% or more would a close-out and a loss of 10% (+ buffer) have occurred on that trading day.

This argumentation comes with a slight uncertainty, however: The prices given are actual **trading prices**, and it is not necessarily a given that the options were traded at exactly the moments when the $S(t)$ was at that day's lowest. A more reliable approach would be to use the maximum bid//ask prices, which are quoted at every moment, but unfortunately these are not available as historical data. To avoid a possible bias here (which could lead to too infrequent close-out assumptions), our program also offers a second "safer" alternative:

Whenever the (larger) difference between the maximum $P_1(t)$ on a trading day and the minimum $P_2(t)$ on that trading day induces a loss of 10% or more, we assume a close-out. Even this variant does not provide a 100% guarantee that every close-out situation is actually "detected", but it tends to overestimate the number of close-outs and therefore tends to lead to more defensive estimates with regard to the performance of the analysed strategies.

The question at what SPX level approximately a certain loss will occur is, of course, again dependent on several factors, namely, on the remaining period to expiration and on the evolution of the implied volatility. How the corresponding analysis is to be performed was already discussed in detail in Volume I Section 4.40, and so we will not go into it again here.

3.5.3 The Various Parameter Choices and Selection of Some Analysis Results

In the following we give, for a few select parameter sets, a small (!) selection of backtest performance results for time periods between early 1990 and late 2010 from the JAI article as well as for periods between early 2010 and late 2018, which were created using our new software. For much more detailed discussions of the results, we refer to the JAI article and to [5] and encourage readers to use the available software to run their own tests. (Section 3.5.5 provides an introduction to the use of the software.) In addition, the author's research group also offers intensive seminars for users interested in implementing such trading strategies on their own.

For the **choice of the parameter K_1**, the analyses in the JAI article, in [5], and in our software offer, among others, the following alternatives:

Alternative 1 *K_1 below S_0 by a certain absolute percentage x.*

Alternative 2 *K_1 below S_0 by a certain percentage that is given by a factor x times the implied volatility (VIX) at the time of trading.*

Alternative 3 *K_1 below S_0 by a certain percentage that is given by a factor x times the historical volatility (hv) at the time of trading (calculated from daily SPX data of the trading month prior to the time of trading). (In the software, the length of the time period from which the hv is calculated can be selected freely.)*

Alternative 4 (Currently Only in the JAI Article) *K_1 (on the assumption that K_2 has also already been chosen as a function of K_1) such that a certain minimum premium (specified as a percentage of the existing investment I_{act}) is earnt.*

(Method 4 for choosing strike K_1 may appear to be based on a similar principle as the one used for choosing K_1 as a function of the VIX. However, in this case we work specifically with the implied volatility of the traded option and not with the VIX, which in a way averages the volatilities of all the options.)

For the **choice of the parameter K_2**, the following alternatives are offered:

Alternative 1 (In the JAI Article) *K_2 is determined as a certain percentage of K_1 (e.g. $K_2 = 0.95 \cdot K_1$)*

Alternative 2 (In the Software) *K_2 is selected at a fixed absolute distance from K_1 (e.g. $K_2 = K_1 - 75$)*

For the **choice of the exit strategy**, the following alternatives are offered:

Alternative 1 *All positions are closed as soon as the SPX falls below K_1 or a certain percentage of K_1 (e.g. $1.01K_1$).*

Alternative 2 *All positions are closed when a certain maximum loss (expressed as a percentage of I_{act}) is reached.*

Alternative 3 *A combination of alternatives 1 and 2 (e.g. All positions are closed as soon as the SPX falls below a certain percentage of K_1 or a certain maximum loss is exceeded).*

As a small sample of backtest results, we now take a table from the JAI article, which lists the best results over the entire test period from 1990 to 2010 in terms of annual returns.

Following that, we give a brief overview of some backtest results for the time period from the beginning of 2010 to the end of 2018, which were generated using our software.

All the results were generated taking into account transaction costs (but no management fees or profit sharing or taxes).

The best strategies in terms of return per annum from the beginning of 1990 to the end of 2010 are given in Table 3.14**:**

Table 3.14 The best strategies with regard to the per-annum return from 1990 to 2010

Rank	Choice of short strike	Choice of long strike	Exit strategy	Return p.a.	Sharpe ratio
1	$K_1 = S_0$	$K_2 = 0.97 \cdot K_1$	At 5% loss or at $0.98 \cdot K_1$	84.62%	1.59
2	$K_1 = (1 - 0.1 \cdot VIX) \cdot S_0$	$K_2 = 0.97 \cdot K_1$	At 15% loss or at $0.98 \cdot K_1$	79.6%	1.54
3	$K_1 = (1 - 0.1 \cdot VIX) \cdot S_0$	$K_2 = 0.97 \cdot K_1$	At 20% loss or at $0.98 \cdot K_1$	76.65%	1.34
4	$K_1 = (1 - 0.1 \cdot hv) \cdot S_0$	$K_2 = 0.97 \cdot K_1$	At 15% loss or at $0.98 \cdot K_1$	76.01%	1.35
5	$K_1 = (1 - 0.1 \cdot VIX) \cdot S_0$	$K_2 = 0.95 \cdot K_1$	At 10% loss or at $0.98 \cdot K_1$	73.2%	1.84
6	$K_1 = (1 - 0.1 \cdot VIX) \cdot S_0$	$K_2 = 0.97 \cdot K_1$	At 10% loss or at $0.98 \cdot K_1$	72.36%	1.58
7	$K_1 = (1 - 0.1 \cdot hv) \cdot S_0$	$K_2 = 0.95 \cdot K_1$	At 10% loss or at $0.98 \cdot K_1$	72.33%	1.67
8	$K_1 = (1 - 0.1 \cdot VIX) \cdot S_0$	$K_2 = 0.95 \cdot K_1$	At 15% loss or at $0.98 \cdot K_1$	71.98%	1.6
9	$K_1 = (1 - 0.1 \cdot hv) \cdot S_0$	$K_2 = 0.97 \cdot K_1$	At 20% loss or at $0.98 \cdot K_1$	71.74%	1.14
10	$K_1 = (1 - 0.1 \cdot hv) \cdot S_0$	$K_2 = 0.97 \cdot K_1$	At 10% loss or at $0.98 \cdot K_1$	70.77%	1.41

A sample of some select strategies for the period from the beginning of 2010 to the end of 2018 is given in Table 3.15:

(Here "s" stands for "S_0", "h" stands for "hv", and "v" stands for "VIX" and an exit at, for example, $10\% + 2\%$ means that positions are closed at a loss of 10% but assuming an actual loss of 12% (buffer $= 2\%$). The long strike K_2 is always chosen as $K_2 = K_1 - 75$:

The results are clearly positive for most strategies. Regarding the data for the period from 1990 to 2010, we will take a closer look at the critical year 2008 further below. Regarding the results for the period from 2010 to 2018, it is worth noting that 2018 was a very negative year, and losses were incurred in the range of 30–60% in this year. In [5] you can also find results for the chaotic period in spring 2020 during the first COVID wave. As emphasized, these are only a few sample results; we leave it to the readers to interpret the results and/or run further tests.

3.5.4 Variant: 2-M Strategy

We explain here another variant of the put-write strategies, the so-called 2-M strategy (the analysis of which was commissioned by the asset management company involved). "2-M" stands for "2 months". Here, options with 2 months to expiration (instead of 1 month) are traded according to the otherwise same system. However, in this case, we do not invest (or use as margin) the entire available investment for 2 months; instead, we proceed in an overlapping manner, i.e. trading 50% of the investment each month over a period of 2 months.

Whereas in the examples tabulated above for the 1-month strategy, we gave only the respective results for the exit strategy "$10\% + 2\%$", and for the 2-month strategy, we give only the results for the exit strategy "$20\% + 2\%$". The reason for doing so is that this exit strategy (unlike the above) mostly delivered the better results for the 2-month strategy.

Again, we see clearly positive results in Table 3.16, with heavy losses again in the course of 2018.

3.5.5 Instructions for the Use of the Analysis Program on the Website

This section was created by the main developer of the analysis program, Alexander Brunhuemer.

We will now take a look at how the analyses described above can be performed using the software. The program can be executed under the following link, provided that you have activated access to the restricted area: https://app.lsqf.org/book/put-write-strategy.

When you open the program, you first see the input window. Explanations of the various input parameters are given in the following table (Table 3.17).

Table 3.15 Sample of strategies for the time span from 2010 to 2018

Short strike	$0.9 \cdot s$	$0.93 \cdot s$	$0.95 \cdot s$	$0.98 \cdot s$
Exit	10% + 2%	10% + 2%	10% + 2%	10% + 2%
2010	−9.98%	10.23%	39.14%	136.22%
2011	15.67%	−1.38%	−3.36%	41.91%
2012	23.57%	55.62%	66.11%	22.43%
2013	9.82%	25.72%	27.76%	56.75%
2014	−2.71%	−5.34%	6.36%	−24.09%
2015	6.55%	−11.36%	−8.67%	−39.14%
2016	23.86%	58.51%	3.92%	61.03%
2017	5.86%	17.78%	37.34%	96.18%
2018	−38.94%	−46.85%	−49.68%	−37.81%
Total (per annum)	1.98%	6.94%	8.76%	24.16%
Total return	19.29%	83.08%	113.11%	602.26%

Short strike	$(1 - 0.4 \cdot h) \cdot s$	$(1 - 0.3 \cdot h) \cdot s$	$(1 - 0.2 \cdot h) \cdot s$	$(1 - 0.1 \cdot h) \cdot s$
Exit	10% + 2%	10% + 2%	10% + 2%	10% + 2%
2010	15.42%	35.68%	82.33%	92.45%
2011	−9.09%	7.77%	9.84%	46.12%
2012	35.44%	65.58%	10.37%	11.17%
2013	17.79%	14.81%	43.90%	98.46%
2014	−16.75%	−4.14%	−26.25%	−11.38%
2015	−15.17%	−16.16%	−43.51%	−49.49%
2016	11.82%	4.37%	36.46%	43.59%
2017	69.31%	67.24%	145.41%	59.17%
2018	−38.96%	−46.79%	−38.62%	−53.32%
Total (per annum)	3.90%	9.61%	10.75%	15.37%
Total return	41.14%	128.47%	150.94%	262.63%

Short strike	$(1 - 0.4 \cdot v) \cdot s$	$(1 - 0.3 \cdot v) \cdot s$	$(1 - 0.2 \cdot v) \cdot s$	$(1 - 0.1 \cdot v) \cdot s$
Exit	10% + 2%	10% + 2%	10% + 2%	10% + 2%
2010	−4.77%	−9.24%	22.85%	72.73%
2011	−4.37%	15.18%	24.09%	35.82%
2012	52.36%	57.65%	13.07%	29.01%
2013	21.33%	22.82%	29.47%	89.29%
2014	2.10%	−1.08%	−15.51%	−17.22%
2015	45.51%	−33.04%	−45.49%	−50.26%
2016	37.32%	46.87%	28.89%	40.64%
2017	41.38%	71.07%	52.72%	113.65%
2018	−31.69%	−44.37%	−48.74%	−56.59%
Total (per annum)	16.99%	5.69%	1.46%	12.39%
Total return	310.97%	64.56%	13.97%	186.28%

Table 3.16 Sample of strategies for the period from 2010 to 2018

Short strike	$0.85 \cdot s$	$0.88 \cdot s$	$0.9 \cdot s$	$0.93 \cdot s$	$0.95 \cdot s$	$0.98 \cdot s$
Exit	20% + 2%	20% + 2%	20% + 2%	20% + 2%	20% + 2%	20% + 2%
2010	9.34%	19.23%	−3.03%	10.01%	2.18%	4.74%
2011	12.73%	−8.36%	−3.56%	−5.29%	28.91%	4.17%
2012	13.45%	23.99%	33.01%	59.35%	−5.49%	22.06%
2013	6.02%	11.34%	16.62%	32.35%	55.32%	57.76%
2014	5.94%	10.67%	3.40%	15.82%	−4.76%	−16.08%
2015	3.21%	−6.80%	−14.32%	−15.10%	−7.07%	−0.48%
2016	12.73%	21.52%	31.96%	54.47%	56.85%	125.16%
2017	4.05%	7.88%	12.44%	24.54%	23.68%	67.47%
2018	−0.46%	−3.20%	−22.54%	−34.72%	−46.21%	−46.61%
Total (per annum)	8.41%	10.58%	4.54%	10.09%	7.59%	15.06%
Total return	108.30%	149.28%	49.69%	139.47%	94.30%	257.59%

Short strike	$(1-0.6 \cdot h) \cdot s$	$(1-0.5 \cdot h) \cdot s$	$(1-0.4 \cdot h) \cdot s$	$(1-0.3 \cdot h) \cdot s$	$(1-0.2 \cdot h) \cdot s$	$(1-0.1 \cdot h) \cdot s$
Exit	20% + 2%	20% + 2%	20% + 2%	20% + 2%	20% + 2%	20% + 2%
2010	−6.03%	−0.50%	−7.49%	0.83%	12.11%	24.75%
2011	6.41%	15.38%	−23.41%	3.63%	−26.87%	−17.17%
2012	31.73%	23.24%	−5.69%	3.65%	18.00%	29.38%
2013	37.49%	49.27%	41.96%	62.18%	54.12%	49.57%
2014	5.65%	11.50%	−12.46%	−10.79%	2.26%	−2.26%
2015	−21.50%	−17.36%	−9.74%	−2.46%	−9.78%	−18.00%
2016	52.31%	70.42%	89.11%	77.98%	108.62%	151.60%
2017	34.09%	43.42%	53.94%	68.08%	89.03%	112.13%
2018	−21.42%	−38.24%	−34.72%	−28.51%	−35.31%	−54.46%
Total (per annum)	11.91%	15.83%	7.27%	8.15%	19.37%	14.22%
Total return	177.97%	280.05%	89.15%	103.72%	399.58%	234.51%
Short strike	$(1-0.6 \cdot v) \cdot s$	$(1-0.5 \cdot v) \cdot s$	$(1-0.4 \cdot v) \cdot s$	$(1-0.3 \cdot v) \cdot s$	$(1-0.2 \cdot v) \cdot s$	$(1-0.1 \cdot v) \cdot s$
Exit	20% + 2%	20% + 2%	20% + 2%	20% + 2%	20% + 2%	20% + 2%
2010	−1.05%	24.96%	0.36%	11.99%	6.06%	0.21%
2011	12.58%	3.78%	13.13%	9.31%	9.85%	0.70%
2012	30.79%	42.72%	39.21%	−6.97%	2.95%	24.41%
2013	25.61%	35.39%	50.69%	50.05%	72.45%	44.92%
2014	8.38%	14.60%	−8.28%	−0.62%	−3.03%	−23.25%
2015	18.48%	−18.78%	−12.46%	−4.15%	7.88%	−17.30%
2016	33.05%	46.92%	66.70%	91.95%	128.09%	145.92%
2017	10.79%	33.39%	27.74%	41.65%	62.20%	98.14%
2018	−14.70%	−26.70%	−21.00%	−43.01%	−22.14%	−54.50%
Total (per annum)	16.12%	15.95%	17.54%	12.60%	22.17%	10.63%
Total return	288.54%	283.45%	334.14%	193.75%	516.55%	150.41%

Table 3.17 Input parameters of the LSQF software

Buttons "1-month strategy" and "2-month strategy"	These buttons allow you to switch between the 1- and 2-M strategies described above
Capital	The capital available for investment (denoted by I_0 above) in US dollars
Start period	The first month in which the strategy is to be executed (on the third Friday). Entered as *YYYYMM*
End period	The last month in which the strategy is to be executed (on the third Friday). Entered as *YYYYMM*
Days for computing reference value of underlying	Number of days used to compute the reference value of the underlying asset (see next parameter) Subsequently denoted by "n"
Reference value for underlying	Function dependent on the S_0, $S_{\max}$, and $S_{\min}$ to be used for computing the value s in the formula for K_1 These values denote: $S_0 \ldots$ value of the underlying on the trading day (closing price) $S_{\max} \ldots$ highest closing price in the last n days $S_{\min} \ldots$ lowest closing price in the last n days S_0, $S_{\max}$, and $S_{\min}$ are always entered as s_0, $s_{\max}$, and $s_{\min}$
Formula for computing short strike (K_1)	Formula for computing K_1 with possible parameters s, v, h which denote the following values: $s \ldots$ reference value for the underlying (computed according to the formula for the underlying) $v \ldots \frac{\text{closing price of the VIX on the trading day}}{100}$ $h \ldots$ historical volatility based on the last 20 trading days Formulas for the short strike could be, for example: $0.95 \cdot s$ $(1 - 0.4 \cdot v) \cdot s$ $(1 - 0.3 \cdot h) \cdot s$
Formula for computing long strike (K_2)	Formula for computing K_2 as a function of K_1. Examples: $0.95 \cdot K_1$ $K_1 - 75$
Exit threshold upon loss (in % of invested capital)	Specifies at what percentage of the invested capital the strategy is to be terminated within 1 month
Additional loss buffer (%)	This value in % of the invested capital is used as an additional loss parameter if the exit threshold (exit strategy) is reached
Transaction cost per contract	This is the transaction cost per traded contract in US dollars that is used in the calculations

(continued)

Table 3.17 (continued)

Bid//ask distance	Specifies at what (absolute) distance to the midpoint between bid and ask prices the trade is executed
Upper and lower bound for short strike relative to the underlying	Specifying these boundaries can make sense if one uses a formula based on the volatility to compute K_1 and wants to restrict the trading range. Example: $(1 - 0.4 \cdot v) \cdot s$, but in any case within the range $[0.9 \cdot s, 0.97 \cdot s]$
Defensive performance verification	When activated, the process for verifying the success of the monthly strategy does not compare the daily high prices of the options, but instead compares the maximum of P_1 and minimum of P_2 (see Sect. 3.5.2)

Once you have entered the desired parameters, you can calculate the success of the strategy. You get the calculated total return (per annum and in total) as well as an overview of the respective trading periods including a graphical illustration of the profit analysis.

3.5.6 Closing Remarks on Operational Risks and Strategy Variants

Some concluding remarks on the results, the (operational) risks of the strategies, and on further possible variants of these put-write strategies are in order.

We have carried out backtests here using historical price data and need to add a caveat—one that you will often come across in investment fund marketing brochures and fact sheets too: Backtesting results must be viewed with caution in regard to their reliability, and they do not allow conclusions to be drawn about the future performance of strategies.

Still, the results of our analyses do point to a certain overpricing of put options. The premiums generated by selling put options lead to a significant outperformance compared to the SPX.

It could be argued here that in the strategies presented here, put options are both sold and bought, and therefore any such overpricing should balance out. In fact, systematic observation of volatility smiles suggests that the further out of the money the option is, the more it is overpriced. That is, the purchased put options with strike K_2 that are further out of the money are more overpriced on average than the sold put options with strike K_1. However, this is to be seen as relative, i.e. as a percentage of the option price. Let us illustrate this with the first numerical example that we gave in Sect. 3.5.2:

The price of the short position was 10.40, and the price of the long position was 5.70. The premium received was therefore 4.70 USD. Let us suppose that the short position was overpriced by 10% and that the long position was even more overpriced, say, by 15%. The actual fair price of the short position would thus be $\frac{10.40}{1+0.10} = 9.45$ USD, and the actual fair price of the long position would be

$\frac{5.70}{1+0.15} = 4.95$ USD. The fair difference would then be 4.50 USD. The actually collected premium of 4.70 is thus higher than the fair premium despite the stronger (relative) overpricing of the long position.

But why do we even work with combinations of short and long positions in this strategy? Wouldn't it be more efficient to trade solely short positions (so-called naked positions)? In doing so we would save ourselves the expensive purchase of those long positions!

There are at least two reasons for working with combinations instead of naked positions:

Firstly

When trading naked positions, a loss may exceed the available margin in the worst case. In this case, the investor has to re-margin (i.e. deposit additional funds as margin) and bear the entire loss even beyond the investment amount.

Of course, one could argue here that exit strategies have been provided to prevent higher losses. On the one hand, that is correct, but on the other hand, an exit strategy is not an absolute guarantee. Large overnight gaps or an exchange that closes (and reopens at possibly much lower prices) have certainly happened in the past and can thwart the best intentions for closing positions in time, thus resulting in enormous losses. A loss in excess of the investment is not possible when using short/long positions as described above.

In trading for clients, in some legal constellations, even the theoretical possibility of a margin call must be excluded in principle, effectively prohibiting the intensive use of naked positions.

Secondly

When working with combinations, it is usually possible to trade a much higher number of contracts than when working with naked positions. As an example, we will again take the first trade given in Volume I Section 2.15: At a difference of 75 points between the strikes K_1 and K_2, we were able to trade 13 combination contracts on an investment of 100,000 USD. As we have seen, this translates into a premium income of over 6000 USD. In the case of naked positions, the margin rules described in Volume I Section 2.15 come into play. Essentially (without going into exact detail), these rules require a margin of $100 \times 15\%$ of S_0 per option contract. At a value for the S_0 of 2950 points, the required margin is 44,250 USD. Thus, only two option contracts can be traded on the 100,000 USD investment. Considering the option prices, this means a premium income of just over 2000 USD. On the other hand, a planned exit upon reaching a certain loss level is then generally triggered much later (in case of a much stronger price decline in the SPX) than when trading a larger number of contracts. Or, put differently: The amount of loss sustained upon closing positions when the SPX falls below a certain threshold is generally significantly lower than when trading combinations. In their article "A Comparison of Different Families of Put-Write Option Strategies" in the ACRN Journal of Finance and Risk Perspectives, L. Del Chicca and the author [6] demonstrate that the performance of put-write strategies is almost consistently significantly better when using combinations than when using naked positions.

Let us now return to the issue of potential operational risks. To illustrate, we will cite an example from a real trading setting:

Probably the most extreme conditions for trading options specifically on the SPX occurred on 10 October 2008.

In fact, the trading strategies that the author implemented in his asset management role around 10 October 2008 required that all positions be closed as soon as the SPX were to fall below the threshold of 900 points. The short position to be closed out was a put option with strike 900 and expiration on 17 October 2008.

At the close of trading on 9 October, the SPX had closed at 909 points. Until then, no close-out had taken place. Figure 3.39 shows how the SPX moved on 10 October 2008. Immediately upon opening of the exchange, the price of the SPX had already fallen to 840. The order to close out the options, in particular to buy back the short position with strike 900, had been placed immediately when trading opened.

In Fig. 3.40 we see the price movement of this 900 put option from 9 October 2008 to 10 October 2008 and on 10 October 2008.

Instead of being able to close the option as anticipated, that is, at a price of about 40 dollars when the SPX fell below the 900 threshold, the prices that were actually obtained fluctuated uncontrollably in a range averaging about 70 dollars.

As a result of this unfortunate coincidence of the close-out requirement directly at the opening of trading on 10 October 2008 with the most extreme trading conditions

Fig. 3.39 SPX movement from 9 October 2008 to 10 October 2008 and on 10 October 2008 (source: Bloomberg)

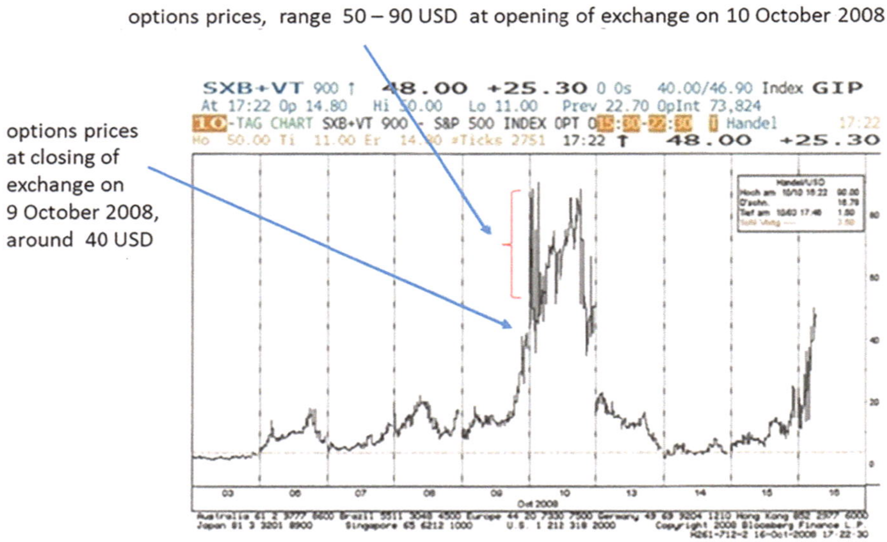

Fig. 3.40 Price movement of the strike 900 put option from 9 October 2008 to 10 October 2008 and on 10 October 2008 (source: Bloomberg)

at the height of the financial crisis, immense losses were incurred in the executed put-write strategies that would never have occurred under normal circumstances.

Note that the tests presented in the JAI article included such potentially extreme trading conditions and also strong overnight gaps!

The parameter choices described above are relatively static in nature. In the actual executions, more flexible procedures were developed, with a particular view to optimizing the choice of the short strike K_1 where possible. In particular, certain systematic waves in the SPX price movement as well as the SPX movement in the days immediately preceding the opening of the position were taken into account in choosing K_1.

By integrating the parameters $S_{\min}$ and $S_{\max}$ into our analysis program, such considerations can be partially factored in when using this program.

3.6 Case Study VI: Valuation of an Asian Option on the Basis of a Foreign Currency Loan Within the Scope of an Expert Witness Report

The following case relates to an expert witness report prepared in connection with a court case in which an Investor A had sued Bank B after a transaction that the Investor had entered into with the Bank in 2006 for a 5-year term until 2011 had gone substantially awry. Investor A claimed that Bank B had not sufficiently informed them about the structure and risks of the product, in particular about the fact that the product's underlying instrument was an **Asian** option and not an option of the usual type (and also lack of information about what this meant for the product's performance potential).

In the following, we will take a closer look at the product in question and two of the commissioned expert reports.

3.6.1 Overall Product Structure

The **product** consisted of two main components:

Component 1: A CHF foreign currency loan with an interest rate cap
Component 2: A commodity index guarantee bond issued by Bank *B*

The foreign currency loan was used exclusively to finance the commodity index guarantee bond.

The **foreign currency loan** (Component 1) was structured as follows:

Loan amount: 1,000,000 EUR
Loan period: 3 April 2006 until 2 April 2011
Interest rate: CHF 1-month LIBOR at payment due date rounded up to $\frac{1}{8} + 0.75\%$
(This results in an initial interest rate of 2.25%.)
Payment in 20 quarterly instalments

Interest rate is capped at 3% per annum.
The cost of this cap is 0.42% per annum.

The loan amount was paid out in CHF on 3 April 2006. On 3 April 2006, the EUR CHF exchange rate was 1.571092. Therefore, an amount of 1,571,092 CHF was paid out.

Thus repayment on 2 April 2011 also had to be effected as a retransfer of CHF 1,571,092.

The **commodity index guarantee bond** (Component 2) was structured as follows:

Nominal amount: 1,000,000 EUR
Term to maturity: 3 April 2006 until 2 April 2011
Annual coupon: 0.80% (paid annually on 2 April)
Repayment at maturity: 100% + maximum of 0% and "profit share" (as defined below)

Profit share:
The profit share is based on the performance of the "S&P GSCI Index" (hereinafter referred to as GSCI). This Goldman Sachs Commodity Index, which was developed by Goldman Sachs in 1991 (back-calculated to 1970) and taken over by Standard

Fig. 3.41 Movement of the GSCI from the start of back-calculation at the beginning of 1970 to when the arrangement was entered into on 3 April 2006

& Poor's in 2007, is one of the best-known benchmarks for commodities. Its performance over time from the beginning of 1970 until 3 April 2006, when the arrangement was entered into, is shown in Fig. 3.41.

The GSCI value $S(0)$ on 3 April 2006 was 399.17 points.

The profit share is calculated as follows:

The relative increase in the GSCI from 3 April 2006 is measured on the following reference dates (closing prices):

2 July 2006, 2 October 2006, 2 January 2007, 2 April 2007
2 July 2007, 2 October 2007, 2 January 2008, 2 April 2008
2 July 2008, 2 October 2008, 2 January 2009, 2 April 2009
2 July 2009, 2 October 2009, 2 January 20108, 2 April 2010
2 July 2010, 2 October 20107, 2 January 2011, 2 April 2011

The arithmetic mean is then taken over these increases. This mean is the profit share.

The investor took out a foreign currency loan to buy a product that was partly a bond (with a fixed coupon that was low at the time) and partly an Asian option on a commodity index.

The risks to which the investor is thus exposed are, on the one hand, payment of loan interest of up to a maximum of 3.42% per annum (including cap costs) against a fixed-coupon income of only 0.8% from the bond and, on the other hand, the exchange rate risks of a foreign currency loan. In principle, however, both risks can also translate into profit opportunities for the investor. The Asian option does not present any risk of investment loss to the investor.

In the investor's lawsuit against the bank, the investor argued that while they had been aware of the foreign currency risk, the bank had not informed them about the low-profit prospects from the Asian option and had especially not informed them about the option being Asian style.

Two questions or mandates—among others—that the court put to the court-appointed expert were:

1. *"Preparation of an ex-ante risk assessment from the perspective of 2006 for the bond in question as regards realistic earnings prospects when using debt financing".*
2. *"Could the offered option component conceivably have been understood to be European style, given the present pricing of component 2 of the product?"*

3.6.2 Fair Price of the Guarantee Bond in Asian and European Style

We start our analyses with an attempt to answer the second question, that is, the question as to the fair price of the bond from the perspective of 3 April 2006, once for the actual Asian style and once for a potential European style.

The approach to use here is relatively clear: We model the GSCI in a risk-neutral manner using the Wiener model. A mean-reverting Ornstein-Uhlenbeck model would also be conceivable, but mean reversion does not seem to be present for the commodity index in this case. So, we only need an estimate of the volatility for the index. In the absence of a truly liquid option market on the GSCI, we will estimate historical volatilities of the GSCI and use those to determine a specific range for the volatility to be used.

We discount the payment expected from the option component and the payments due from the bond component (coupon and principal repayment at maturity) using appropriate discount factors.

We begin by determining the discount factors and do this by proceeding as set out in Sect. 3.4. We denote the discount factors for the annual coupon payments by d_1, d_2, d_3, d_4, d_5 from the perspective of 2 April 2006. For this we need the interest rates $f_{0,1}, f_{0,2}, f_{0,3}, f_{0,4}, f_{0,5}$ from the perspective of 2 April 2006. We use these either directly or interpolate them linearly from the swap rates on 3 April 2006, i.e. from:

$$S_1 = 0.03294$$
$$S_2 = 0.03536$$
$$S_5 = 0.03782$$

This gives us the interpolated values:

$$S_3 = 0.03617$$
$$S_4 = 0.03701$$

We then obtain the discount rates from this with $dt = \frac{1}{(1+f_{0,t})^t}$. The discount rates that are then finally used are as follows:

$d_1 = 0.9681$
$d_2 = 0.9329$
$d_3 = 0.8946$
$d_4 = 0.8675$
$d_5 = 0.8338$

The payments to be expected from the actual bond component therefore had the following value from the perspective of 3 April 2006:

$$8000 \times (0.9681 + 0.9329 + 0.8946 + 0.8675 + 0.8338) + 1,000,000 \times 0.8338$$
$$= 869{,}820\,\text{EUR}.$$

To estimate a reliable volatility range from the perspective of 2 April 2006, we calculate annualized historical volatilities from daily GSCI data for various time periods up to 3 April 2006.

This gives us the following percentage values (in a sequential order) for the annualized historical volatilities:

2 April 2005 to 2 April 2006
2 April 2004 to 2 April 2006
2 April 2003 to 2 April 2006
...
2 April 1970 to 2 April 2006:
16.94, 16.58, 15.59, 15.03, 14.63, 14.37, 13.96, 13.65, 13.25, 13.06, 12.70, 12.42, 12.10, 11.78, 11.53, 11.87, 11.66, 11.67, 11.60, 11.54, 11.56, 11.43, 11.35, 11.38, 11.38, 11.62, 11.91, 11.97, 12.00, 12.11, 12.42, 12.87, 13.42, 13.44, 13.33, 13.25

Furthermore, we obtain for all historical 5-year periods, in the following order:

2 April 2001 to 2 April 2006
2 April 2000 to 2 April 2005
2 April 1999 to 2 April 2004
...
2 April 1970 until 2 April 1975

The following annualized historical volatility values, as percentages, are as follows: 14.63, 13.81, 12.76, 12.33, 11.59, 11.26, 10.34, 9.86, 9.08, 8.53, 7.6, 9.82, 9.61, 10.47, 11.08, 11.58, 10.51, 10.58, 10.14, 10.51, 10.7, 11.85, 13.83, 14.46, 14.6, 15.25, 15.94, 17.18, 19.67, 19.85, 19.03, 17.49

If as our estimator we only use volatilities that occurred in 5 years before the reference date, we would work with a value of 14.63%. In both number sequences, the majority of the values are somewhere between 10 and 15%. We will therefore calculate the value of the Asian option component using volatilities σ in the range between 10% and 15%. This means: We will calculate the value of the option component for $\sigma = 10\%$, 12.5%, and 15%.

As the risk-free interest rate r for the 5-year option, we choose the 5-year swap rate from the perspective of 3 April 2006, that is, $r = 3.782\%$.

The risk-neutral model for the GSCI therefore has the form

$$S(t) = 399.17 \cdot e^{\left(r-\frac{\sigma^2}{2}\right)\cdot t+\sigma\cdot W_t} = 399.17 \cdot e^{0.03282\cdot t+0.1\cdot W_t} \text{ or}$$
$$S(t) = 399.17 \cdot e^{0.02657\cdot t+0.15\cdot W_t} \text{ or } S(t) = 399.17 \cdot e^{0.03\cdot t+0.125\cdot W_t}$$

depending on whether $\sigma = 10\%$, 15%, or 12.5% is selected. Here, W_t again denotes a standard Brownian motion.

We now simulate 10,000 paths for each of the three cases based on weekly step lengths (i.e. $dt = \frac{1}{52}$ and thus 260 steps each) and evaluate on a quarterly basis (i.e. after every 13 steps) the returns accumulated up to that point, which are then averaged to determine the payoff for each path.

The payoffs are then discounted using the 5-year discount factor $d_5 = 0.8338$. Taking the average over all the discounted payoffs, we obtain three estimators (expressed as a percentage of the principal (nominal value)) for the fair price of the option component (for three different volatility estimates).

In total therefore, the guarantee bond has approximately the following fair price, depending on the assumed volatility (compared to Fig. 3.42 to obtain the associated factors in the calculation):

If $\sigma = 10\%$, then $869{,}820 + 94{,}089 = 963{,}909$ EUR.
If $\sigma = 12.5\%$, then $869{,}820 + 100{,}568 = 970{,}388$ EUR.
If $\sigma = 15\%$, then $869{,}820 + 102{,}689 = 972{,}509$ EUR.

So, we get an estimated range for the fair price of the guarantee bond of approximately 964,000–972,500 EUR. The 1,000,000 EUR purchase price that was actually asked can therefore be considered (taking into account fees and margins for the bank) as quite reasonable.

Let us now estimate the fair price of the option component and subsequently of the guarantee bond on the assumption that the option portion is European style. We proceed exactly as before, except that the payoff is merely determined from the values up to maturity on 2 April 2011 and that the index values at earlier points are not relevant. The convergence behaviour and values of the European option component are shown in Fig. 3.43.

In total therefore, if the option component were European, the guarantee bond **would have** approximately the following fair price, depending on the assumed volatility:

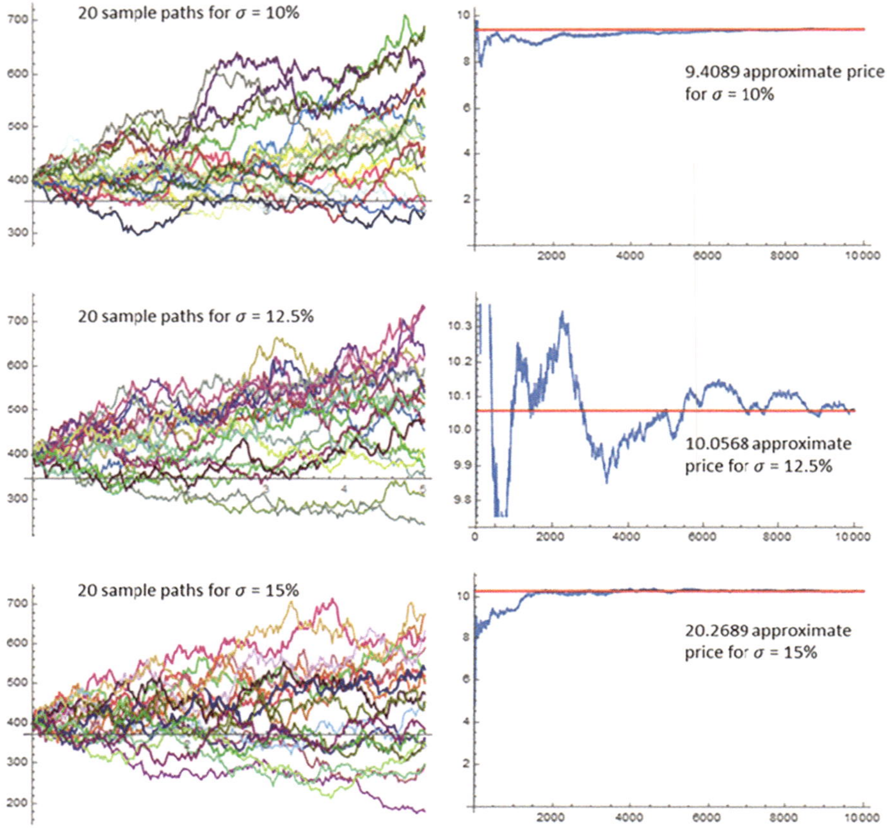

Fig. 3.42 Twenty scenarios each (left) of price paths and convergence behaviour (right) from 10,000 scenarios using Monte Carlo approximation for the Asian option component with different volatilities

If $\sigma = 10\%$, then $869{,}820 + 178{,}238 = 1{,}048{,}058$ EUR.
If $\sigma = 12.5\%$, then $869{,}820 + 178{,}560 = 1{,}048{,}380$ EUR.
If $\sigma = 15\%$, then $869{,}820 + 185{,}673 = 1{,}055{,}493$ EUR.

So, if the option component had been European, we would get an estimated range for the fair price of the guarantee bond of approximately 1,048,000–1,055,500 EUR. Therefore, the 1,000,000 EUR purchase price that was actually asked could not have been the price for a guarantee bond with a European option component. However, because the difference between the fair price and the actual price is not large, one cannot say that an investor not versed in financial mathematics should immediately recognize that, at the price of EUR 1,000,000, the present guarantee bond could not have a European-style option component.

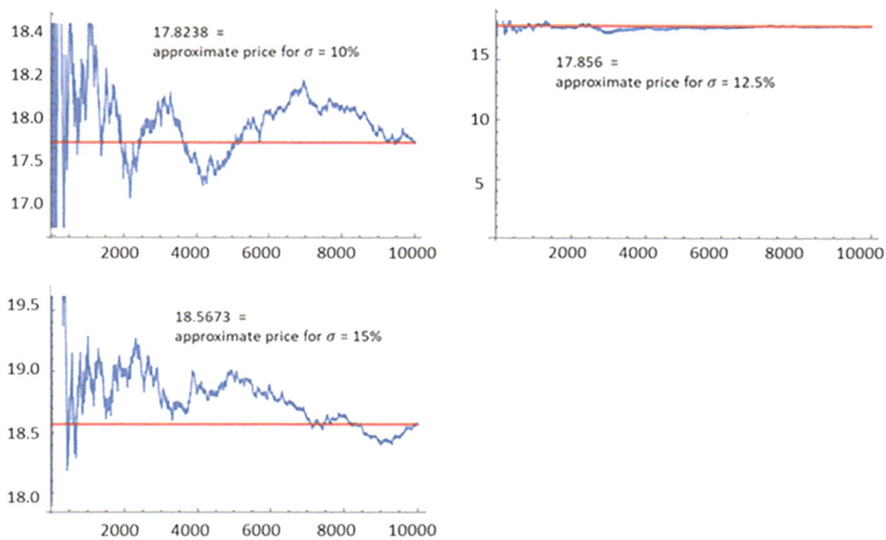

Fig. 3.43 Convergence behaviour in the case of 10,000 scenarios using Monte Carlo approximation for a European-style option component with different volatilities

3.6.3 Preparation of an Ex Ante Risk Assessment from the Perspective of Realistic Earnings Prospects When Using Debt Financing

We now turn to the first of the expert's mandates, that is, "preparation of an ex-ante risk assessment from the perspective of 2006 for the bond in question as regards realistic earnings prospects when using debt financing".

To do this, we first need to be clear what exactly we want to calculate or estimate here, what exactly is being asked for, and what information the court can and will actually expect from the expert. The background to the question posed by the court was as follows: Is the offer to invest in this product bundle consisting of a guarantee bond and a foreign currency loan to be qualified as absurd in principle, or did the investor stand reasonably realistic chances to make a profit from this product combination? So, the task here is neither about determining a fair price for the product combination nor about calculating an "expected profit".

The former would be irrelevant in the present case: The investor did not intend to actively manage and hedge the product, but to buy it and hold it to maturity. When initiating the investment, the investor was not interested in a possible fair price, but rather in whether there were realistic (!) (not only theoretical) chances of making a profit.

The latter—determination of the expected profit from this product combination—is simply not possible. To do that, we would need a reliable estimate for the future price moves of the GSCI index (from the perspective of April 2006). However, we cannot realistically presume to make such a reliable estimate.

We can therefore approach the task as follows:

We define various realistic scenarios for possible price moves of the financial variables underlying the product combination (GSCI index, EUR CHF development, CHF 1-month LIBOR). We then use Monte Carlo simulation to determine the expected profit on the assumption that these price moves actually occur. In other words, we will not (only) determine the expected profit but also an approximate probability distribution for different profit levels. After that we will discuss the obtained profit probabilities.

We start by arguing a few such parameter choices (again, the point here is not to choose and argue **probable** parameters, but realistically **possible** parameters).

The **first realistic starting point** would be to assume that both the lending rates and the EUR CHF exchange rate remain essentially unchanged over the product combination's life. This would mean a loan interest rate of essentially 2.25% plus the cost of the cap of 0.42%, i.e. of 2.67% per annum, and an essentially unchanged EUR CHF exchange rate of 1.577 throughout the product combination's life.

As regards an assumption for the volatility of the GSCI, we rely on the analyses performed above, which suggest a volatility in the range of 10% to 15% per annum. A realistic possible assumption would therefore be a volatility in the order of 12.5%.

For these realistic basic assumptions, we now run Monte Carlo simulations for different trend assumptions for the GSCI to create a profit probability distribution. For this, we proceed as follows: We again generate 10,000 scenarios each of possible GSCI price trends in the period from April 2006 to April 2011. For each scenario, we determine the profit (or loss) of the product combination discounted to April 2006. This gives us a set of 10,000 possible discounted profits from which we can create a historical probability distribution.

Before doing so, we give some thought to the question as to which possible trends we want to use for the simulations in terms of realistic trend assumptions. For this purpose, we proceed in the same way as above, when we estimated realistic GSCI volatilities. Specifically, for all 5-year periods since the (back)-calculation of the GSCI, we will determine the per-annum trend in these periods, which will give us some indication of **possible** trends (because they occurred in the past). Furthermore, we obtain for all historical 5-year periods, in the following order:

2 April 2001 to 2 April 2006
2 April 2000 to 2 April 2005
2 April 1999 to 2 April 2004
...
2 April 1970 to 2 April 1975

The following annualized historical volatility values, as percentages, are as follows: 6.68, 6.47, 6.45, −4.05, −10.44, −6.51, −3.33, −5.82, −1.29, 3.42, 2.66, −0.17, 1.96, 5.63, 7.69, 12.13, 11.17, 6.1, 3.67, 0.8, −5.38, −8.05, −4.3, −2.98, 2.47, 1.58, 5.81, 6.89, 8.55, 13.34, 19.68, 20.87

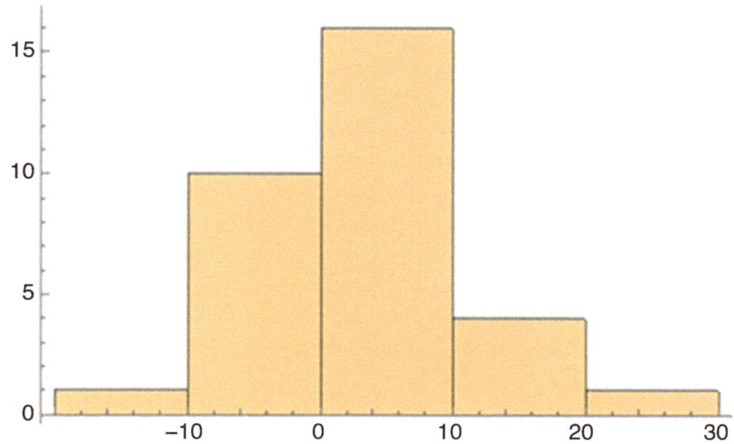

Fig. 3.44 Histogram of frequencies of the 32 5-year per-annum trends of the GSCI

They ranged from -10.44% to 20.87%. The mean of these 5-year per-annum trends was 3.18%. A histogram for this is shown in Fig. 3.44.

So in principle, based on this histogram, 5-year trends of the GSCI in the range between 0% and 10% are definitely realistic.

Now let us generate the profit probability distributions for GSCI trends in this value range.

The profits or losses from the product combination consist of the discounted fixed payments in the foreign currency loan (due to the parameter assumptions, these are indeed fixed payments) and in the bond component on the one hand (which we can determine exactly) and the discounted payoffs of the Asian option on the other hand.

From the perspective of 3 April 2006, the annual payments of the 0.8% coupon from the bond had the discounted value that was already (implicitly) calculated above of

$$8000 \times (0.9681 + 0.9329 + 0.8946 + 0.8675 + 0.8338) = 35{,}975 \text{ EUR}.$$

The quarterly loan interest payments in the amount of $\frac{2.67}{4} = 0.6675\%$ each, on the other hand, had a discounted value of 121,759 EUR. (We calculated the discount factors here from the EUR interest rates of 3 April 2006, which we interpolated from the 3-month LIBOR, the 6-month LIBOR, and the Euro swap rates that we already used above. The quarterly discount factors used were therefore 0.993142, 0.985368, 0.977058, 0.96811, 0.959596, 0.950878, 0.941964, 0.932862, 0.923558, 0.914076, 0.904425, 0.89461, 0.887492, 0.880606, 0.873948, 0.867513, 0.859101, 0.850684, 0.842265, and 0.833845.)

A comment is called for here with regard to the use of discount factors based on EUR interest rates instead of CHF interest rates. After all, the loan repayments are made in CHF and not in EUR. However, it is important to keep in mind that our task is not to calculate the fair price of the product in financial mathematical terms but the value of the product to the investor based on that investor's assumptions and approach: We assume here that the amount of future payments **in EUR** is already known at time 0 (based on the assumption made) and that the investor will raise these **EUR amounts** on the respective future payment dates, change them into CHF, and thus make the payments due. So, for this reason, the EUR amounts to be raised need to be discounted using discount factors based on EUR interest rates.

The fixed reciprocal payments thus had a discounted value of $35{,}975 - 121{,}759 = -85{,}784$ from the investor's perspective on 3 April 2006. If we combine this value of fixed payments with the discounted payoffs under 10,000 Monte Carlo scenarios, we obtain the simulation values illustrated in the following graphs for various selected trends.

Assuming the long-term average per-annum **trend of 3.18%**:
average discounted profit of **17,200 EUR**
probability of making a **profit: 43.02%**
Histogram 3.45 of **profit amounts**:
Assuming a per-annum **trend of 0%**:
average discounted profit of **−28,173 EUR**
probability of making a **profit: 25.3%**
Histogram 3.46 of **profit amounts**:
Assuming a per-annum **trend of 5%**:
average discounted profit of **54,270 EUR**
probability of making a **profit: 54.67%**
Histogram 3.47 of **profit amounts**:

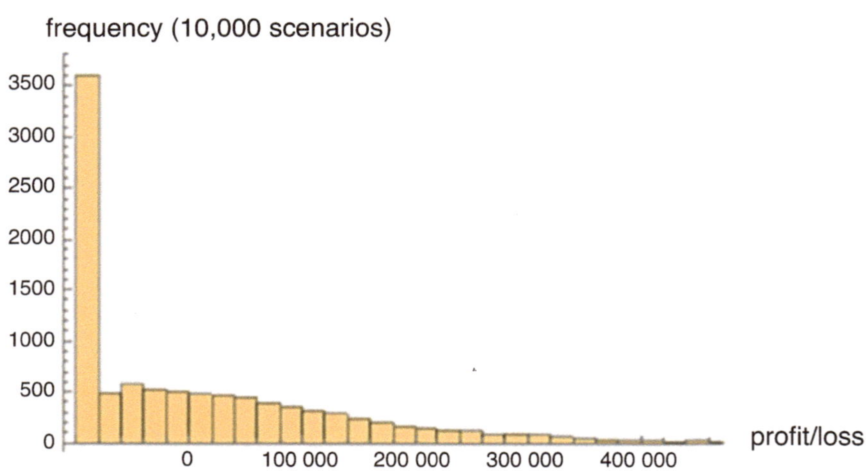

Fig. 3.45 Realistic assumptions, part 1: frequencies of discounted profits/losses assuming GSCI trend = 3.18%

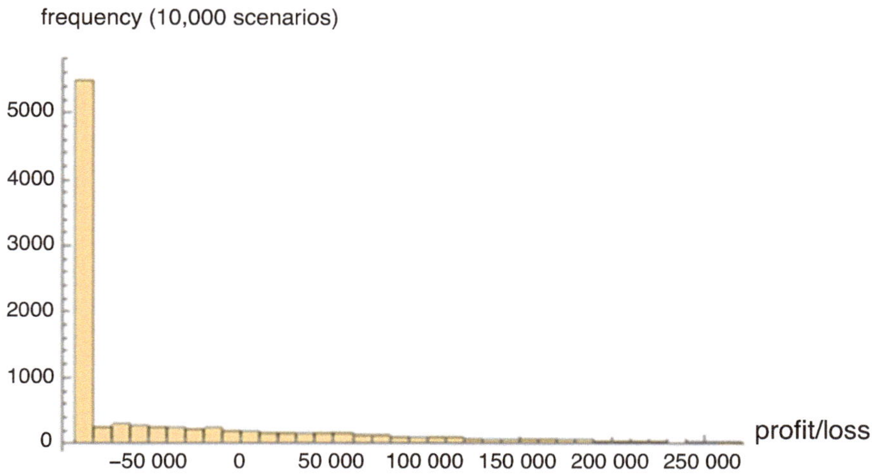

Fig. 3.46 Realistic assumptions, part 1: frequencies of discounted profits/losses assuming GSCI trend = 0%

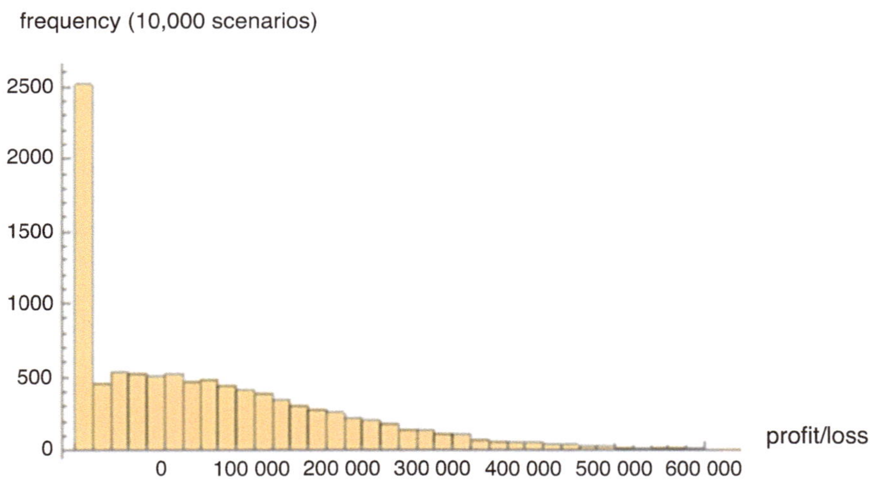

Fig. 3.47 Realistic assumptions, part 1: frequencies of discounted profits/losses assuming GSCI trend = 5%

A second realistic starting point would be to assume that the lending rates remain essentially unchanged over the product combination's life. This would again mean a loan interest rate of essentially 2.25% plus the cost of the cap of 0.42%, so 2.67% per annum. We recall that due to the cap, the total interest charge can increase to no more than 3.42% at most; therefore the assumption of a fixed interest rate of 2.67% is a rather cautious (even pessimistic) estimate from the investor's point of view.

Region G10 | Währung Euro / Schweizer Franken - Cros | Zum 06/01/06

EURCHF

Spot 1.56

Q1 06 Tatsächl.

1.58

Q1 06 Prognose

.00

	Q3 06	Q4 06	Q1 07	Q2 07	2007	2008	2009	2010
Median	1.55	1.55	1.55	1.54	1.54	1.53	1.51	1.51
Mean	1.55	1.55	1.55	1.53	1.54	1.54	1.53	1.52
High	1.59	1.59	1.59	1.59	1.60	1.64	1.63	1.62
Low	1.51	1.50	1.46	1.44	1.48	1.49	1.46	1.43
Forward price	1.56	1.55	1.54	1.54	1.53	1.51	1.50	1.48

Datenanbieter (25)	Zum	Q3 06	Q4 06	Q1 07	Q2 07	2007	2008	2009	2010
1) Credit Suisse Group	06/01/06	1.55	--	--	--	--	--	--	--
2) Danske Bank	06/01/06	1.54	1.53	1.53	1.52	--	--	--	--
3) Morgan Stanley	06/01/06	1.55	1.55	1.55	1.52	1.52	1.49	1.46	1.43
4) Rabobank	06/01/06	1.55	1.55	1.54	1.54	1.53	--	--	--
5) UBS	05/31/06	1.56	1.52	--	--	1.50	--	--	--
6) Commonwealth Ban	05/26/06	1.54	1.52	1.46	1.44	1.59	1.63	1.63	1.63
7) Deutsche Bank	05/26/06	1.56	1.56	1.56	--	1.56	1.56	--	--
8) Saxo Bank	05/26/06	1.55	1.57	1.57	--	--	--	--	--
9) Canadian Imperial B	05/25/06	1.59	1.59	1.59	1.59	1.58	--	--	--
10) Citigroup	05/25/06	1.54	1.54	1.54	--	1.53	1.53	1.53	1.53
11) Royal Bank of Scotla	05/25/06	1.56	1.56	1.57	1.60	1.57	1.57	1.58	1.57
12) Scotiabank	05/25/06	1.51	1.50	1.48	--	--	--	--	--
13) Bank of America Me	05/23/06	1.56	1.55	1.54	1.52	1.53	1.50	--	--
14) Brown Brothers Harr	05/22/06	1.55	1.54	1.53	--	--	--	--	--

Australia 61 2 9777 8600 Brazil 5511 2395 9000 Europe 44 20 7330 7500 Germany 49 69 9204 1210 Hong Kong 852 2977 6000
Japan 81 3 3201 8900 Singapore 65 6212 1000 U.S. 1 212 318 2000 Copyright 2014 Bloomberg Finance L.P.
SN 343859 EST GMT-5:00 H194-2470-0 13-Dec-2014 13:56:21

Fig. 3.48 Forecasts by international banks on future EUR CHF exchange rate trends from the perspective of 1 June 2006 (source: Bloomberg)

With regard to EUR CHF exchange rate forecasts, we resort to a Bloomberg chart from 1 June 2006 (see Fig. 3.48).

We assume here that these forecasts are roughly in line with the forecasts as seen at the beginning of April 2006. What we are looking at here are realistic trends from the perspective of April 2006 (not probable trends or worst-case trends). A realistic forecast (perhaps more realistic than the assumption of a constant exchange rate) would be to use the average ("mean") forecast of the banks for the EUR CHF exchange rate in the years following 2006, which can be taken from Fig. 3.48.

One could summarize this mean forecast in simplified terms as follows:

Forecast for the remainder of 2006:	1.55
Forecast for 2007 and 2008:	1.54
Forecast for 2009:	1.53
Forecast for 2010:	1.52

And possibly a continued forecast for 2011 of 1.51

How would such an estimation of the EUR CHF exchange rate trend affect possible opportunities and probabilities of making a profit? This assumed change in the exchange rate would affect the loan interest payments and the repayment of the loan amount, since they are all to be paid in Swiss francs and increase due to the stronger franc.

With this in mind, let us now take another look at the payments due from the loan in detail and as they actually accrue:

The loan amount was 1,571,092 CHF. Of this amount, 0.6675%, i.e. 10,487 CHF, was payable in interest on a quarterly basis (i.e. 20 times). After 5 years, repayment of the principal was due at 1,571,092 CHF. If we now convert the payments due using the above EUR CHF exchange rate forecasts, we obtain:

Two payments of 6766 EUR each in 2006
Four payments of 6810 EUR each in 2007
Four payments of 6810 EUR each in 2008
Four payments of 6854 EUR each in 2009
Four payments of 6899 EUR each in 2010
Two payments of 6945 EUR each in 2011

as well as a repayment in the amount of 1,040,458 EUR in April 2011.

If we discount these payments with the appropriate discount factors (given above), we see that we have to make discounted payments on the loan (minus the 1,000,000 EUR we receive from the bond at maturity) amounting to the equivalent of 124,825 EUR in interest payments and 33,736 EUR due to the loss incurred from the changed exchange rate when repaying the principal (nominal) amount. The total discounted payment for the loan is therefore the equivalent of 158,561 EUR (instead of the 121,759 EUR in the first assumed "realistic scenario").

Netted, the discounted fixed payments from the bond and loan therefore amount to $35{,}975 - 158{,}561 = -122{,}586$ EUR.

Now, in analogy to our approach in the first realistic case, we again run simulations (this time based on these fixed payments) for the Asian option component under different trend assumptions and obtain the following results:

Assuming the long-term average per-annum **trend of 3.18%**:
average discounted profit of **−20,370 EUR**
probability of making a **profit: 33.77%**
Histogram 3.49 of **profit amounts**:
Assuming a per-annum **trend of 0%**:
average discounted loss of **−66,137 EUR**
probability of making a **profit: 18.2%**
Histogram 3.50 of **profit amounts**:
Assuming a per-annum **trend of 5%**:
average discounted profit of **15,857 EUR**
probability of making a **profit: 45.2%**
Histogram 3.51 of **profit amounts**:
So we see:

Under quite realistic (and not at all overly optimistic) assumptions, there was a clearly positive probability for making profits with the product combination from the perspective of 2006.

It cannot be said therefore that making a profit would have had to be more or less ruled out from the outset. However, we emphasize once again: This statement does not provide any information or estimation as to whether it "made sense" to invest in such a product, and the calculations also do not provide any estimation as to the

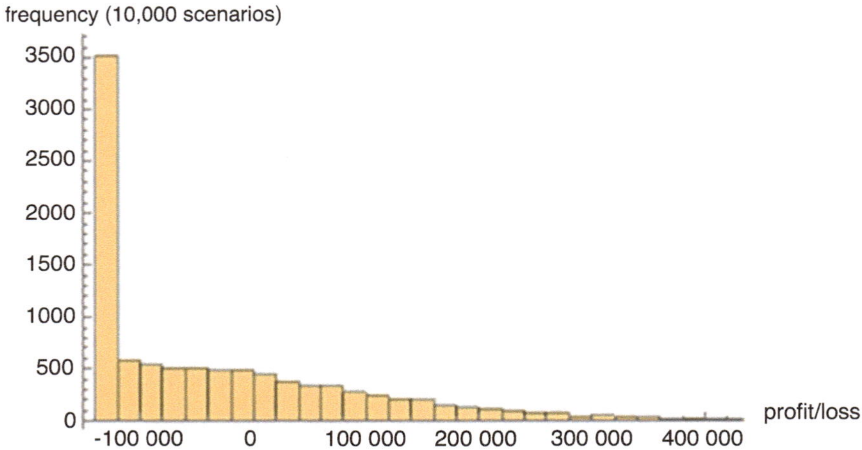

Fig. 3.49 Realistic assumptions, part 2: frequencies of discounted profits/losses assuming GSCI trend = 3.18%

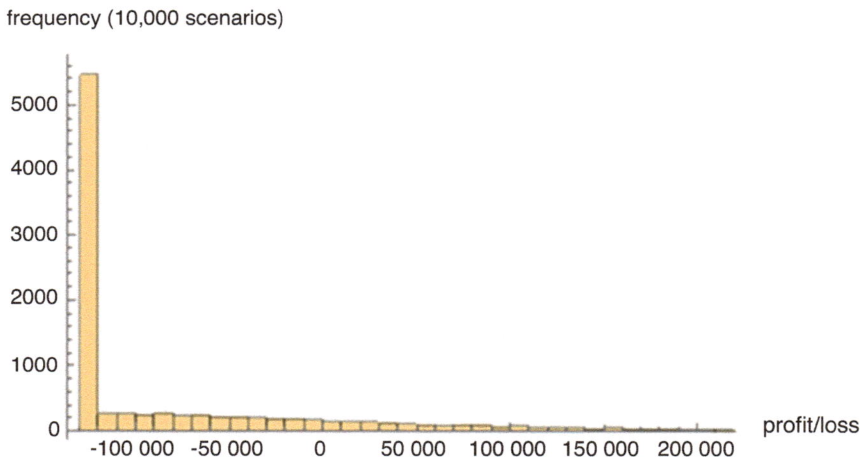

Fig. 3.50 Realistic assumptions, part 2: frequencies of discounted profits/losses assuming GSCI trend = 0%

amount of an average expected profit nor do they provide any risk estimation for the product bundle.

This answers the court's questions to the expert.

Let's now take a look—just out of interest—at the product combination's actual performance from April 2006 to April 2011. Presumably it wasn't favourable, since otherwise the investor would hardly have brought this lawsuit against the bank.

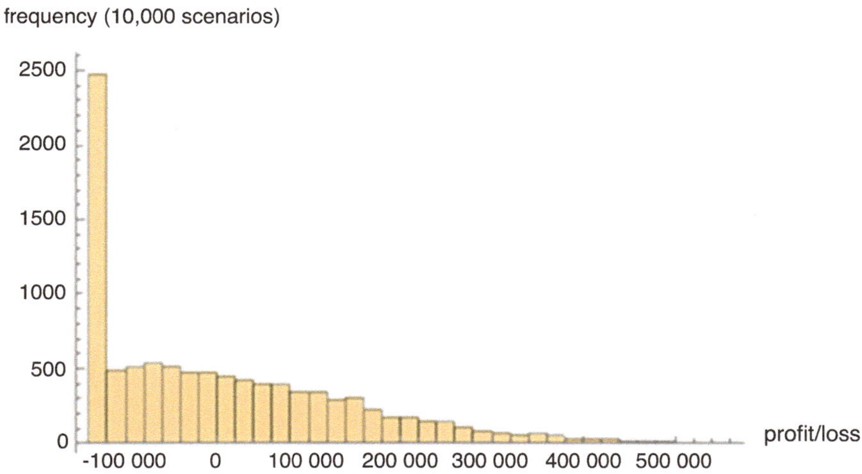

Fig. 3.51 Realistic assumptions, part 2: frequencies of discounted profits/losses assuming GSCI trend = 5%

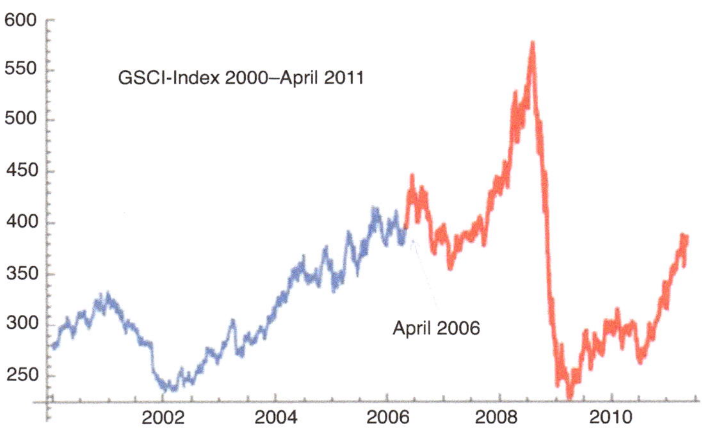

Fig. 3.52 GSCI index from 2000 to April 2011

3.6.4 Actual Performance of the Product Combination

For this purpose, let us first look at the actual performance of the three main financial variables underlying the product (GSCI index given in Fig. 3.52, EUR CHF exchange rate in Fig. 3.53, CHF 1-month LIBOR in Fig. 3.54) before April 2006 and subsequently until April 2011.

The chart clearly shows that the EUR CHF exchange rate moved in a way that was fatal to the investor. The exchange rate was 1.3147 at maturity on 2 April 2011.

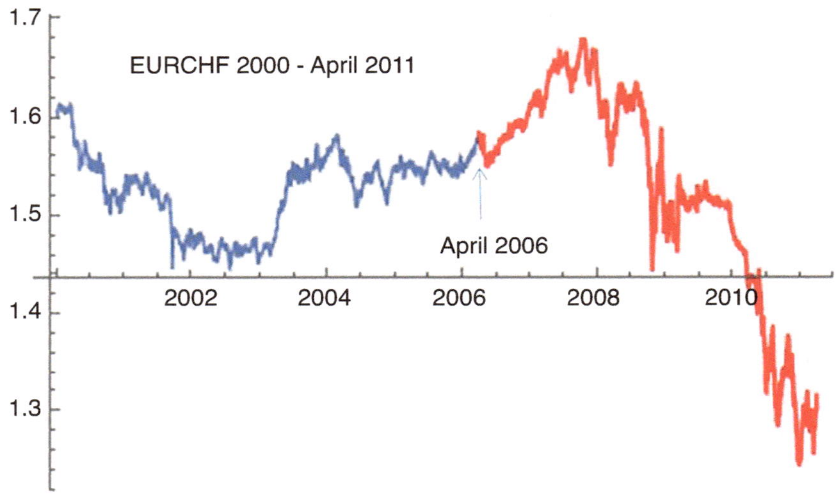

Fig. 3.53 EUR CHF exchange rate from 2000 to April 2011

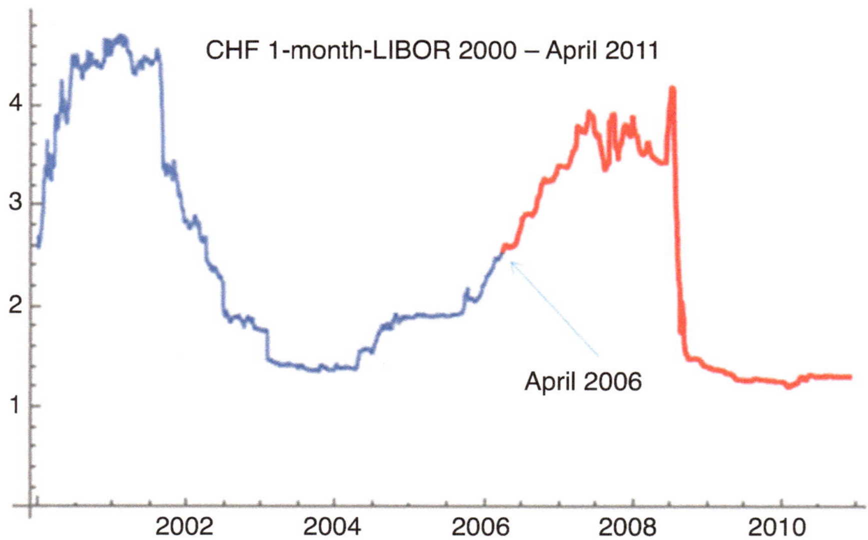

Fig. 3.54 CHF 1-month LIBOR from 2000 to April 2011

The required repayment was therefore $\frac{1{,}571{,}092}{13{,}147} = 1{,}195{,}019$ EUR, i.e. 195,019 EUR more than what the bond component of the guarantee bond paid out.

One glance at the GSCI index over the period from 2006 to 2011 already suffices to understand that the average performance of the index on the 20 sampling dates was just short of positive, meaning that no payoff resulted from the option component.

Finally, one look at the lending rates in Fig. 3.54 shows that, although they fell dramatically from 2008 onwards, they were nevertheless (including cap costs) always above the bond's coupon of 0.8%.

This means that, for the investor, the regular fixed payments from the bond on the one hand and the loan on the other hand were slightly negative in total. Thus, for the investor, this resulted in total losses from the product combination of just over 200,000 EUR.

3.7 Case Study VII: Expert Opinion on the EUR CHF Stop-Loss Order Fiasco in January 2015

The following case study is again about a dispute in a court case. On 15 January 2015, the exchange rate of the EUR against the CHF dropped sharply within seconds. Having remained very constant at around 1.20 for about three and a half years, the exchange rate suddenly dropped to below 0.90 on 15 January practically without warning—as said: within seconds—and then levelled off at about 1.05 (see Fig. 3.55).

A large number of investors in the euro zone were still invested in CHF loans at the time, and for many of these investors, stop-loss orders had been placed by their banks to hedge the risks of these currency loans. These stop-loss orders were essentially as follows:

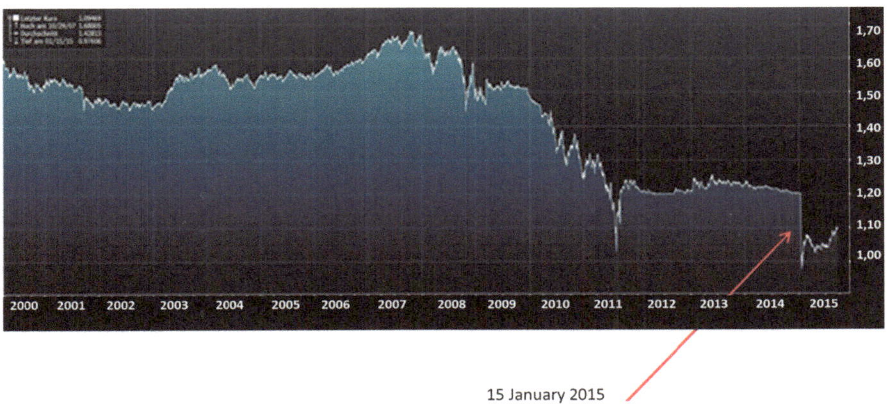

Fig. 3.55 EUR CHF exchange rate before and around 15 January 2015 (source: Bloomberg)

"As soon as the CHF exchange rate falls below (for example) 1.15, immediately buy a certain volume of Swiss francs for the customer to hedge the customer and their loan (which is to be repaid in CHF) against a further strengthening Swiss franc".

All of these stop-loss orders were triggered on 15 January 2015. This resulted in a massive problem: Due to the extreme plunge, a large part of the automated (!) orders could not be executed at the specified threshold of, for example, 1.15 and were then executed when the price had already dropped to 1.05, or 1.00, or even (for a short time) 0.90. This resulted in significantly higher than anticipated losses for a number of customers. And in the aftermath, a large number of these customers filed lawsuits against their banks. The general tenor of these complaints was as follows: *"By setting these stop-loss orders, the banks took inappropriate risk management measures".*

Two questions that arose during subsequent court proceedings were:

1. *Shouldn't the banks have been aware, through careful study of comparable situations, that such stop-loss orders are inappropriate risk management measures?*
2. *Would it have been possible, in this specific case, to take other, more efficient, more risk-reducing measures against this form of currency risk?*

We will address these two questions below from the perspective of a private expert opinion commissioned by one of the defendant banks. The questions are therefore dealt with below in a manner that, while being factually entirely objective, will tentatively support the bank's position.

3.7.1 Analysis of Comparable Events: GBP DEM September 1992

Probably the most dramatic event to occur between highly developed stable currencies until 15 January 2015 happened on 16 September 1992, the "Black Wednesday", between the British pound and the German mark. We will not go into the details that led to the massive crisis of the British pound at the time; we will only cite the facts that are relevant for the following analysis.

In October 1990, the British government signed an agreement guaranteeing that the exchange rate between the British pound (GBP) and other ERM member currencies (including the German mark (DEM)) would not be allowed to change by more than 6%. When the agreement was signed, the GBP DEM exchange rate was 2.95 (i.e. one British pound could be traded for 2.95 German marks). The agreement ensured that this rate could not fall below a value of 2.773. Due to massive speculation against the GBP by investor George Soros and others, the British government was forced to announce its withdrawal from the agreement on the evening of 16 September 1992, despite massive attempts to intervene in the interest rate market.

The impact on the GBP DEM exchange rate is illustrated in Fig. 3.56: a relatively stable rate between 2.80 and 3.00 in the period before September 1992 and a fast

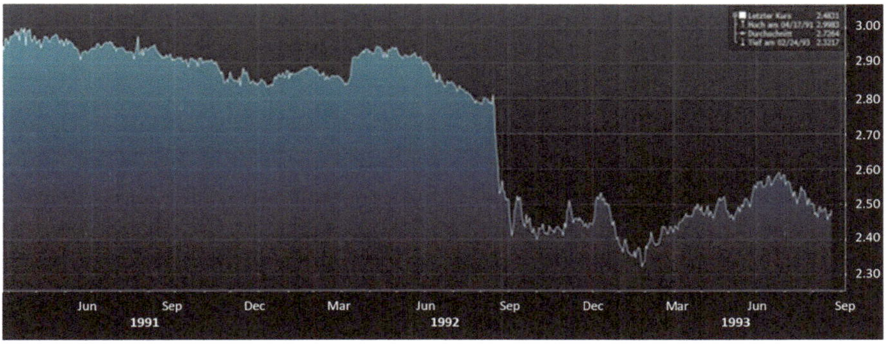

Fig. 3.56 GBP DEM exchange rate from March 1991 to September 1993 (source: Bloomberg)

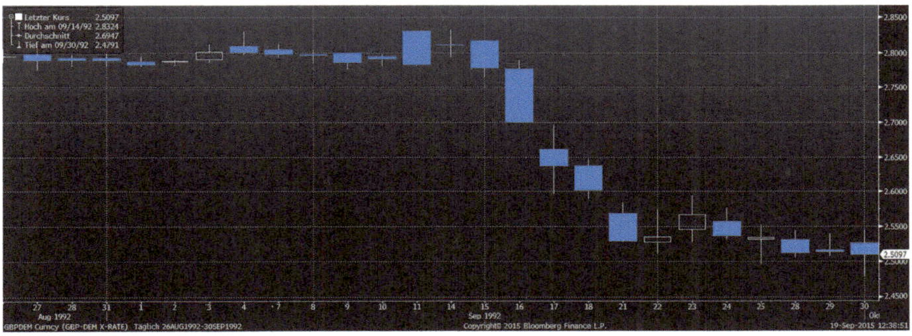

Fig. 3.57 GBP DEM exchange rate from 27 August 1992 to 30 September 1992 (source: Bloomberg)

decline to a new level in the range of approximately 2.40–2.50 as from 16 September 1992.

For the analysis of possible hedging strategies for future comparable events, the speed of the decline in this dramatic situation is of course essential. The following graphic is very telling in this regard:

The bar Chart 3.57 of the exchange rate in the period from 27 August 1992 to 30 September 1992 clearly shows that there was a sharp drop in the exchange rate immediately following the announced withdrawal on 16 September to a value of around 2.70, declining further on 17 September to a low of around 2.60, and that the further decline to a range of around 2.50 only took place in the course of the following days.

In that extremely critical situation, a stop-loss order on the GBP DEM exchange rate at a value of 2.77, for example, would in all probability have had the following effect in the worst case (from the perspective of a British citizen with the GBP as the base currency): In the worst-case scenario, the stop loss would have been exercised at the GBP DEM exchange rate low on 16 September 1992, hence, in the worst case, at a value of approximately 2.70. The stop-loss mechanism would therefore have resulted in a loss of approximately 2.5% in the worst case.

Placing the stop loss was essential here: The termination of the agreement was announced on the evening of 16 September 1992 at 7:00 p.m. GMT (i.e. at 8:00 p.m. CET). In the vast majority of cases, no new actions (i.e. new market orders) would have been placed by the banks at this time. No transactions would therefore have occurred until the next day, 17 September 1992. The exchange rates on this following day were clearly lower than on 16 September, as can be seen from the chart. So, the exchange rates would in any case have been worse than those achieved with the stop loss on 16 September.

What alternative hedging strategies (again, from the perspective of a British citizen with GBP as the base currency) would have been available in the situation at the time?

It would have been possible to definitively secure the rate of 2.77 for 1 year, respectively, by buying (American) put options with strike 2.77 and about 1 year to expiration (one unit of such an option entitles the investor to sell one GBP at the price of DEM 2.77 at any time during the coming year, that is, to buy one DEM at the price of $\frac{1}{2.77} = 0.361$ GBP). The price of such a put option reflects the hedging costs in this method. Since we do not have historical prices of GBP DEM put options for the period in question nor historical prices for implied volatilities for the GBP DEM exchange rate over this period, we will calculate theoretical estimates for such put options based on the Black-Scholes model and on historical volatilities in the following.

Specifically, we will use the Black-Scholes model adapted for FX rates that we provided in Volume II Section 2.2 and that we state again here:

Theorem 3.1 *For the fair price $C(t)$ of a call option with expiration T and strike K at time $t \in [0, T]$ on a foreign currency XXX with price $X(t)$ which follows a Wiener model with parameters μ and σ, we have*

$$C(t) = e^{-r_f \cdot (T-t)} \cdot X(t) \cdot \mathcal{N}\left(\tilde{d}_1\right) - e^{-r \cdot (T-t)} \cdot K \cdot \mathcal{N}\left(-\tilde{d}_2\right)$$

with

$$\tilde{d}_1 = \frac{\log\left(\frac{X(t)}{K}\right) + \left(r - r_f + \frac{\sigma^2}{2}\right)(T-t)}{\sigma\sqrt{T-t}}$$

and

$$\tilde{d}_2 = \frac{\log\left(\frac{X(t)}{K}\right) + \left(r - r_f - \frac{\sigma^2}{2}\right)(T-t)}{\sigma\sqrt{T-t}}$$

and $\mathcal{N}$ being the distribution function of the standard normal distribution and r_f the risk-free interest rate of the foreign currency.

Theorem 3.2 *For the fair price $P(t)$ of a put option with expiration T and strike K at time $t \in [0, T]$ on a foreign currency XXX with price $X(t)$ which evolves according to a Wiener model with parameters μ and σ, we have*

$$P(t) = e^{-r\cdot(T-t)} \cdot K \cdot \mathcal{N}\left(-\tilde{d}_2\right) - e^{-r_f\cdot(T-t)} X(t) \cdot \mathcal{N}\left(-\tilde{d}_1\right),$$

with $\tilde{d}_1$ und $\tilde{d}_1$ as in the theorem above and $\mathcal{N}$ being the distribution function of the standard normal distribution.

So, how do we get the price data needed to apply these formulas?

Figures 3.58, 3.59, and 3.60 show the historical volatility of the GBP DEM exchange rate (from daily data on a 1-month basis) as well as the 6-month GBP LIBOR and the 6-month DEM LIBOR.

As an example and to give you an idea of the order of magnitude, we will approximate the cost of hedging the product combination with put options for the period from October 1990 (signing of the agreement) to October 1992 at a level of 2.77. We assume the purchase of two types of 1-year put options for the periods October 1990 to October 1991 and October 1991 to October 1992.

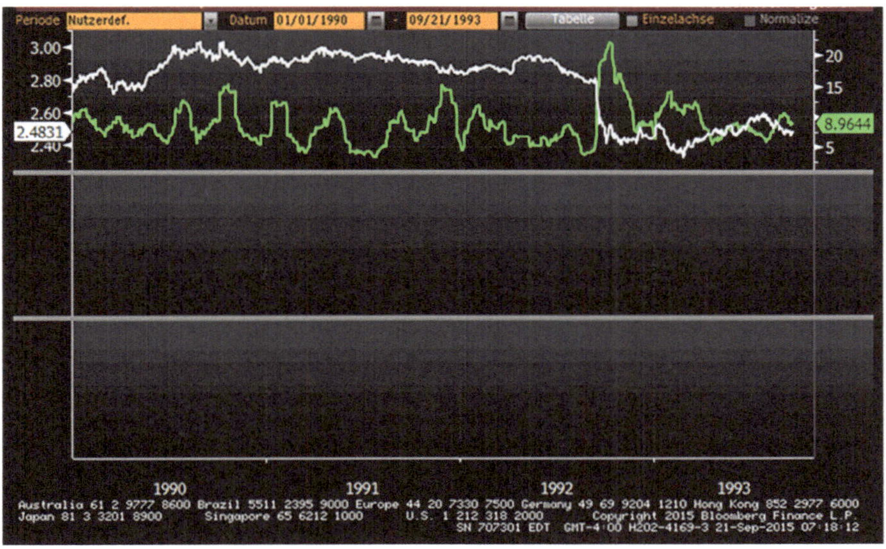

Fig. 3.58 GBP DEM exchange rate (white) and historical volatilities on a monthly basis of the exchange rate in the period 1990–1993 (source: Bloomberg)

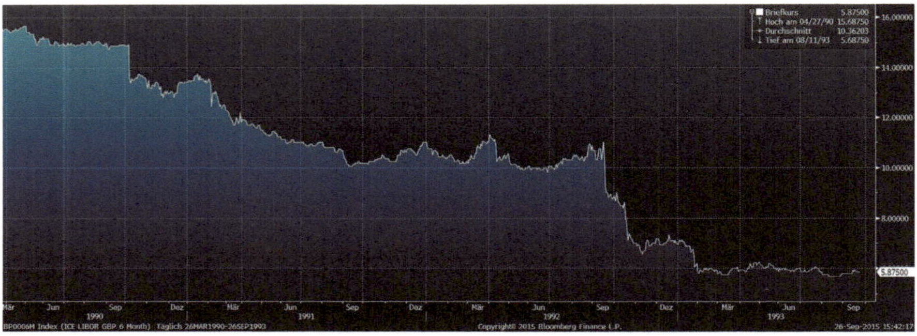

Fig. 3.59 Six-month GBP LIBOR from March 1990 to September 1993 (source: Bloomberg)

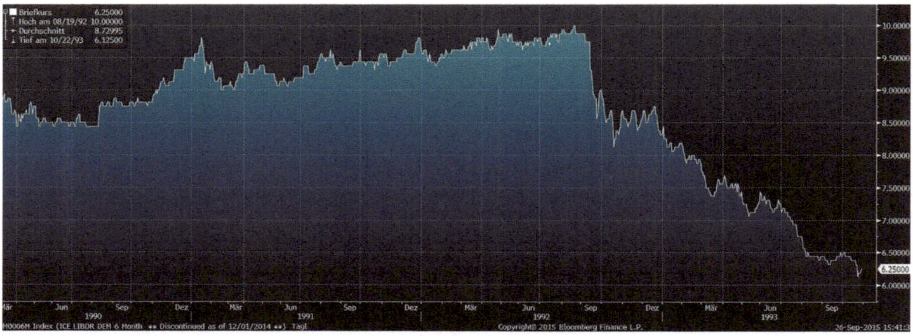

Fig. 3.60 Six-month DEM LIBOR from March 1990 to September 1993 (source: Bloomberg)

Note, however, that if the strong GBP devaluation had occurred at a later date, the cost of further put options for subsequent time periods would have greatly increased the overall cost of such a hedge.

The thing here is that, in order to achieve a definitive hedge, it would be necessary to either buy put options expiring when the loan matures (often 10 years, 20 years, or more) from the outset or to buy 1-year put options annually. The latter method has the advantage that, if an exchange rate event occurs earlier, it is not necessary to hedge the investment over the entire term of the loan, but the disadvantage is that it may be considerably more expensive than the first method. Therefore, in addition to our calculations for the 2 years preceding the sudden GBP exchange rate drop in 1992, we will run the analogous approximations for longer-running hedges (10 years) below.

3.7.2 The GBP DEM Case: Comparison of Stop Loss Versus Hedging with Put Options in the Period October 1990 to October 1992

Above we gave the formulas for computing the fair prices of foreign currency options. For American put options, which would be the more relevant type for hedging purposes here, the price of the hedge in this example and in all of the following examples is slightly higher. However, since the difference is not significant, we will limit the following discussion to the simpler analysis of European options. Volatility over the period under review was mostly in a range between 5% and 13% (as shown in the chart). For our calculations, we therefore assume volatility of approximately 9% when the hedges were entered into.

In the following, the interest rate r denotes the DEM 6-month LIBOR (DEM denotes the base currency of our analyses), and r_f denotes the GBP 6-month LIBOR.

r was about 8.5% in October 1990 and about 9.5% in October 1991.
r_f was about 13.5% in October 1990 and about 10.5% in October 1991.

The GBP DEM exchange rate was approximately 2.90 in both October 1990 and October 1991.

With these parameters, using the above formula, we obtain prices for the two put options that were used for hedging of:

DEM 0.096 for the option opened in October 1990 and expiring in October 1991
DEM 0.053 for the option opened in October 1991 and expiring in October 1992.

To evaluate the consequences of the two hedging types, we analyse the following example (to be able to compare the results with the EUR CHF situation—which is what interests us here—we will again proceed from the perspective of a British citizen with GBP as the base currency):

In October 1990, a DEM loan in the amount of 100,000 GBP is taken out at an exchange rate of 2.90. The loan period is 2 years, that is, the loan is to be repaid in October 1992.

We consider three possible approaches:

(a) No hedging
(b) Hedging using a stop-loss order at 2.77 (assuming delayed actual exercise at 2.70)
(c) Hedging using above put options

Since the interest payments are the same in all three cases, they can be ignored in the analysis. We can focus exclusively on the effects of exchange rate fluctuations.

Case (a)
In October 1990, a loan is taken out in the amount of 290,000 DEM (= 100,000 GBP). In October 1992, the exchange rate was approximately 2.52. The cost of purchasing the 290,000 DEM to repay the loan is therefore approximately 115,079 GBP. **The loss in case (a) (no hedging) is therefore 15.08%.**

Case (b)
In October 1990, a loan is taken out in the amount of 290,000 DEM (= 100,000 GBP). On 16 September 1992, the stop-loss order is exercised at the worst-case price of 2.70 (as described above). The cost of purchasing the 290,000 DEM to repay the loan is therefore approximately 107,407 GBP. **The loss in case (b) (hedging using a stop-loss order) is therefore 7.41%.** (The remaining 1-month loan period can be neglected in this analysis.)

Case (c)
In October 1990, a loan is taken out in the amount of 290,000 DEM (= 100,000 GBP). For hedging purposes, a total of 104,693 (= 290,000/2.77) put options are purchased in October 1990 at a price of 0.096 DEM each, that is, at a total price of 10,050 DEM = 3465 GBP. In addition, in October 1991 a total of 104,693 (= 290,000/2.77) put options are purchased for hedging purposes at a price of 0.053 DEM each, that is, at a total price of 5549 DEM = 1913 GBP.

On 16 September 1992, the options are exercised at a price of 2.77. The cost of purchasing the 290,000 DEM to repay the loan is therefore approximately 104,693 GBP. The total cost (loan repayment + hedging costs) was therefore 104,693 + 3465 + 1913 = 110,071 GBP. **The loss in case (c) (hedging using put options) is therefore 10.07%.** (The remaining 1-month loan period can be neglected in this analysis.)

In this example, it would therefore have been clearly preferable to hedge the investment by a stop-loss order. It should also be noted that if the event (massive devaluation of the GBP) had occurred at a later date, additional options would have had to be purchased, which would have increased the cost in case (c) even more and to an extent that could not have been computed in advance. In contrast, the costs of a stop-loss order are independent of when (and if) the order is actually executed and are therefore known in advance.

Based on this analysis of one of the most massive exchange rate events between stable currencies in modern financial markets, we can conclude in principle that hedging through stop-loss orders is undoubtedly sensible and economically viable.

3.7.3 The GBP DEM Case: Hedging with Put Options for 10-Year Loans as from October 1990

Since we do not have historical data for GBP 10-year and GBP 20-year interest rates, nor for DEM 10-year and DEM 20-year interest rates for the relevant periods, we give possible ranges for the hedging cost in these cases (10-year and 20-year loan periods) below, assuming realistic parameter ranges. As a reference for October 1990, we have the respective 6-month LIBOR values of 8.5% for the DEM and of 13.5% for the GBP. Assuming a range of 4–10% for DEM 10-year interest rates and of 8–15% for GBP 10-year interest rates and assuming a negative difference of less than or equal to 2% and of greater than or equal to −6% between the DEM 10-year interest rate and the GBP 10-year interest rate and keeping the other parameters unchanged, the above put price formula results in put option prices in the range of:

0.195 DEM
assuming a DEM 10-year interest rate of 10%
and assuming a GBP 10-year interest rate of 12%
to 0.79 DEM
assuming a DEM 10-year interest rate of 9%
and assuming a GBP 10-year interest rate of 15%

For our example above, this means: In October 1990, a loan is taken out for a period of 10 years in the amount of 290,000 DEM (= 100,000 GBP). To hedge the investment for the entire period, a total of 104,693 (= 290,000/2.77) put options are purchased in October 1990 at a minimum price of 0.195 DEM each, that is, at a total price of 20,415 DEM = 7040 GBP, and a maximum price of 0.79 DEM each, that is, at a total price of 82,707 DEM = 28,520 GBP.

On 16 September 1992, the options are exercised at a price of 2.77. The cost of purchasing the 290,000 DEM to repay the loan is therefore approximately 104,693 GBP. The total cost (loan repayment + hedging) was therefore between 104,693 + 7,040 = 111,733 GBP and 104,693 + 28,520 = 133,213 GBP.

The loss in the case of hedging using 10-year put options is therefore between 11.73% and 33.21%

3.7.4 Alternative Hedging Method for the EUR CHF Exchange Rate Using Put Options

As stated above, based on the analysis of one of the most massive exchange rate events between stable currencies in modern financial markets, hedging by means of stop-loss orders is a sensible and viable approach—especially in cases like the one presented above.

Still, in the following we want to examine which alternative hedging methods using put options would have been possible at what cost in the present EUR CHF exchange rate case. For this analysis, we proceed in the same way as in the analysis of the option hedging strategies in the previous subsection, except that this time, we take the perspective of an investor with EUR as the base currency. As in the case of the GBP DEM analysis, we will again need the volatilities of the EUR CHF exchange rate as well as the risk-free EUR and CHF interest rates in the relevant period to calculate the required option prices. The relevant data for the relevant period can be seen in the following Charts 3.61, 3.62, 3.63 and 3.64:

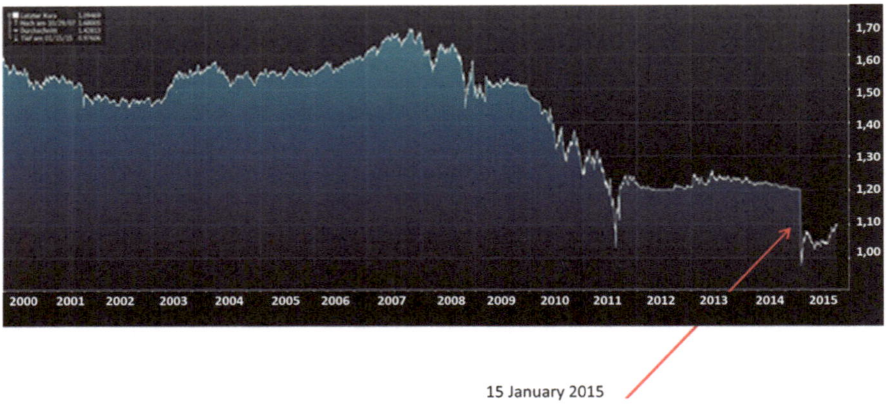

Fig. 3.61 EUR CHF exchange rate from 2000 to 2015 (source: Bloomberg)

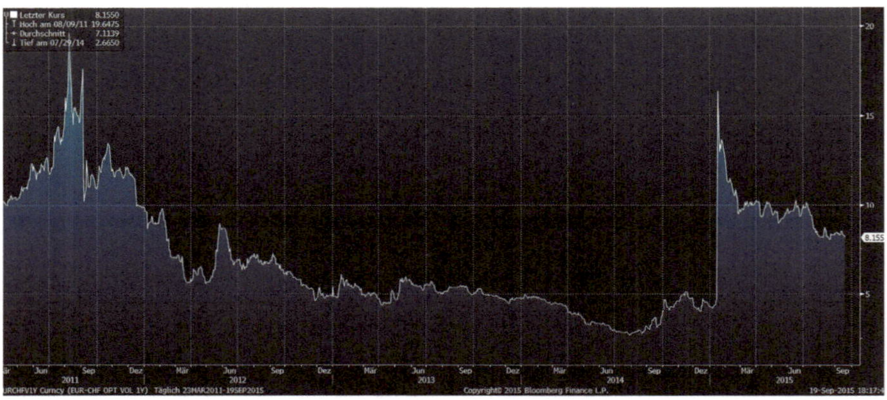

Fig. 3.62 Implied (1Y) volatility (at the money) of the EUR CHF exchange rate 2011–2015 (source: Bloomberg)

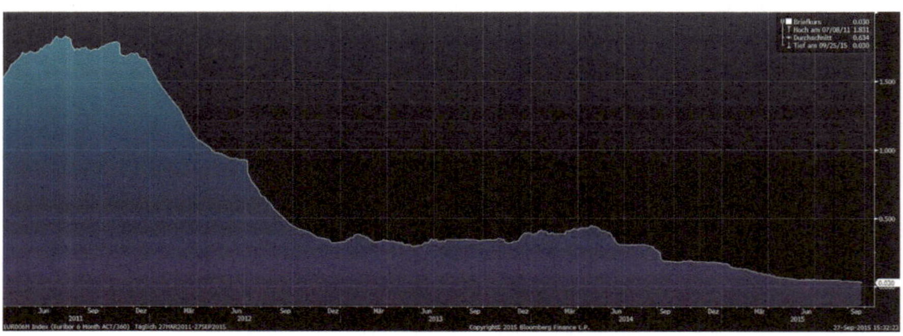

Fig. 3.63 Six-month LIBOR EUR 2011–2015 (source: Bloomberg)

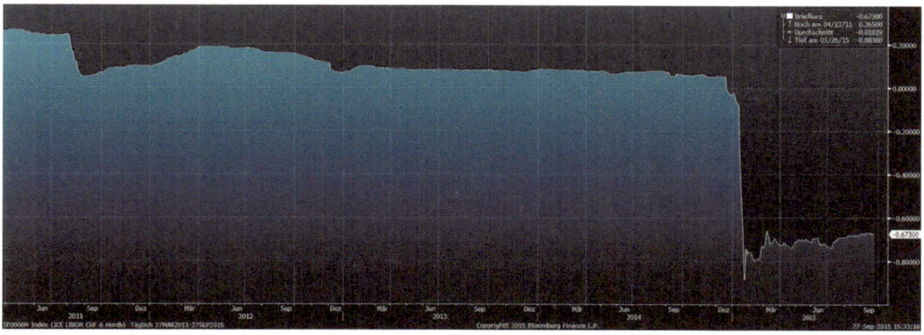

Fig. 3.64 Six-month LIBOR CHF 2011–2015 (source: Bloomberg)

In the following, we will again approximate the costs of a hedging strategy using put options and compare them with the executed stop-loss order variant. For this purpose, we take as a starting point a point in time shortly after the 1.20 CHF barrier was set for the EUR CHF exchange rate, that is, at the end of January 2012.

We further assume (Case 1) that hedging will again be done using put options with a strike of 1.20 and for a period of 1 year, starting the end of January 2012, January 2013, and January 2014, respectively.

Again, we need to point out, just like we did for the analysis of the GBP DEM exchange rate event in September 1992 that in order to achieve a definitive hedge, it is necessary to either buy put options expiring when the loan matures (often 10 years, 20 years, or more) or to buy 1-year put options annually. So, in addition to our calculations for the 3 years preceding the sudden CHF hike in January 2015, we will run analogous calculations for longer-running hedges for the loans (10 years, 20 years) in the following.

Again, we are going to use the valuation methodology according to the option pricing formulas given above and will focus exclusively on the analysis of European options (calculations with American options would yield only slightly different but in any case higher hedging prices).

The implied volatility (1-year volatility, at the money) of the EUR CHF exchange rate on the following dates was:

31 January 2012 ... 9.375
31 January 2013 ... 5.583
31 January 2014 ... 4.780

The EUR CHF exchange rate on the following dates was:

31 January 2012 ... 1.204
31 January 2013 ... 1.236
31 January 2014 ... 1.222

The EUR 6-month LIBOR rate on the following dates was:

31 January 2012 ... 1.418
31 January 2013 ... 0.378
31 January 2014 ... 0.396

The CHF 6-month LIBOR rate on the following dates was:

31 January 2012 ... 0.121
31 January 2013 ... 0.094
31 January 2014 ... 0.080

Using the option price formula, we get the following put option prices:
The put option price on the following dates was:

31 January 2012 ... 0.051 CHF
31 January 2013 ... 0.014 CHF
31 January 2014 ... 0.015 CHF

To evaluate the consequences of this type of hedging using put options, we analyse the following example:

On 31 January 2012, a CHF loan in the equivalent amount of 100,000 EUR is taken out at an exchange rate of 1.204. The loan period is 3 years, that is, the loan's principal is due for repayment in January 2015. Thus, a loan in the amount of 120,400 CHF (= 100,000 EUR) is taken out on 31 January 2012.

For hedging purposes, a total of 100,333 (= 120,400/1.20) put options are purchased on 31 January 2012 at a price of 0.051 CHF each, that is, at a total price of 5117 CHF = 4250 EUR.

Then, on 31 January 2013, a total of 100,333 (= 120,400/1.20) put options are purchased at a price of 0.014 CHF each, that is, at a total price of 1405 CHF = 1136 EUR.

Then, on 31 January 2014, a total of 100,333 (= 120,400/1.20) put options are purchased at a price of 0.051 CHF each, that is, at a total price of 1505 CHF = 1232 EUR.

On 15 January 2015, the options are exercised at a price of 1.20. The cost of purchasing the 120,400 CHF to repay the loan is therefore approximately 100,333 EUR. Thus, the total cost (loan repayment + hedging) was $100,333 + 4,250 + 1,136 + 1,232$ EUR $= 106,951$ EUR.

The loss in the case of hedging using 1-year put options is therefore 6.95%. (The remaining loan period of half a month can be neglected in this analysis.)

However, the reason why the hedging costs were so low is that the exchange rate event occurred just 3 years into the loan period. If the event had occurred at a later point during the loan period, the hedging cost would have been much higher (see also the next section).

3.7.5 Hedging Using Put Options for 10-Year or 20-Year Loans as from January 2012

We will again look at hedging strategies for loans taken out at the end of January 2012 for a period of 10 years and 20 years, respectively, using put options with the same maturity as the respective loan.

For this purpose, we use the following parameters as of 31 January 2012:

The implied volatility (10-year volatility, at the money) of the EUR CHF exchange rate on the following date was:

31 January 2012 ... 13.31 (see also Fig. 3.65)

(Historical data for the implied volatility for the 20-year period is not available, so we use the 10-year volatility data to price the 20-year options too.)

The EUR CHF exchange rate on the following date was:

31 January 2012 ... 1.204

The 10-year EUR swap rate on the following date was:

31 January 2012 ... 2.2395% (see also Fig. 3.66)

The 20-year EUR swap rate on the following date was:

31 January 2012 ... 2.5835% (see also Fig. 3.67)

The 10-year CHF swap rate on the following date was:

31 January 2012 ... 1.0735% (see also Fig. 3.68)

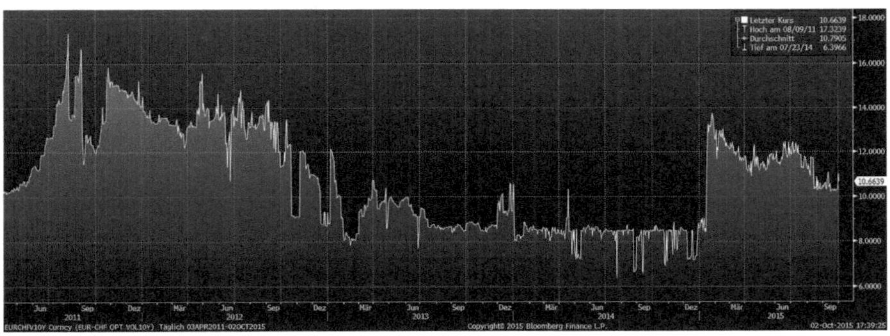

Fig. 3.65 Implied (10Y) volatility (at the money) of the EUR CHF exchange rate 2011–2015 (source: Bloomberg)

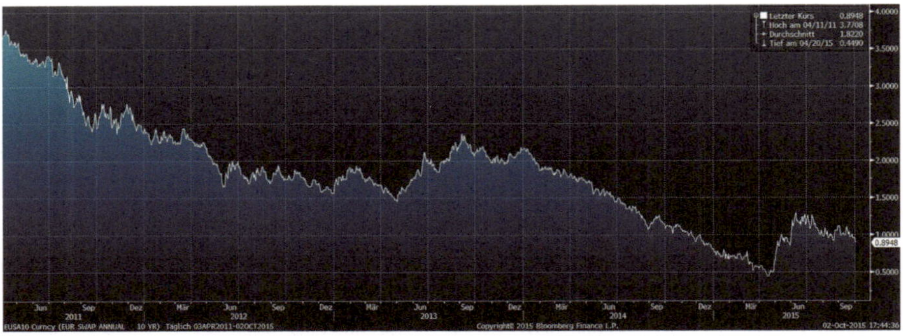

Fig. 3.66 Ten-year Euro swap rate 2011–2015 (source: Bloomberg)

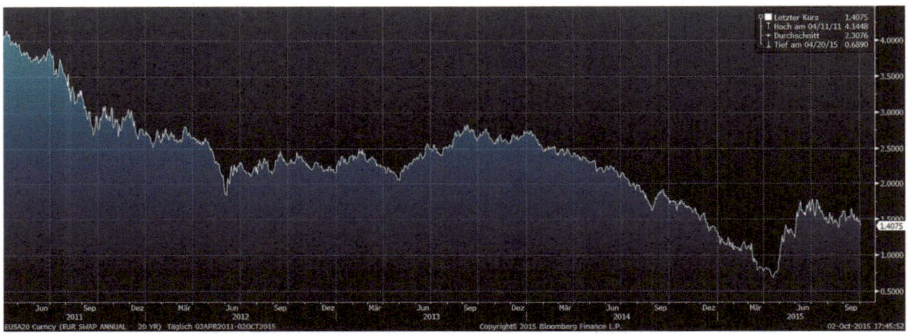

Fig. 3.67 Twenty-year Euro swap rate 2011–2015 (source: Bloomberg)

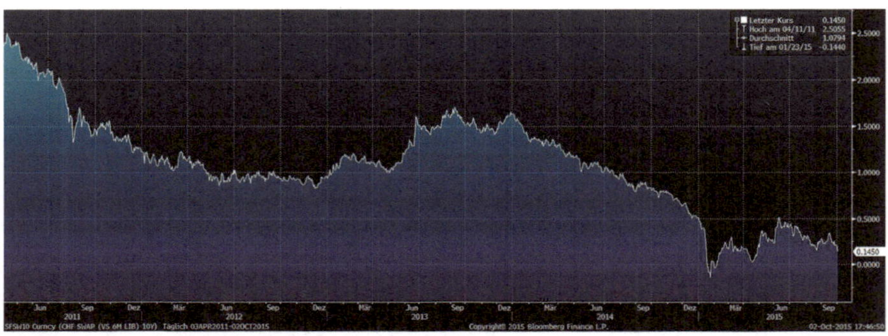

Fig. 3.68 Ten-year CHF swap rate 2011–2015 (source: Bloomberg)

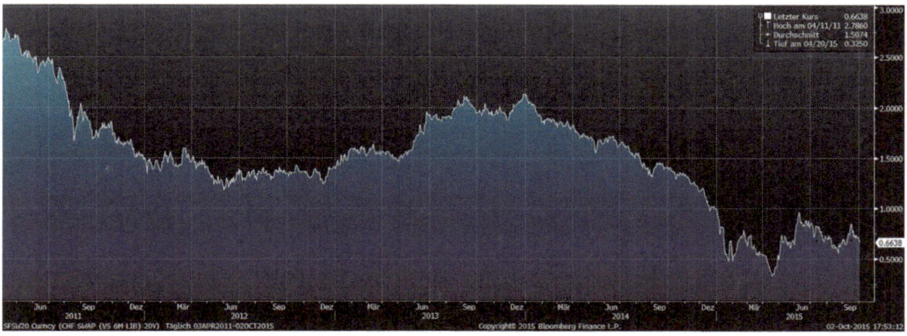

Fig. 3.69 Twenty-year CHF swap rate 2011–2015 (source: Bloomberg)

The 20-year CHF swap rate on the following date was:

31 January 2012 ... 1.3710% (see also Fig. 3.69)

Using the above option price formula, we get the following put option prices: The put option price on the following dates was:

31 January 2012 for the 10-year put option ... 0.2365 CHF
31 January 2012 for the 20-year put option ... 0.3025 CHF

To evaluate the consequences of this type of hedging strategy using put options with the same maturity as the loan, we analyse the following example:

Ten-Year Loan
On 31 January 2012, a CHF loan in the equivalent amount of 100,000 EUR is taken out at an exchange rate of 1.204.

So, actually, a loan in the amount of 120,400 CHF (= 100,000 EUR) is taken out on 31 January 2012.

The loan period is 10 years, that is, the loan's principal is due for repayment in January 2022.

To hedge the loan over the entire loan period, a total of 100,333 (= 120,400/1.20) 10-year put options are purchased on 31 January 2012 at a price of 0.2365 CHF each, that is, at a total price of 23,729 CHF = 19,708 EUR.

On 15 January 2015, the options are exercised at a price of 1.20. The cost of purchasing the 120,400 CHF to repay the loan is therefore approximately 100,333 EUR. The total cost (loan repayment + hedging) was therefore 100,333 + 19,708 EUR = 120,041 EUR.

The loss in the case of hedging a 10-year loan using 10-year put options is therefore 20.04%.

Twenty-Year Loan

On 31 January 2012, a CHF loan in the equivalent amount of 100,000 EUR is taken out at an exchange rate of 1.204. So, actually, a loan in the amount of 120,400 CHF (= 100,000 EUR) is taken out on 31 January 2012.

The loan period is 20 years, that is, the loan's principal is due for repayment in January 2032.

To hedge the loan over the entire loan period, a total of 100,333 (= 120,400/1.20) 20-year put options are purchased on 31 January 2012 at a price of 0.3025 CHF each, that is, at a total price of 30,351 CHF = 25,208 EUR.

On 15 January 2015, the options are exercised at a price of 1.20. The cost of purchasing the 120,400 CHF to repay the loan is therefore approximately 100,333 EUR. The total cost (loan repayment + hedging) was therefore 100,333 + 25,208 EUR = 125,541 EUR.

The loss in the case of hedging a 20-year loan using 20-year put options is therefore 25.54 %.

3.7.6 Comparison with the Stop-Loss Approach

Let us now return to the starting point of our analyses, that is, to the lawsuits referred to above, and ask what losses would have resulted if the simple stop-loss strategy at an exchange rate of 1.15 had been applied in the example discussed above, that is, for a CHF loan taken out on 31 January 2012 in the amount of EUR 100,000 at an exchange rate of 1.204.

So, in that example, a loan in the amount of 120,400 CHF (= 100,000 EUR) was taken out on 31 January 2012.

On 15 January 2015, the stop-loss order would have become effective.

The required 120,400 CHF would have had to be purchased at prices between 1.15 (best case) and 0.90 (worst case).

In the best-case scenario, the cost of the CHF purchase would have been only 120,400/1.15 = 104,696 EUR. The additional cost incurred ("hedging cost") would have been 4696 EUR.

In the worst case, however, the cost of the CHF purchase would have been 120,400/0.90 = 133,778 EUR. The additional cost incurred ("hedging cost") would have been 33,778 EUR.

For illustration, let us summarize the cost incurred in all hedging variants in Table 3.18:

Keep in mind, however, as noted before, that the hedging variant consisting of 1-year put options is relatively cheap here only because the hedge became effective already after 3 years. If the sharp drop in the exchange rate and the corresponding exercise of the options had occurred at a later date, more 1-year options would have had to be purchased, possibly resulting in even significantly higher cost.

We can thus draw the following **conclusion**:

Table 3.18 Costs of the different hedging variants

Hedging variant	Cost
One-year put options as from 2012	6.951 EUR
Ten-year put options	20,041 EUR
Twenty-year put options	25,541 EUR
Stop loss (best case)	4696 EUR
Stop loss (worst case)	33,778 EUR

In principle, the drop in the euro against the Swiss franc could have been much more dramatic and led to even higher losses for CHF borrowers. It is therefore necessary to plan risk mitigation measures. The possible alternative to placing stop-loss orders would be to use put options. But especially if stop-loss orders weren't executed so early as in this case but became effective only at a later time, it cannot be assumed that using put options will give better results. This is borne out both by the analysis of what was probably the most massive event prior to January 2015 (GBP DEM drop in September 1992) and the analysis of the EUR CHF event in January 2015.

3.8 Case Example VIII: Analyses of the Risk of Individual Higher-Quality Bonds Versus Portfolios of Lower-Quality Bonds

The following case study was again taken from a court case and a privately commissioned expert opinion in connection with that court case. The background was as follows:

An asset manager had agreed with a client to put together a portfolio of investment-grade bonds, that is, of bonds with a Standard & Poor's rating of "BBB" or better or a Moody's rating of "Baa" or better.

The bond portfolio that had been invested in for that client subsequently suffered heavy losses. The client then sued the asset manager on the following grounds: Contrary to their agreement, the bond portfolio had also contained (a small amount of) bonds with a rating below investment grade.

The asset manager now instructed the private expert to build up a line of defence for the trial in which the inclusion of these lower-rated bonds was to be defined and justified as a deliberate decision intended to minimize risk through suitable diversification.

The following is not a reproduction of the entire expert opinion, but merely explains the basic structure on which the line of argumentation was based. In the method, we use Credit Metrics, the credit risk management system presented in Chap. 1. Therefore, rather than perform the calculations explicitly in the following, we will refer to the techniques that can be looked up in Chap. 1 instead.

Again, we have kept the somewhat tendential tone of a privately commissioned expert report.

3.8.1 Two Simple Hypothetical Examples for Illustration

In the following, we compare the loss probabilities of higher-rated individual bonds with the loss probabilities of portfolios of lower-rated bonds with low correlations. We can limit our analyses here to **losses related to credit quality** (i.e. due to defaults or rating changes), since all other influencing factors affect all bonds essentially to the same extent.

Since this is a broad topic and space is limited, it can only be dealt with by way of examples here. Yet it will still become clear from those examples that diversified portfolios of low-correlation bonds, even if they are non-investment grade, **can** be significantly less likely to experience major price declines than individual bonds, even if the latter are investment grade.

To clarify the principle underlying the following analyses, we will first present a **trivial, hypothetical example** to illustrate the fact that a **well-diversified combination of lower-rated bonds can have a lower default risk than a better-rated individual bond**. After that, we will illustrate the fact, again using a **simple numerical example**, that a **well-diversified combination of lower-rated bonds can actually have a higher expected return at a lower risk than a better-rated individual bond**, before we proceed with the second part of this analysis and carry out more extensive calculations with more complex real data:

(a) For the **first hypothetical (!) example**, let us envision the following artificial, very simple "bond model":

- In this model there are bonds with three different ratings A, B, and D.
- Let an A-rated bond have value 2.
- A B-rated bond has value 1.
- A D-rated bond has value 0 (D = default).
- A bond with rating A defaults with probability $\frac{1}{3}$ in the course of a certain time period $[0, T]$ (i.e. new rating D)
(with probability $\frac{2}{3}$ the rating remains A).
- A bond with rating B defaults with probability $\frac{1}{2}$ in the course of a certain time period $[0, T]$ (i.e. new rating D)
(with probability $\frac{1}{2}$ the rating remains B).
This second bond thus has a higher risk.

Let us now compare a **bond portfolio $P1$** consisting of only **one bond X rated A** with a **bond portfolio $P2$** consisting of **two individual unrelated bonds $Y1$ and $Y2$ that are each rated B**.

The total **default probability** for the first portfolio with the **better-rated single bond X** is $\frac{1}{3}$**, i.e. 33.3%**.

The total **default probability** for the second portfolio with the **two lower-rated bonds $Y1$ and $Y2$** is $\frac{1}{2} \times \frac{1}{2} = \frac{1}{4}$**, so 25%**.

Table 3.19 Transition probability matrix in the analysed time period

	Rating A	Rating B	Rating D
Rating A	50%	25%	25%
Rating B	33.3%	33.3%	33.3%

The second bond portfolio, consisting of two B-rated bonds, thus has a lower (total) default probability than the A-rated single bond.

(b) For the **second hypothetical example**, let us envision the following artificial, slightly more complex "bond model":

- In this model there are bonds with three different ratings A, B, $and D$.
- Let an A-rated bond have value 2.
- A B-rated bond has value 1.
- A D-rated bond has value 0 (D = default).
- The **transition probability matrix** in the analysed time period $[0, T]$ is as given in Table 3.19:
  (This means that, for example, a bond with an A rating at time 0 has a 50% probability of being rated A at time T, a 25% probability of being rated B at time T, etc.)

Now let us again compare a **bond portfolio $P1$** consisting of only **one bond X rated A** with a **bond portfolio $P2$** consisting of **two individual unrelated bonds $Y1$ and $Y2$ that are each rated B**.

At the current time 0, both bond portfolios have the same value 2.

At time T (our trading horizon), the portfolio $P1$ will have:

Value 2 with probability $0.5(= 50\%)$
Value 1 with probability $0.25(= 25\%)$
Value 0 with probability $0.25(= 25\%)$

The **expected value E for the price of the bond portfolio $P1$ at time T** is therefore $\boldsymbol{E} = 0.5 \times 2 + 0.25 \times 1 + 0.25 \times 0 = \mathbf{1.25}$.

At time T (our trading horizon), the portfolio $P2$ will have:

- Value 4 with probability $\frac{1}{9}$
  (viz. if both—currently B-rated bonds—have an A rating at time T, which will happen (because they are independent individual bonds) with probability $\frac{1}{3} \times \frac{1}{3}$)
- Value 3 with probability $\frac{2}{9}$
  (viz. if at time T either the first of the two bonds in the portfolio $P2$ has an A rating and the second one has a B rating or vice versa, which will happen (again because they are independent individual bonds) with probability $\frac{1}{3} \times \frac{1}{3} + \frac{1}{3} \times \frac{1}{3}$)
- Value 2 with probability $\frac{3}{9} = \frac{1}{3}$
  (viz. if at time T either the first of the two bonds in the portfolio $P2$ has an A rating and the second one has a D rating or vice versa, or if both bonds are again rated B at time T, which will happen (again because they are independent individual bonds) with probability $\frac{1}{3} \times \frac{1}{3} + \frac{1}{3} \times \frac{1}{3} + \frac{1}{3} \times \frac{1}{3}$)

- Value 1 with probability $\frac{2}{9}$
  (viz. if at time T either the first of the two bonds in the portfolio $P2$ has a B rating and the second bond has a D rating or vice versa, which will happen (again because they are independent individual bonds) with probability $\frac{1}{3} \times \frac{1}{3} + \frac{1}{3} \times \frac{1}{3}$)
- Value 0 with probability $\frac{1}{9}$
  (viz. if both—currently B-rated bonds—have a D rating at time T, which will happen (because they are independent individual bonds) with probability $\frac{1}{3} \times \frac{1}{3}$).

The **expected value E for the price of the bond portfolio $P2$ at time T** is therefore $\boldsymbol{E} = \frac{1}{9} \times 4 + \frac{2}{9} \times 3 + \frac{3}{9} \times 2 + \frac{2}{9} \times 1 + \frac{1}{9} \times 0 = \mathbf{2}$.

So, the second **portfolio $P2$**, which consists of lower-rated bonds, **has a greater expected return than the portfolio $P1$ consisting of one single better-rated bond**.

The **probability of a complete default** for the second **portfolio $P2$**, which consists of lower-rated bonds, is $\mathbf{\frac{1}{9} = 11.1\%}$, whereas for the **portfolio $P1$** consisting of one single higher-rated bond, that probability is $\mathbf{\frac{1}{4} = 25\%}$.

The **probability** for the second **portfolio $P2$**—consisting of lower-rated bonds—to sustain a **loss** by time T is $\frac{2}{9} + \frac{1}{9} = \frac{1}{3} = \mathbf{33.3\%}$, whereas for the **portfolio $P1$** consisting of one single better-rated bond, the probability of making a loss is $\frac{1}{4} + \frac{1}{4} = \mathbf{50\%}$.

So, in this model, the portfolio consisting of independent lower-rated bonds has a higher expected value, a lower loss probability, and a lower total probability of default than a better-rated individual bond.

3.8.2 Analyses Based on Realistic Data

The above examples are only intended to illustrate the calculations that we are going to run below for realistic bond prices and bond parameters. The calculations are done on the same basis as above.

Instead of the three rating classes used in the illustrative example, we now have the nine Moody's rating classes Aaa, Aa, A, Baa, Ba, B, Caa, Ca-C, and Default (the WR class, which stands for "withdrawn rating", will not be considered for subsequent analyses; furthermore, for the following calculations, we will combine the classes Caa and Ca-C into one class C, and the probabilities will be renormalized after eliminating the WR class; the result of the calculations will be only marginally affected by this).

The average rating transition probabilities (averages from 1920 to 2012) for these classes are given in the subsequent table (Table 3.20).

The simplified and normalized values—as described above—are then given in Table 3.21:

To determine the current (fair) value of a bond (to reduce the amount of technical detail, we will focus on EUR bonds only in the following) with a certain rating and a certain residual maturity and a certain coupon, we are going to follow—as

Table 3.20 Rating transition probabilities (averaged over the time span 1920–2012)

From/to	Aaa	Aa	A	Baa	Ba	B	Caa	Ca-C	WR	Default
Aaa	86.407%	7.927%	0.851%	0.159%	0.033%	0.001%	0.001%	0.000%	4.620%	0.000%
Aa	1.164%	83.428%	7.502%	0.835%	0.183%	0.040%	0.006%	0.005%	6.767%	0.070%
A	0.076%	2.782%	84.069%	5.692%	0.739%	0.122%	0.029%	0.009%	6.384%	0.098%
Baa	0.039%	0.276%	4.219%	81.533%	4.936%	0.776%	0.126%	0.014%	7.802%	0.277%
Ba	0.007%	0.081%	0.461%	5.855%	73.411%	6.835%	0.596%	0.065%	11.421%	1.268%
B	0.006%	0.045%	0.144%	0.565%	5.589%	71.436%	5.597%	0.497%	12.553%	3.569%
Caa	0.000%	0.018%	0.025%	0.166%	0.713%	8.367%	63.516%	3.624%	11.854%	11.715%
Ca-C	0.000%	0.025%	0.108%	0.058%	0.500%	3.053%	7.990%	50.776%	12.344%	25.146%

Table 3.21 Simplified and normalized values of the previous probabilities

From/to	Aaa	Aa	A	Baa	Ba	B	C	Default
Aaa	90.59	8.31	0.89	0.16	0.03	0.001	0.001	0
Aa	1.24	89.48	8.04	0.89	0.19	0.04	0.01	0.07
A	0.08	2.90	89.80	6.08	0.79	0.13	0.04	0.10
Baa	0.04	0.30	4.57	88.43	5.35	0.84	0.15	0.30
Ba	0.01	0.09	0.52	6.61	82.88	7.72	0.75	1.43
B	0.01	0.05	0.16	0.65	6.39	81.69	6.97	4.08
Caa	0	0.02	0.08	0.13	0.69	6.49	71.60	20.99

Table 3.22 Recovery rates for various seniority classes by Moody's

Seniority class	Recovery rate (statistics for the period 1987–2012)
Loans	80.6%
Senior secured bonds	63.7%
Senior unsecured bonds	48.6%
Subordinated bonds (includes senior subordinated, subordinated, and junior subordinated bonds)	28.5%

Source: Moody's: Average Corporate Debt Recovery Rates Measured by Ultimate Recoveries, 1987–2012

announced above—the J. P. Morgan's credit risk management approach known as the "Credit Metrics" system.

So in the spirit of keeping the technical details to a minimum, the following analyses will focus on bonds with annual coupons and whose coupons have just been paid (i.e. with integral residual maturity).

Furthermore, we will limit the analyses in the following—again to keep the technical details to a minimum—to observation periods of 1 year. For this, we will also need a bond's (fair) value 1 year from now (time 0) in the following. This value will obviously depend on what rating the bond will have in 1 year's time. Since this rating is not yet known at time 0, we will be dealing with nine different values that the bond might have in 1 year.

A special case in determining a bond's value 1 year into the future is when the rating in 1 year is D, i.e. when the bond defaults in the course of the period under review. The average value of a bond at default (recovery rate) depends on the bond's seniority class and is generally not equal to 0.

Moody's maintains the following statistics on average recovery rates for bonds in various seniority classes (Table 3.22).

In our examples, we will assume bonds in medium seniority classes (senior unsecured bonds) and equate the value at default with the average recovery rate of 48.6%.

We will now present a typical return/risk behaviour of bond portfolios versus individual bonds using a real-life example and based on the above data and approach.

Example 3.3 We have a euro bond portfolio $P1$ that consists of only one single Baa-rated bond X (i.e. an investment-grade bond, equivalent to a Standard & Poor's BBB rating) and compare it with a portfolio $P2$ consisting of several Ba-rated euro bonds $Y1, Y2, Y3, \ldots$ (i.e. non-investment grade, equivalent to a Standard & Poor's BB rating).

In this example, to minimize technical details, we assume zero-coupon bonds with the same maturity of 2 years.

Let the rating transitions of two bonds in the $P2$ portfolio be mutually independent in each case.

To determine the values of these bonds in the portfolios $P1$ and $P2$ at time 0 and the potential values of these bonds after 1 year (according to Credit Metrics), we need the interest rates ${f_{0,1}}^{(Z)}$ and $f_{0,2}^{(Z)}$ for all rating classes Z. These can then be used to calculate the required implied interest rates $f_{1,2}^{(Z)}$.

At the moment when the privately commissioned expert report was prepared, the relevant values were:

$$
\begin{aligned}
{f_{0,1}}^{(Aaa)} &= 0.12\\
{f_{0,1}}^{(Aa)} &= 0.50\\
{f_{0,1}}^{(A)} &= 0.73\\
{f_{0,1}}^{(Baa)} &= 1.73\\
{f_{0,1}}^{(Ba)} &= 2.43\\
{f_{0,1}}^{(B)} &= 2.81\\
{f_{0,1}}^{(C)} &= 3.34
\end{aligned}
$$

$$
\begin{aligned}
{f_{0,2}}^{(Aaa)} &= 0.18\\
{f_{0,2}}^{(Aa)} &= 0.89\\
{f_{0,2}}^{(A)} &= 0.89\\
{f_{0,2}}^{(Baa)} &= 1.78\\
{f_{0,2}}^{(Ba)} &= 2.68\\
{f_{0,2}}^{(B)} &= 3.03\\
{f_{0,2}}^{(C)} &= 4.87
\end{aligned}
$$

These values can be used to calculate the implied interest rates

$$
\begin{aligned}
{f_{1,2}}^{(Aaa)} &= 0.23\\
{f_{1,2}}^{(Aa)} &= 1.38\\
{f_{1,2}}^{(A)} &= 1.06\\
{f_{1,2}}^{(Baa)} &= 1.83\\
{f_{1,2}}^{(Ba)} &= 2.95\\
{f_{1,2}}^{(B)} &= 3.26\\
{f_{1,2}}^{(C)} &= 6.94
\end{aligned}
$$

(For the record, the following analyses apply in principle even for interest rates of other comparison periods.)

Using the above figures and based on the "Credit Metrics" method, we get, for the current value $A_0(X)$ of the bond X and the value $A_1(X, Y)$ of the bond in 1 year, if its rating at that point is Y:

$$A_0(X) = 96.6\%$$

and

$A_1(X, Aaa) = 99.77\%$
if X is rated Aaa in 1 year (which will happen with probability 0.04%)
$A_1(X, Aa) = 98.64\%$
if X is rated Aa in one year (which will happen with probability 0.30%)
$A_1(X, A) = 98.45\%$
if X is rated A in 1 year (which will happen with probability 4.58%)
$A_1(X, Baa) = 98.20\%$
if X is rated Baa in 1 year (which will happen with probability 88.43%)
$A_1(X, Ba) = 97.13\%$
if X is rated Ba in 1 year (which will happen with probability 5.35%)
$A_1(X, B) = 96.84\%$
if X is rated B in 1 year (which will happen with probability 0.84%)
$A_1(X, C) = 93.51\%$
if X is rated C in 1 year (which will happen with probability 0.15%)
$A_1(X, D) = 48.6\%$
if X is rated D in 1 year (which will happen with probability 0.30%)

For the mutually identical bonds $Y1, Y2, Y3, \ldots$ (we denote one of them by Y), analogous calculations give us:

$$A_0(Y) = 95.31\%$$

and the potential values of the bond Y in 1 year are given by:

$A_1(Y, Aaa) = 99.7\%$
if Y is rated Aaa in 1 year (which will happen with probability 0.008%)
$A_1(Y, Aa) = 98.64\%$
if Y is rated Aa in 1 year (which will happen with probability 0.09%)
$A_1(Y, A) = 98.45\%$
if Y is rated A in 1 year (which will happen with probability 0.52%)
$A_1(Y, Baa) = 98.20\%$
if Y is rated Baa in 1 year (which will happen with probability 6.61%)
$A_1(Y, Ba) = 97.13\%$
if Y is rated Ba in 1 year (which will happen with probability 82.88%)
$A_1(Y, B) = 96.84\%$

if Y is rated B in 1 year (which will happen with probability 7.72%)
$A_1(Y, C) = 93.51\%$
if Y is rated C in 1 year (which will happen with probability 0.75%)
$A_1(Y, D) = 48.6\%$
if Y is rated D in 1 year (which will happen with probability 1.43%)

From the above figures, it follows, for example:

Conclusion 1
The probability that the value of the bond X, i.e. of the portfolio $P1$, drops from 96.63% to an average of 48.6%, i.e. suffers a loss of approximately 50%, is 0.3%.

The probability that the value of the $P2$ portfolio will suffer a corresponding loss to a price value of about 50% can now be easily calculated from the above values (due to the assumed independence of the bonds $Y_1, Y_2, Y_3, \ldots$) and becomes extremely small very quickly as n grows, i.e. as the number of bonds in the portfolio increases. The values are:

If $n = 1$, then probability = 1.43%.
If $n = 2$, then probability = 0.000204%.
If $n = 3$, then probability = 0.0000003%.
If $n = 4$, then probability = 0.000000000418%.

So already at a portfolio size of two independent bonds, the probability of a loss of about half of the investment in this example is incomparably lower than for a (better rated) single bond.

The probability of a loss of around 50% obtained by a portfolio of one (or more) Baa-rated bonds is given below:

Conclusion 2
The probability that the value of the bond X, i.e. of the portfolio $P1$, drops from 96.63% to an average of 48.6%, i.e. suffers a loss of approximately 49.70%, is—as seen in Table 3.23—0.3%.

Table 3.23 Probability of making a loss of approximately 50%

Portfolio	Probability of a loss of approximately 50%
Portfolio of one **Baa**-rated bond	0.3%
Portfolio of **one** Ba-rated bond	1.43%
Portfolio of **two** Ba-rated bonds	0.000204%
Portfolio of **three** Ba-rated bonds	0.0000003%
Portfolio of **four** Ba-rated bonds	0.000000000418%

Table 3.24 Maximum loss that the portfolio is expected to incur with 0.3%

Portfolio	Amount of loss with probability 0.3%
Portfolio of one **Baa**-rated bond	49.7%
Portfolio of **two** Ba-rated bonds	23.6%
Portfolio of **three** Ba-rated bonds	15.4%
Portfolio of **four** Ba-rated bonds	10.8%
Portfolio of **five** Ba-rated bonds	9.1%
Portfolio of **ten** Ba-rated bonds	3.6%
Portfolio of **20** Ba-rated bonds	3.4%

We now calculate the maximum loss that the portfolio Y is expected to incur with a probability of 0.3%.

The results are easily determined based on the above values and are listed in the following table (Table 3.24):

These numbers tend to be borne out even if the bonds in the $P2$ portfolio are not strongly positively correlated, for example, if the correlation between two bonds each in the $P2$ portfolio does not exceed the value of, for example, 0.2.

3.9 Case Study IX: Portfolio Selection Based on Sustainability Parameters

The following analyses were conducted on behalf of a fund data provider that specialized, among others, in providing sustainability parameters for funds. The task was to create portfolios of funds in the most efficient manner—in the sense of the Markowitz portfolio selection theory presented in Chap. 2—yet with the proviso that the portfolios meet certain minimum sustainability requirements. The corresponding strategy paper, presented in an abridged version below, was prepared by the author in cooperation with L. Del Chicca.

3.9.1 The Problem

The objective of this article is to generalize the Markowitz portfolio selection theory to the case where the selected portfolio has to meet certain sustainability criteria. We assume that the contents of Chap. 2 are essentially known but will occasionally restate some of it briefly. In the following we proceed from the assumption that we can invest in n financial products $A_1, \ldots, A_n$, for which we have an estimate for the expected trends $\mu_1, \ldots, \mu_n$ and for the expected volatilities $\sigma_1, \ldots, \sigma_n$ for the relevant time period $[0, T]$.

We further assume that we have estimates for the correlations $\rho_{i,j}$ between the returns of the ith and the jth product A_i resp. A_j.

In addition to these assumptions, which are also made in the classic Markowitz model, we now assume that we have a sustainability metric available for each of the products $A_1, \ldots, A_n$. That is, for each of the products A_i, sustainability is valued by a parameter $\tau_i \in [0, 1]$: $\tau_i = 0$ indicates the lowest sustainability and $\tau_i = 1$ indicates the highest sustainability.

We invest a (normalized) investment of 1 in the n products. In each case, let an amount of x_i be invested in the product A_i. This gives us the relation

$$x_1 + \cdots + x_n = 1.$$

For the expected portfolio return μ_P, we get

$$\mu_P = x_1\mu_1 + \cdots + x_n\mu_n$$

and for the volatility σ_P, based on the known formula for the volatility of a sum of dependent random variables

$$\sigma_P = \sigma_P(x_1, \ldots, x_n) = \sqrt{\sum_{i=1}^{n} x_i^2\sigma_i^2 + \sum_{i=1}^{n} \sum_{j=1, j\neq i}^{n} \rho_{i,j}\sigma_i\sigma_j x_i x_j}.$$

We also assume linearity for the sustainability parameter τ_P of the total portfolio, hence:

$$\tau_P = x_1\tau_1 + \cdots + x_n\tau_n.$$

In the following, it will again be necessary to distinguish whether shorting the products under consideration is not possible, possible within bounds, or possible without bounds (unbounded short selling). In the following we consider the case of unbounded short selling, i.e. arbitrary $x_i \in \mathbb{R}$, and the case in which short selling is not allowed, i.e. $x_i \in [0, 1]$ for all i. The individual products A_i can again be represented in the risk-return diagram (with the product's expected volatility on the $x-$axis and the product's expected return on the $y-$axis).

To illustrate the techniques developed in this article, we give an example with the following parameters: We choose five products with expected returns 3%, 6%, 2%, 5%, and 7% and volatilities 2%, 4%, 5%, 5%, and 7%, which are to be uncorrelated in this example. For the sustainability parameters of the five products in this example, we set 0.5, 0.2, 0.7, 0.6, and0.7.

The five individual products are depicted in the risk-return diagram in Fig. 3.70 (portfolios with a larger sustainability parameter are shown in greener shades, and portfolios with a lower sustainability parameter are more reddish):

The "opportunity set" of the products again refers to the set of all possible portfolios that can be created from these products (at a normalized investment of 1) as well as their representation in the risk-return diagram. For our illustrative example, Fig. 3.71 shows the opportunity set in the case without short selling

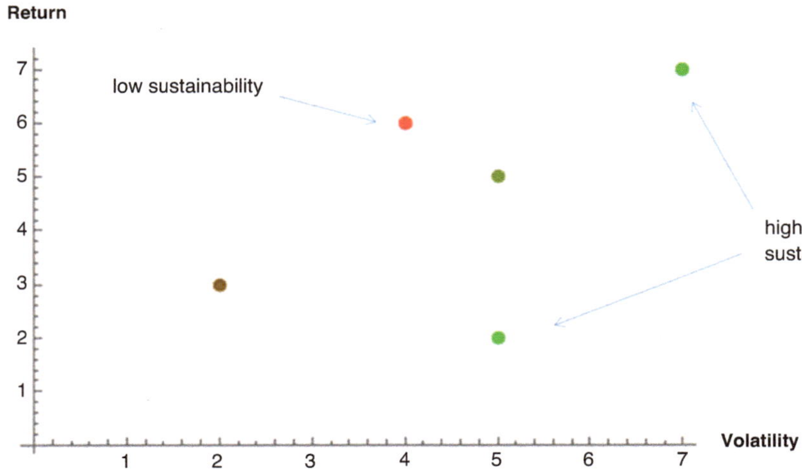

Fig. 3.70 Products with different sustainability parameters

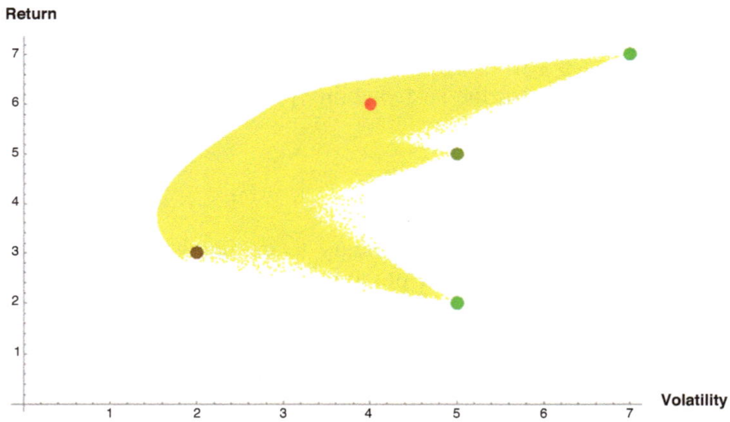

Fig. 3.71 Opportunity set in the case without short selling

(the irregularities in the shape are due to approximation through Monte Carlo simulation), Fig. 3.72 illustrates the case with unbounded short selling, and Fig. 3.73 superimposes both opportunity sets for comparison. The opportunity sets are approximations using Monte Carlo simulation.

Naturally, preference will be given to portfolios that can be expected to generate the highest possible return at the lowest possible risk (i.e. the lowest possible volatility). In both cases, these portfolios lie on the opportunity set's upper left boundary, the "efficient border".

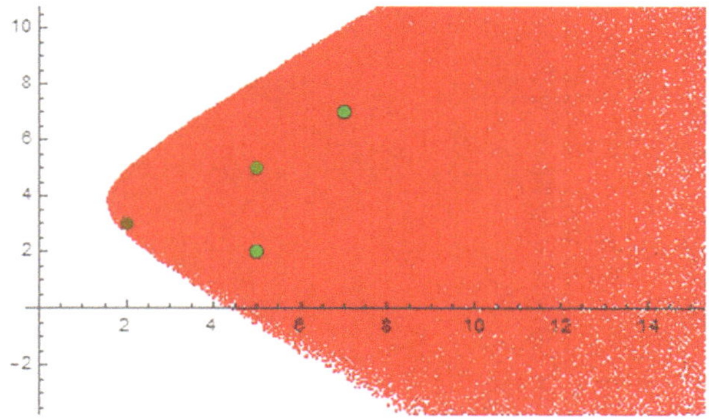

Fig. 3.72 Opportunity set in the case of unbounded short selling

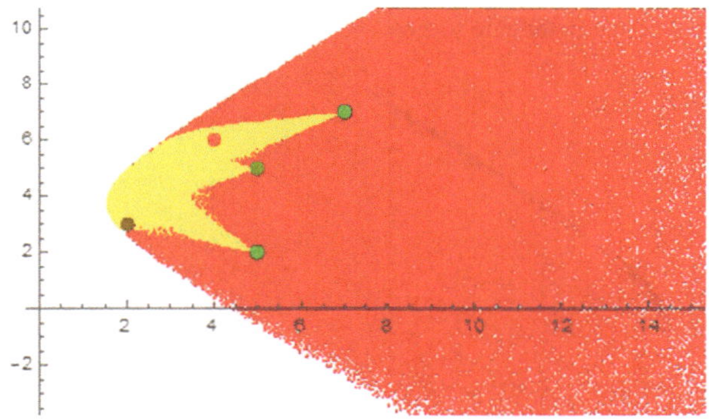

Fig. 3.73 Combination of both opportunity sets

Figures 3.74 and 3.75 show the efficient borders in the two cases under consideration. In the case without short selling, the efficient border was again approximated using Monte Carlo simulation (an exact calculation in explicit analytical terms is not possible here), while in the case with unbounded short selling, the explicit analytical representation was used. It stands to reason therefore that only portfolios with a risk-return value on this efficient border should be selected. The "market portfolio" among the efficient-border portfolios, that is, the one with maximum Sharpe ratio, has been marked in the following diagrams.

For both cases, the portfolio with maximum Sharpe ratio is marked by a blue dot on the efficient border in Figs. 3.76 and 3.77. (The market portfolio in the case without short selling was again determined through Monte Carlo simulation, and the market portfolio in the case with short selling was determined analytically.) Geometrically, the market portfolio is obtained by placing a tangent from the point $(0, r)$ of the risk-return diagram to the respective opportunity set. The point of contact between the tangent and the opportunity set represents the market portfolio.

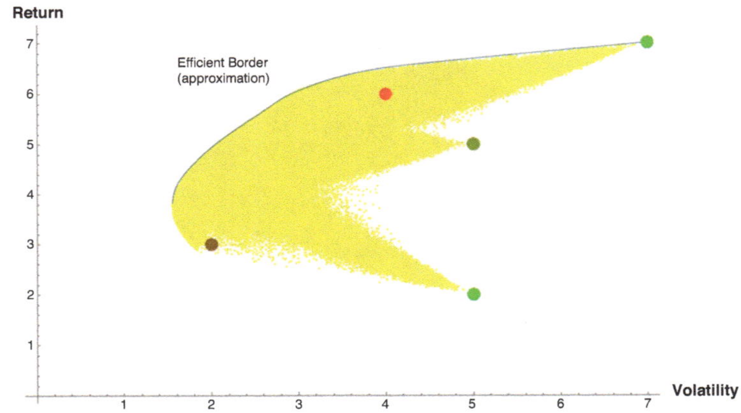

Fig. 3.74 Efficient border in the case without short selling

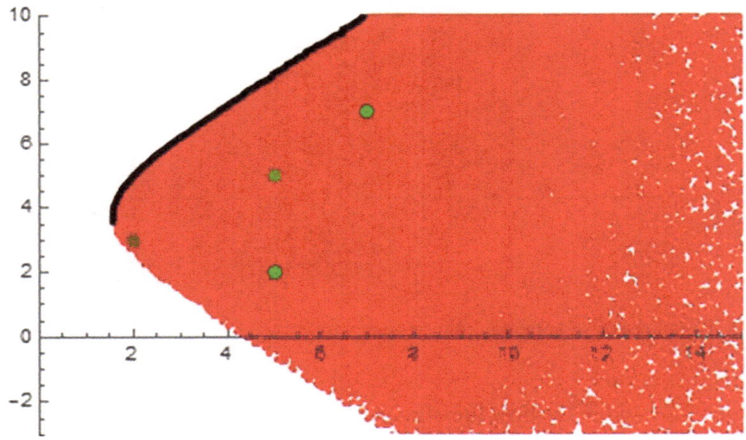

Fig. 3.75 Efficient border in the case with unbounded short selling

What interest us in this article are the following questions:

(a) What do opportunity sets, efficient borders, and market portfolios look like if the portfolios must satisfy a certain predefined minimum sustainability value?
(b) How can the efficient borders and market portfolios be determined under this constraint?
(c) Is there a suitable "sustainability Sharpe parameter" that defines an optimal "sustainability market portfolio"?

We will discuss these questions in the next subsection and give approximate solutions, using Monte Carlo methods, for the case without short selling. In the

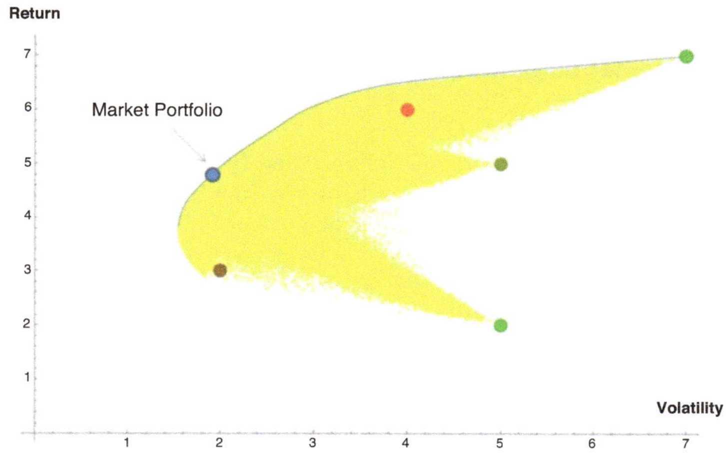

Fig. 3.76 Market portfolio in the case without short selling

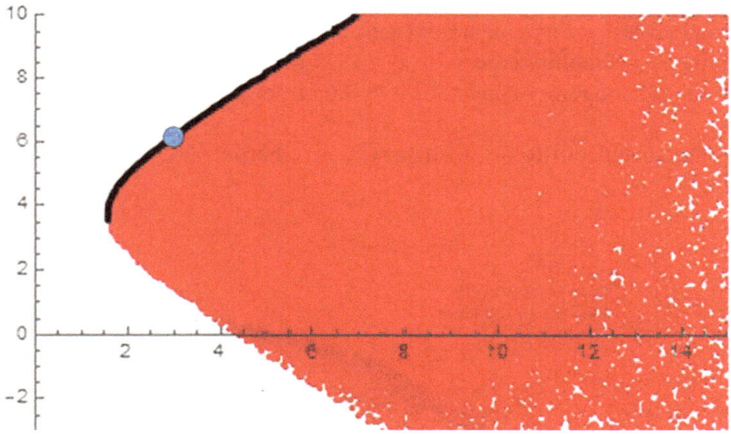

Fig. 3.77 Market portfolio in the case with unbounded short selling

case without short selling, an explicit analytical solution is not generally possible. However, due to the bounds on the x_i to be invested in a product A_i, this case is well suited for Monte Carlo methods. We will then address the case with unbounded short selling and identify explicit analytical solutions to all of the above questions. This is possible in this case (which has limited suitability for Monte Carlo methods).

3.9.2 Monte Carlo Determination of Efficient Frontiers and Market Portfolios Subject to Sustainability Constraints and Optimal Sustainability Sharpe Parameters in the Case *without* Short Selling

The obvious approach to solving questions a) and b) approximately using Monte Carlo simulation is as follows: In each of the simulations performed for our illustrative example, we generate a large number of 500,000 different scenarios (random portfolios) $(x_1, x_2, \ldots, x_n)$ with $0 \leq x_i \leq 1$ for all i and with $x_1 + x_2 + \cdots + x_n = 1$. For each of these random portfolios, we calculate expected return, expected volatility, and the sustainability parameter. All portfolios with sustainability parameters below a certain predefined minimum sustainability threshold τ are discarded. For the remaining portfolios, we again determine their opportunity set. Furthermore, the Sharpe ratio is determined for each of the remaining portfolios, and the portfolio with maximum Sharpe ratio is marked as "τ−market portfolio". For our example we depict, in Fig. 3.78, the three opportunity sets

(a) Without sustainability constraint (yellow)
(b) With the sustainability constraint $\tau \geq 0.55$ (blue)
(c) With the sustainability constraint $\tau \geq 0.6$ (green)

with the associated portfolios with maximum Sharpe ratio (under the respective constraints).

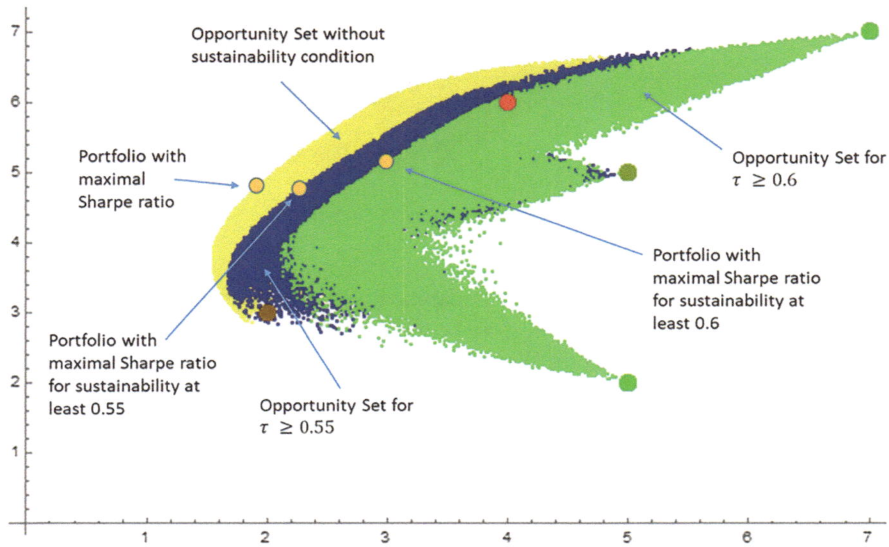

Fig. 3.78 Opportunity sets and efficient frontiers with and without sustainability constraint

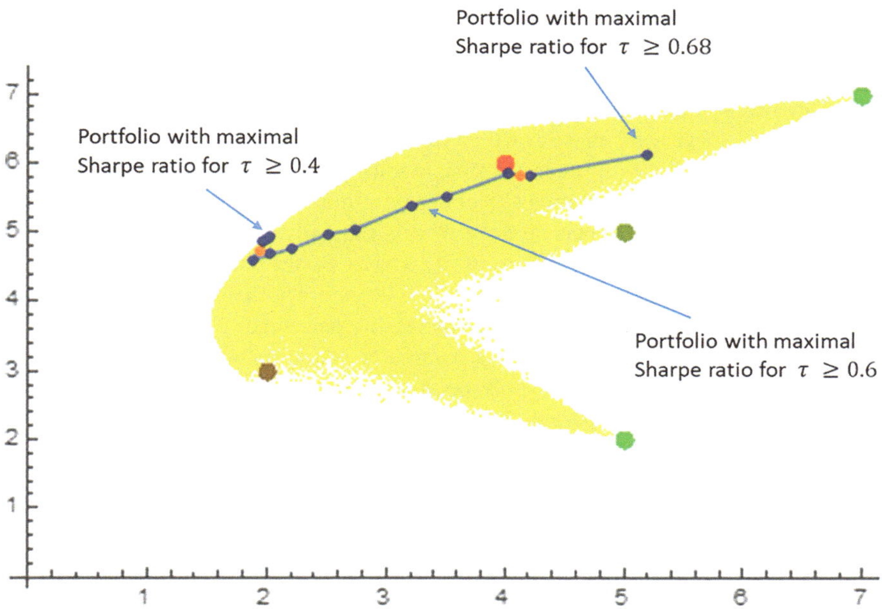

Fig. 3.79 Sustainability Sharpe line (without short selling)

The maximum Sharpe ratios in each of the cases have the values:

(a) 1.455 = maximum Sharpe ratio among all portfolios
(b) 1.215 = maximum Sharpe ratio among all portfolios with sustainability parameter greater than or equal to 0.55
(c) 1.055 = maximum Sharpe ratio among all portfolios with sustainability parameter greater than or equal to 0.6

In Fig. 3.79, we represented the portfolios with maximum Sharpe ratio for a number of other sustainability constraints, connected them, and plotted the resulting line in the original opportunity set. The left endpoint of the line corresponds to the portfolio with globally maximum Sharpe ratio (= 1.455), while the right endpoint of the line corresponds to the portfolio with maximum Sharpe ratio (= 0.795) under the constraint that the sustainability parameter be greater than or equal to 0.68. Note that in our example, the sustainability parameter of a portfolio can be at most 0.7, since the largest sustainability parameter of the five products is 0.7.

Of course, sustainability-minded investors will now wonder which of the portfolios on this "sustainability Sharpe line" (which are the only portfolios that a rational sustainability-minded investor will consider) they should invest in. "Radical" sustainability-minded investors will invest in the portfolio at the right end of the "sustainability Sharpe line" (with a Sharpe ratio of 0.795), while investors who are not sustainability-minded will invest in the portfolio at the left end of the line (with a Sharpe ratio of 1.455).

To answer the question in which of the portfolios on the "sustainability Sharpe line" an investor should invest, it would make sense to define a sustainability parameter γ for every investor, as well as a "sustainability Sharpe ratio" that defines the optimal portfolio with a certain sustainability parameter γ for every investor. Here, the sustainability parameter γ could be a number in the range $[0, \infty)$. The value $\gamma = 0$ would define an investor who is radically not sustainability-minded, and a very large value for γ, such as $\gamma > 10$, would define investors who are extremely to radically sustainability-minded. Investors with a "reasonable rational" sustainability mindset are assigned a sustainability parameter γ in the range of about $0.5 < \gamma < 2$. Such a sustainability parameter valuation would be self-assessed by the investor or could conceivably be done using a suitable test for classifying investors by sustainability parameters. The sustainability Sharpe ratio of a portfolio P as proposed by us is now

$$NSR(\gamma) = \tau^{\gamma} \cdot \frac{\mu_P - r}{\sigma_P},$$

hence, the product of a weighted sustainability factor and a conventional Sharpe ratio. The portfolio for a sustainable investor with sustainability parameter γ should have maximal $NSR(\gamma)$. Of course, for each γ, the portfolio with maximal $NSR(\gamma)$ is on the "sustainability Sharpe line". When $\gamma = 0$, that portfolio is found at the left end of the line, and when γ is large, it is at the right end of the line. Figure 3.80 shows the optimal portfolio in terms of NSR—again determined through Monte Carlo simulation—for an investor with "average" sustainability parameter $\gamma = 1$ and for $\gamma = 2$ in our illustrative example.

The optimal portfolio for the parameter $\gamma = 1$ has a τ value of 0.51 and a Sharpe ratio of 1.35 (thus close to the maximum Sharpe ratio). The optimal portfolio for the parameter $\gamma = 2$ has a τ value of 0.66 and a significantly lower Sharpe ratio of 0.91.

3.9.3 Determination of Efficient Frontiers and Market Portfolios Subject to Sustainability Constraints and Optimal Sustainability Sharpe Parameters in the Case *with* Short Selling

In principle, all the graphs shown in the previous section could also be generated using Monte Carlo simulation in the case with unbounded short selling. However, the method—especially when there is a larger number of underlying products, and if it is to provide reliable results—quickly becomes extremely laborious, since for each parameter x_i, values can now be selected from all $\mathbb{R}$, whereas in the case without short selling, these values are limited to the range $x_i \in [0, 1]$.

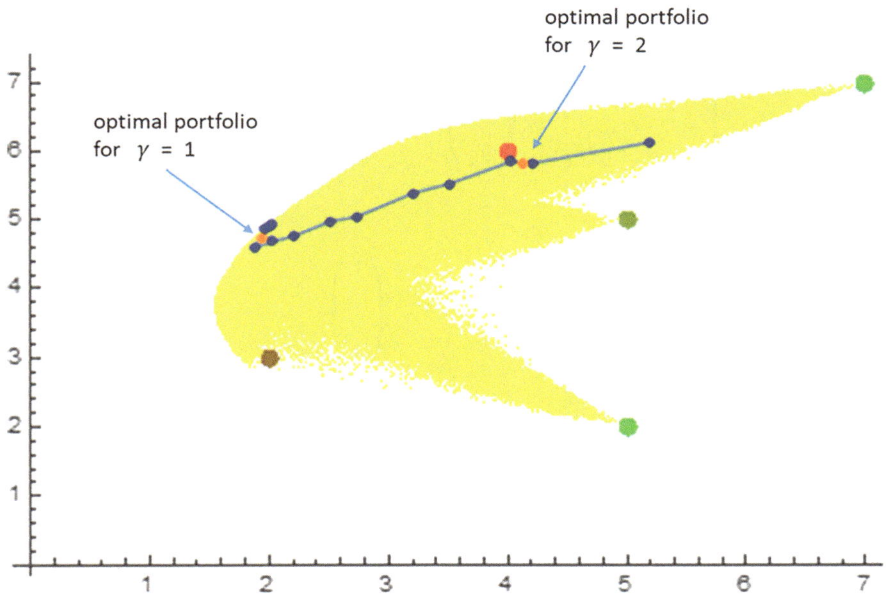

Fig. 3.80 Portfolio with optimal $NSR(\gamma)$ (without short selling)

However, in the case with unbounded short selling, it is possible to give explicit analytical representations for all relevant objects and ratios described in the previous section (efficient borders with and without sustainability constraints, optimal Sharpe ratios with and without sustainability constraints, sustainability Sharpe line, maximal $NSR(\gamma)$), which can be used to represent these objects graphically as well.

In the following, we will only give the resulting formulas for these objects. They can be derived using the Lagrange multiplier method in the same way as was done for the formulas in Sects. 2.9 and 2.10. The only difference is that, for variable τ, there is the additional constraint

$$x_1\tau_1 + x_2\tau_2 + \ldots + x_n\tau_n = \tau$$

that needs to be taken into account. We leave this task as an exercise for the reader.

First, we determine an explicit representation of the efficient border under the constraint that the sustainability parameter τ_P of the considered portfolios must have a given value τ. The parameter representation of such an efficient frontier can be specified for a free parameter $\mu \geq \mu_0$ and a given value τ for the sustainability parameter of the considered portfolios in the following way (here and in the following, we use the notations defined in the previous sections):

$$\text{efficient border} = (\sigma_P(x_1(\mu, \tau), \ldots, x_n(\mu, \tau)), \mu), \quad \mu \geq \mu_0$$

where

$$\begin{pmatrix} x_1(\mu,\tau) \\ \vdots \\ x_n(\mu,\tau) \end{pmatrix} = \frac{\lambda}{2}\begin{pmatrix} Y_1 \\ \vdots \\ Y_n \end{pmatrix} + \frac{\beta}{2}\begin{pmatrix} Z_1 \\ \vdots \\ Z_n \end{pmatrix} + \frac{\kappa}{2}\begin{pmatrix} W_1 \\ \vdots \\ W_n \end{pmatrix}$$

where

$$\begin{pmatrix} Y_1 \\ \vdots \\ Y_n \end{pmatrix} = A^{-1}\begin{pmatrix} 1 \\ \vdots \\ 1 \end{pmatrix}$$

$$\begin{pmatrix} Z_1 \\ \vdots \\ Z_n \end{pmatrix} = A^{-1}\begin{pmatrix} \mu_1 \\ \vdots \\ \mu_n \end{pmatrix}$$

$$\begin{pmatrix} W_1 \\ \vdots \\ W_n \end{pmatrix} = A^{-1}\begin{pmatrix} \tau_1 \\ \vdots \\ \tau_n \end{pmatrix}$$

Here, A is the covariance matrix of the underlying assets $A_1, \ldots, A_n$. The λ, β, and κ values are defined as a function of μ and τ, as follows: Let

$$Y := \sum_{i=1}^{n} Y_i, \quad \tilde{Y} := \sum_{i=1}^{n} \mu_i Y_i, \quad \bar{Y} := \sum_{i=1}^{n} \tau_i Y_i$$

$$Z := \sum_{i=1}^{n} Z_i, \quad \tilde{Z} := \sum_{i=1}^{n} \mu_i Z_i, \quad \bar{Z} := \sum_{i=1}^{n} \tau_i Z_i$$

$$W := \sum_{i=1}^{n} W_i, \; \tilde{W} := \sum_{i=1}^{n} \mu_i W_i, \; \bar{W} := \sum_{i=1}^{n} \tau_i W_i$$

Furthermore, let

D be the determinant of the matrix $\begin{pmatrix} Y & Z & W \\ \tilde{Y} & \tilde{Z} & \tilde{W} \\ \bar{Y} & \bar{Z} & \bar{W} \end{pmatrix}$

D_1 be the determinant of the matrix $\begin{pmatrix} 2 & Z & W \\ 2\mu & \tilde{Z} & \tilde{W} \\ 2\tau & \bar{Z} & \bar{W} \end{pmatrix}$

D_2 be the determinant of the matrix $\begin{pmatrix} Y & 2 & W \\ \tilde{Y} & 2\mu & \tilde{W} \\ \bar{Y} & 2\tau & \bar{W} \end{pmatrix}$

and D_3 be the determinant of the matrix $\begin{pmatrix} Y & Z & 2 \\ \tilde{Y} & \tilde{Z} & 2\mu \\ \bar{Y} & \bar{Z} & 2\tau \end{pmatrix}$

Then $\lambda = \frac{D_1}{D}$, $\beta = \frac{D_2}{D}$ and $\kappa = \frac{D_3}{D}$. The lower bound μ_0 for the free parameter μ in the efficient border parameter representation is given by:

$$\mu_0 = x_1^0(\tau) \cdot \mu_1 + \ldots + x_n^0(\tau) \cdot \mu_n,$$

with

$$\begin{pmatrix} x_1^0(\tau) \\ \vdots \\ x_n^0(\tau) \end{pmatrix} = \frac{\Delta}{2} \begin{pmatrix} Y_1 \\ \vdots \\ Y_n \end{pmatrix} + \frac{\Gamma}{2} \begin{pmatrix} W_1 \\ \vdots \\ W_n \end{pmatrix}$$

where

$$\Delta = \frac{2\tau(\sum_{i=1}^n W_i - \sum_{i=1}^n \tau_i W_i)}{\sum_{i=1}^n W_i \sum_{i=1}^n \tau_i Y_i - \sum_{i=1}^n \tau_i W_i \sum_{i=1}^n Y_i}$$

and

$$\Gamma = \frac{2\sum_{i=1}^n \tau_i Y_i - 2\tau \sum_{i=1}^n Y_i}{\sum_{i=1}^n W_i \sum_{i=1}^n \tau_i Y_i - \sum_{i=1}^n \tau_i W_i \sum_{i=1}^n Y_i}$$

This completes the parameter representation of the efficient border with sustainability constraints.

Figure 3.81 shows the counterpart of Fig. 3.78 for the case with unbounded short selling. The efficient borders were determined using the above explicit formula.

Figure 3.82 depicts the "sustainability Sharpe line" for the case with unbounded short selling. The parameter representation of this line for the free parameter τ can be determined as follows: For given τ, the corresponding point on the sustainability Sharpe line is on the efficient border for this τ. Therefore, we can use the notations and the results of the above calculations to determine these efficient borders. The portfolio values $x_1, \ldots, x_n$ again have the representation

$$\begin{pmatrix} x_1(\mu,\tau) \\ \vdots \\ x_n(\mu,\tau) \end{pmatrix} = \frac{\lambda}{2} \begin{pmatrix} Y_1 \\ \vdots \\ Y_n \end{pmatrix} + \frac{\beta}{2} \begin{pmatrix} Z_1 \\ \vdots \\ Z_n \end{pmatrix} + \frac{\kappa}{2} \begin{pmatrix} W_1 \\ \vdots \\ W_n \end{pmatrix}.$$

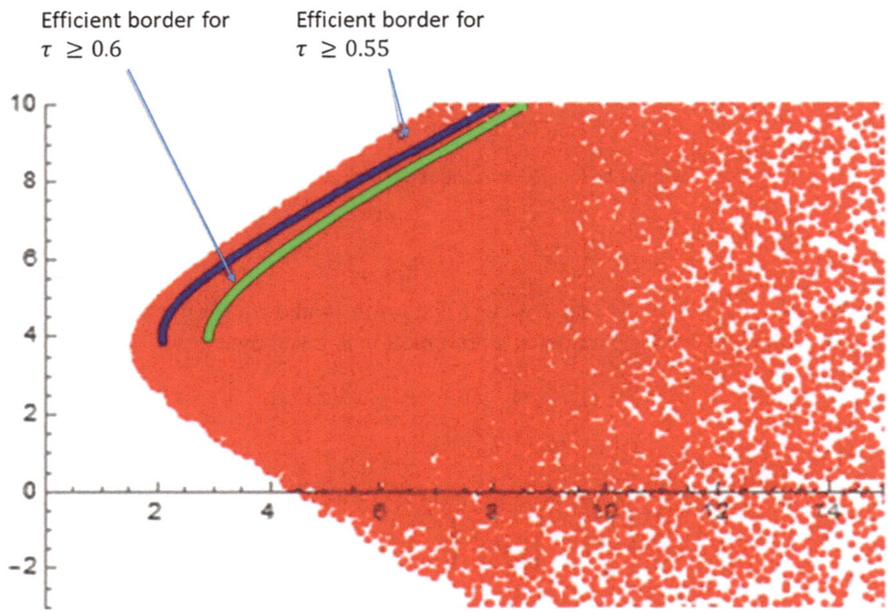

Fig. 3.81 Efficient borders with respect to various sustainability parameters

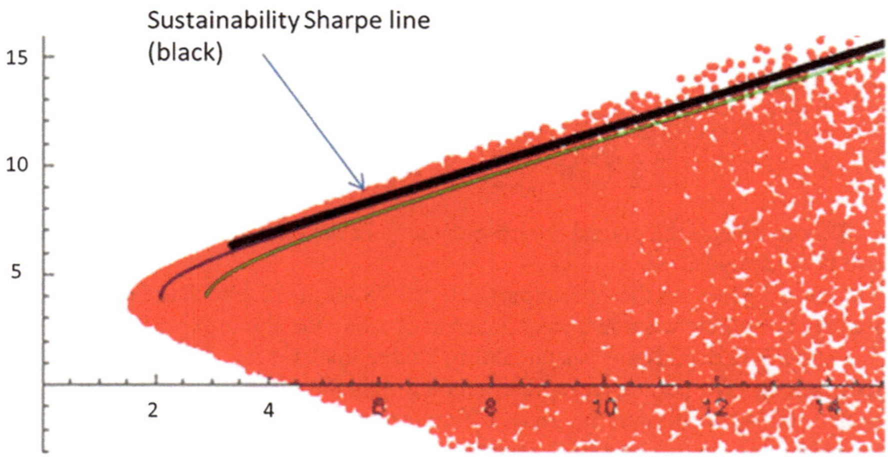

Fig. 3.82 Sustainability Sharpe line (with short selling)

By arranging the terms, we get representations for the coordinates of the form

$$x_i = a_i + \mu b_i + \tau c_i,$$

where the explicit definition of the values a_i, b_i, c_i results, now that the terms have been arranged, from the earlier representation.

All that is left to do now is determine the value μ for given τ such that we get a maximum Sharpe ratio. Using extreme value calculations again with the Lagrange multiplier, we obtain the following value for this optimal μ:

$$\mu = -\frac{Fr + 2G}{F + 2ER}$$

with

$$E = \sum_{i=1}^{n} \sigma_i^2 b_i^2 + \sum_{i,j=1,i\neq j}^{n} \sigma_{i,j} b_i b_j,$$

$$F = \sum_{i=1}^{n} \sigma_i^2 \left[2a_i b_i + 2\tau b_i c_i\right] + \sum_{i,j=1,i\neq j}^{n} \sigma_{i,j} \left[a_i b_j + a_j b_i + \tau(b_i c_j + c_i b_j)\right],$$

and

$$G = \sum_{i=1}^{n} \sigma_i^2 \left[2\tau a_i c_i + a_i^2 + \tau^2 c_i^2\right] +$$

$$+ \sum_{i,j=1,i\neq j}^{n} \sigma_{i,j} \left[a_i a_j + \tau(a_i c_j + c_i a_j) + \tau^2 c_i c_j\right].$$

The corresponding σ_P is then given by

$$\sigma_P(\mu) = \sqrt{E\mu^2 + F\mu + G}.$$

In Fig. 3.82 you see the counterpart of Fig. 3.79 for the case with unbounded short selling. The sustainability Sharpe line was determined using the above explicit formula and is plotted as a thick line on the opportunity set and the efficient borders (shown as thin lines) from Fig. 3.81. We see here that, as τ grows, the portfolio with optimal Sharpe ratio (in contrast to the case without short selling) is located far in the upper right, i.e. in the risky area of the opportunity set.

Finally, we derived the formula for determining the portfolio with maximum sustainability Sharpe ratio $NSR(\gamma)$ for given γ. However, determining this portfolio requires solving an eighth-degree polynomial equation with coefficients composed

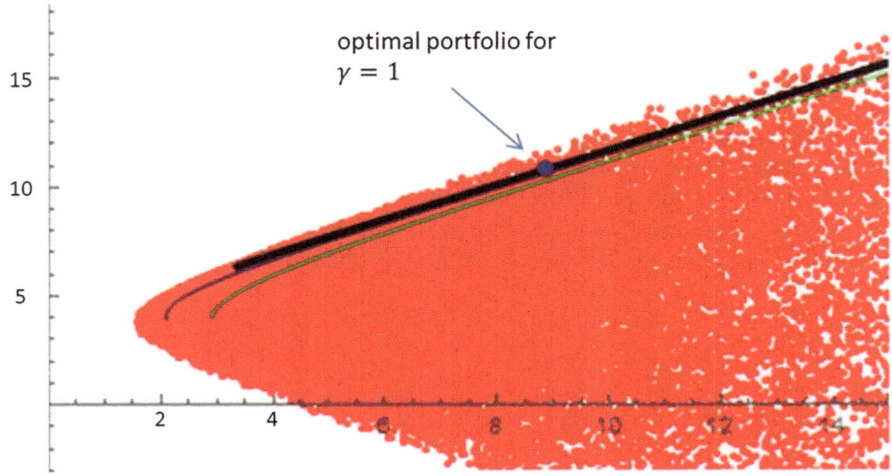

Fig. 3.83 Portfolios with optimal NSR(1) (with short selling)

in a relatively complex way from the initial data. For this reason, we do not give the explicit form of these coefficients and of the polynomial equation here, but only illustrate the case $\gamma = 1$ for our example, that is, the optimal portfolio for an investor with sustainability parameter $\gamma = 1$. For the case $\gamma = 1$, this optimal portfolio is shown on the "sustainability Sharpe line" for the case with unbounded short selling in Fig. 3.83.

3.9.4 Outlook

We have presented a model for integrating sustainability parameters into the classic Markowitz portfolio selection theory, and for the case where unbounded short selling is possible, we have derived explicit representations of efficient borders under sustainability constraints and of portfolios with optimal Sharpe ratios under sustainability constraints, and we have illustrated them using an illustrative example. For this purpose, we used Monte Carlo methods in the case where short selling is not possible and explicit representations in the case where unbounded short selling was possible.

The proposed model is the first approach to such a sustainability portfolio selection theory and is intended to serve as the starting point of further discussions along these lines. What should be discussed in particular are the effects of required minimum sustainability parameter values on short selling (the model here clearly pushes the purchase of sustainable products and of shorting highly non-sustainable products). Another aspect to address is how to appropriately choose the sustainability parameter γ for investors.

The next important step would be to test this model extensively using real-market data and a wide variety of possible scenarios. Initial—incomplete—research in this regard suggests that, under reasonably normal circumstances, requiring a minimum value for sustainability parameters of portfolios affects the Sharpe ratio only marginally and that sustainable portfolios under this model incur a barely significant loss of quality in terms of their risk-return ratio.

3.10 Case Study X: Comparison of Two Basket Derivatives

This article is about a product exchange, the appropriateness (or rather inappropriateness) of which was to be examined in a private expert opinion. Again, we will present and analyse only the essence of this comprehensive issue and core findings.

First: the basic facts of the situation:

On 1 September 2007, Bank B had sold a financial Product $P1$ (which is presented in detail below) running from 1 September 2007 to 31 August 2013 at a price of 100 to a group of investors A (referred to as "Investor A" below). Already at the end of May 2008, this product $P1$ showed a very negative performance, resulting in a price of 74.30 at the end of May (i.e. a loss of 26.70% within 9 months).

As a "goodwill gesture", Bank B offered Investor A the following exchange, at no charge:

The old product $P1$ could be exchanged free of charge for a new product $P2$, which was very similar in design to Product $P1$, but was completely new (without the history of the last 9 months). So the product would start again "from zero", but instead of buying it at a starting price of 100, the investor would buy it by swapping it for Product $P1$, i.e. at a price of 74.30 (= current price of $P1$). Accordingly, $P2$ would mature 1 year later than $P1$ (i.e. it would run from 1 June 2008 to 31 August 2014).

At first sight, the offer seemed tempting (all of the product details will be presented in the next section), and so Investor A accepted it.

At the end of August 2013, the old product $P1$ closed at a price of 160, thus with a handsome profit (contrary to the gloomy outlook in May 2008), while the new product $P2$ closed with a heavy loss at a price of around 40 at the end of August 2014.

Only now did Investor A begin to question the swap that had taken place in June 2008. They retained an expert to analyse the transaction with regard to whether indeed a less valuable (or at least equivalent) product had been exchanged for a more valuable product in this seemingly advantageous exchange in June 2008 or whether they had been deceived.

3.10.1 Detailed Description of the Two Products

As noted above: Both products were of a very similar design, which we will present below.

Design of the (old) product P1:

Contract period: *1 September 2007 to 31 August 2013*
Price at issue: *100*

The performance and the final value of the product depend on the performance of a reference basket (Basket 1) of 30 stocks. We denote these stocks by $X_1, X_2, \ldots, X_{30}$ *in the following. (Approximately one quarter of these stocks were traded on the Tokyo Stock Exchange, approximately one quarter on the New York Stock Exchange and approximately one half on stock exchanges in the euro area.)*
The prices of the stocks $X_1, X_2, \ldots, X_{30}$ *are normalized to 100 as of 1 September 2007, and the subsequent price moves of the stocks are always relative to this initial value of 100. In the following, the "price" of each of these stocks is always deemed to be the respective stock's value relative to this normalized value of 100 (on 1 September 2007).*
A "reference value" is then established at every moment throughout the lifetime of product $P1$*. This reference is determined by taking the mean over the ten stocks with the lowest prices at the respective moment. This gives the* ***reference value*** $\mathbf{R1}(t)$ ***of Basket 1 at time*** t.

On 31 August of each year (starting in 2008), i.e.
on 31 August 2008
on 31 August 2009
on 31 August 2010
on 31 August 2011
on 31 August 2012
on 31 August 2013
the reference value $R1(t)$ *is compared with certain thresholds, with the following consequences:*

– If on ***31 August 2008****,* $R1(t)$ *has a value greater than or equal to* ***88****,*
then the amount ***maximum of 110 and*** $\mathbf{R1}(t)$ *is paid out and the product expires.*

– If not, then:
If on ***31 August 2009****,* $R1(t)$ *has a value greater than or equal to* ***80****,*
then the amount ***maximum of 120 and*** $\mathbf{R1}(t)$ *is paid out and the product expires.*

– If not, then:
If on ***31 August 2010****,* $R1(t)$ *has a value greater than or equal to* ***70****,*
then the amount ***maximum of 130 and*** $\mathbf{R1}(t)$ *is paid out and the product expires.*

– If not, then:
If on ***31 August 2011****,* $R1(t)$ *has a value greater than or equal to* ***60****,*
then the amount ***maximum of 140 and*** $\mathbf{R1}(t)$ *is paid out and the product expires.*

– If not, then:
If on ***31 August 2012****,* $R1(t)$ *has a value greater than or equal to* ***46****,*
then the amount ***maximum of 150 and*** $\mathbf{R1}(t)$ *is paid out and the product expires.*

– If not, then:

*If on **31 August 2013**, $R1(t)$ has a value greater than or equal to **40**,*
*then the amount of **160** is paid out and the product expires.*
*If on **31 August 2013**, $R1(t)$ has a value smaller than **40**,*
then $\mathbf{R1(t)}$ is paid out and the product expires.

While—as said before—the new product $P2$ was very similar in design, it started out almost a year later:

Design of the (new) product P2:
***Contract period:** 1 June 2008 to 31 August 2014*
***Price at issue:** 74.30!! (when exchanged for $P1$)*

The performance and the final value of the product depend on the performance of another reference basket (Basket 2) of 30 stocks. We denote these stocks by $Y_1, Y_2, \ldots, Y_{30}$ in the following. (In this case, approximately one quarter of these stocks was from Asia (Japan, China, Korea), approximately one half was traded on the New York Stock Exchange, and approximately one quarter was traded on stock exchanges in the euro area.)
The prices of the stocks $Y_1, Y_2, \ldots, Y_{30}$ are normalized to 100 as of 1 June 2008, and the subsequent price moves of the stocks are always relative to this initial value of 100. In the following, the "price" of each of these stocks is always deemed to be the respective stock's value relative to this normalized value of 100 (on 1 June 2008).
*A "reference value" is then established at every moment throughout the lifetime of product $P2$. This reference is determined by taking the mean over the ten stocks with the lowest prices at the respective moment. This gives the **reference value $\mathbf{R2(t)}$ of Basket 2 at time t**.*

On 31 August of each year (starting in 2009), i.e.:
On 31 August 2009
On 31 August 2010
On 31 August 2011
On 31 August 2012
On 31 August 2013
On 31 August 2014
The reference value $R2(t)$ is compared with certain thresholds, with the following consequences:

*– If on **31 August 2009**, $R2(t)$ has a value greater than or equal to **100**,*
*then the amount **maximum of 118 and $\mathbf{R2(t)}$** is paid out and the product expires.*

– If not, then:
*If on **31 August 2010**, $R2(t)$ has a value greater than or equal to **95**,*
*then the amount **maximum of 130 and $\mathbf{R2(t)}$** is paid out and the product expires.*

– If not, then:
*If on **31 August 2011**, $R2(t)$ has a value greater than or equal to **90**,*
*then the amount **maximum of 140 and $\mathbf{R2(t)}$** is paid out and the product expires.*

– If not, then:

If on ***31 August 2012****,* $R2(t)$ *has a value greater than or equal to* ***80****,*
then the amount ***maximum of 150 and*** $\mathbf{R2}(\boldsymbol{t})$ *is paid out and the product expires.*

– If not, then:
If on ***31 August 2013****,* $R2(t)$ *has a value greater than or equal to* ***60****,*
then the amount ***maximum of 160 and*** $\mathbf{R2}(\boldsymbol{t})$ *is paid out and the product expires.*

– If not, then:
If on ***31 August 2014****,* $R2(t)$ *has a value greater than or equal to* ***50****,*
then the amount of ***170*** *is paid out and the product expires.*
If on ***31 August 2014****,* $R2(t)$ *has a value smaller than* ***50****,*
then $\mathbf{R2}(\boldsymbol{t})$ *is paid out and the product expires.*

3.10.2 First Brief Comparison of the Two Products

A quick and superficial comparison of the two products reveals the following differences:

- Slightly longer contract period of product $P2$ (slight disadvantage for $P2$)
- Significantly higher thresholds for $P2$ (clear disadvantage for $P2$)
- Significantly higher payouts for $P2$ (clear advantage for $P2$)
- Initial price (= purchase price) of 74.30 for $P2$ instead of 100 for $P1$ (clear advantage for $P2$)

Based on this comparison, the proposed exchange of $P1$ for $P2$ as of 1 June 2008 does not look disadvantageous to Investor A. However, an important point to consider here—which not sufficiently savvy investors may not be aware of—is the quality of the two stock baskets.

In particular, it is necessary to analyse the volatility of the stocks in the two baskets (including for later valuation of these two products). Since implied volatilities were only available for a few of the stocks in the two baskets, historical volatilities were calculated from the daily prices of the stocks in the previous three trading years (so, from 1 September 2004 to 31 August 2007 for Basket 1 and from 1 June 2005 to 31 May 2008 for Basket 2). (The required 3three-year time series were available for all of the stocks.)

A possible objection to using historical volatilities for pricing the two products instead of implied volatilities (which were not sufficiently available here) can be countered as follows: In the following, it is not so much a question of obtaining an exact absolute valuation of the two products, but rather of the relation between the values. And in that regard, i.e. in terms of the relation of the prices to each other, any inaccuracy due to the use of historical instead of implied volatilities is balanced out by the fact that the historical volatility is used for **both** products.

Table 3.25 Historical volatilities of the stocks $X_1, \ldots X_20$ for Basket 1

Stock	X_1	X_2	X_3	X_4	X_5	X_6	X_7	X_8	X_9	X_{10}
Volatility	43.9	27.4	49.7	51.6	58.3	72.9	30.6	26.6	32.5	28.0

Stock	X_{11}	X_{12}	X_{13}	X_{14}	X_{15}	X_{16}	X_{17}	X_{18}	X_{19}	X_{20}
Volatility	48.2	25.6	28.8	21.4	32.9	37.0	20.8	24.6	34.1	23.2

Stock	X_{21}	X_{22}	X_{23}	X_{24}	X_{25}	X_{26}	X_{27}	X_{28}	X_{29}	X_{30}
Volatility	34.0	34.4	30.6	31.4	25.8	38.8	31.1	28.8	44.1	20.6

Table 3.26 Historical volatilities of the stocks $X_1, \ldots X_20$ for Basket 2

Stock	Y_1	Y_2	Y_3	Y_4	Y_5	Y_6	Y_7	Y_8	Y_9	Y_{10}
Volatility	68.2	51.6	57.4	45.6	41.7	32.5	31.9	42.8	88.9	34.5

Stock	Y_{11}	Y_{12}	Y_{13}	Y_{14}	Y_{15}	Y_{16}	Y_{17}	Y_{18}	Y_{19}	Y_{20}
Volatility	39.7	39.0	33.3	32.0	53.1	40.7	69.8	42.5	45.1	37.9

Stock	Y_{21}	Y_{22}	Y_{23}	Y_{24}	Y_{25}	Y_{26}	Y_{27}	Y_{28}	Y_{29}	Y_{30}
Volatility	51.5	42.5	35.8	45.8	38.5	59.1	49.6	34.9	37.7	44.3

Determination of the historical volatilities yielded the following values:

Basket 1—historical volatilities:

The average volatility of the stocks in Basket 1 is 34.59, which follows directly from Table 3.25.

Basket 2—historical volatilities:

The average volatility of the stocks in Basket 2 is 45.60, as one can determine with Table 3.26.

Clearly the stocks in Basket 2 are significantly more volatile than the stocks in Basket 1. What is not yet clear, however, is how this affects the comparative values of $P1$ and $P2$.

Another observation with regard to the two stock prices is also quite interesting and revealing, although it will not be included in the further calculations: Let us compare the analyst ratings of the two stock baskets on 1 June 2008 as per the Global Industry Classification Standard (GICS). These ratings reflect the average analyst opinions of the respective stocks at the respective dates and can be accessed in Bloomberg. The ratings represent scores between 1 and 5. The lowest score is 1 and means "strong sell recommendation", and the best score is 5 and means "strong buy recommendation". We are not going to give the individual scores for each stock here, just the average scores: The average GICS score for the stocks in Basket 1 was 3.93, while the average GICS score for the stocks in Basket 2 was only 3.32. This too indicates a significantly lower quality of the stocks in Basket 2.

3.10.3 Comparison of the Two Product Values on 1 June 2008

The aim now is to determine and compare the values of the two products $P1$ and $P2$ on 1 June 2008.

Let us recall:

According to Bank B, Product $P1$ had a value of 74.30 on 1 June 2008. The bank offered Investor A the possibility to exchange Product $P1$ for Product $P2$, describing this offer as a goodwill gesture. In doing so, the bank implicitly gave the client the impression that Product $P2$ had a value greater than 74.30. Let us now verify these valuations.

We assume a risk-neutral Wiener model for each of the stocks. The stock volatilities that we need for this were already given above. In addition, we need the risk-free interest rate r for the lifetime of the two products. In both cases, we use the 5-year Euro swap rate, which was 4686%. We can assume that a corresponding 6-year swap rate is close to this value.

On 1 June 2008, the stocks of Product $P2$ started at the normalized value of 100. The stocks of Product $P1$, on the other hand, had already run a 9-month course since the product was launched and normalized to 100 on 1 September 2007. To determine the value of Product $P1$, we therefore need the (normalized) values of the stocks in Basket 1 on 1 June 2008. For the stocks $X_1, X_2, \ldots, X_{30}$, these values were 80.17, 76.12, 86.36, 142.67, 118.58, 44.96, 84.35, 109.59, 44.53, 111.78, 95.75, 90.96, 77.74, 104.84, 66.96, 94.11, 100.29, 98.39, 47.72, 84.04, 56.33, 59.58, 98.53, 103.06, 104.14, 103.69, 82.14, 90.01, 72.76, and 114.37.

If we arrange these values by magnitude and average the ten smallest values, we obtain the reference value $R1(0)$ (where we use "0" to denote the reference date of 1 June 2008). The ten smallest values are 44.53, 44.9, 47.72, 56.33, 59.58, 66.97, 72.77, 76.12, 77.75, and 80.17, averaging 62.69.

If we compare this value with the payout thresholds at the respective reference dates for Product $P1$, we see that after a life of 9 months, its value was already well below the first two relevant thresholds of 88% and of 80% and even below the third thresholds of 70%. This explains why the price of Product $P1$ had fallen so much by 1 June 2008. Further, to determine the fair price of the two products, we would also need the correlations between a pair of stocks from Basket 1 and a pair of stocks from Basket 2. These correlations could easily be estimated using the previous 3 years' daily data that we already used to estimate the historical volatilities. However, because this led to certain numerical problems (with regard to the positive definiteness of the correlation matrix), we simplified the procedure as follows: The average correlation of pairs of stocks in Basket 1 and Basket 2 was estimated from the available historical data. In both cases, this average was very close to 0.3. In a simplified process, we then worked with a correlation matrix M in both cases, where all values outside the main diagonal were set equal to 0.3. This matrix M is positive-definite and can be easily decomposed with the Cholesky decomposition method.

The rest of the procedure is exactly the same as set out in Volume II Section 2.31 on pricing multi-asset options using Monte Carlo: A number of scenarios (5000 in this specific case) are simulated. In each scenario, the prices of 30 stocks are to be simulated over a period of 5 years and 3 months (in the case of $P1$) or 6 years and 3 months (in the case of $P2$), and the payoff is to be determined for each scenario.

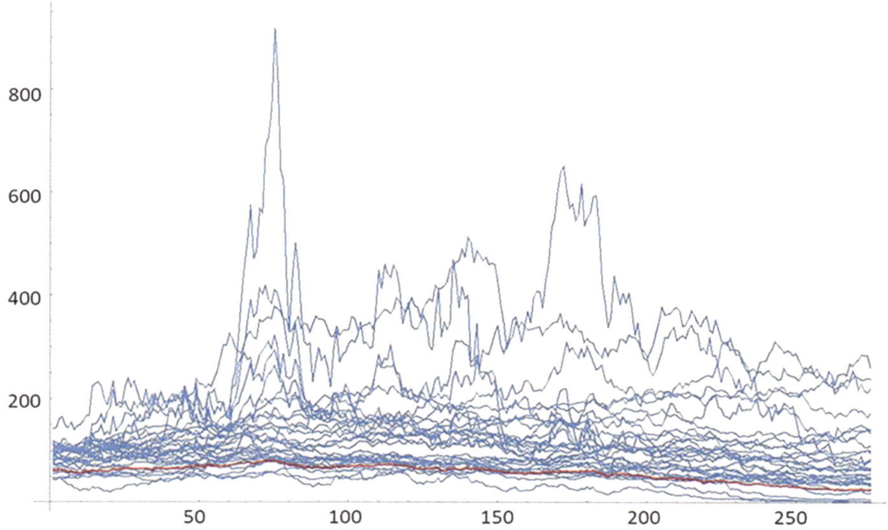

Fig. 3.84 Possible scenario of the performance of the 30 stocks (blue) in Stock Basket 1 from 1 June 2008 to 31 August 2013 (273 weeks) and performance of the reference value $R1(t)$ (red) of each week's ten worst performing stocks

All prices are simulated in weekly steps. This means that we generate 273 resp. 325 steps for each path.

In Fig. 3.84 we see a typical scenario of the paths of 30 stocks and the corresponding reference value R1(t) for Basket 1. We see that, because the basket had already been in existence for some time, the stocks start out with different values.

Let us now take a closer look at the reference value $R1(t)$ in this scenario, including a view to the payoff that this scenario would have led to (see Fig. 3.85).

We see that in this scenario, the reference value $R1(t)$ would have exceeded the relevant threshold (70 in this case) for the first time at the end of August 2010. This would have resulted in a payoff of 130 (since $R1(t)$ was greater than 70 but evidently less than 130).

The payoffs under each scenario are discounted as required. The discount factors resulting from the interest rates at that time for each of the six potential payment dates, i.e. on 31 August in each year from 2008 through 2013, were 0.9839, 0.9433, 0.9044, 0.8671, 0.8307, and 0.7965.

The average over the payoffs of the 5000 scenarios then gives us the approximate value of Product $P1$ on 1 June 2008. In Fig. 3.86 we see the convergence behaviour of this Monte Carlo simulation and get a relatively stable **value for Product $P1$ on 1 June 2008 of 76.89**. Thus, we obtain a slightly higher value than the bank had attributed to the Product $P1$ on 1 June 2008 (74.30).

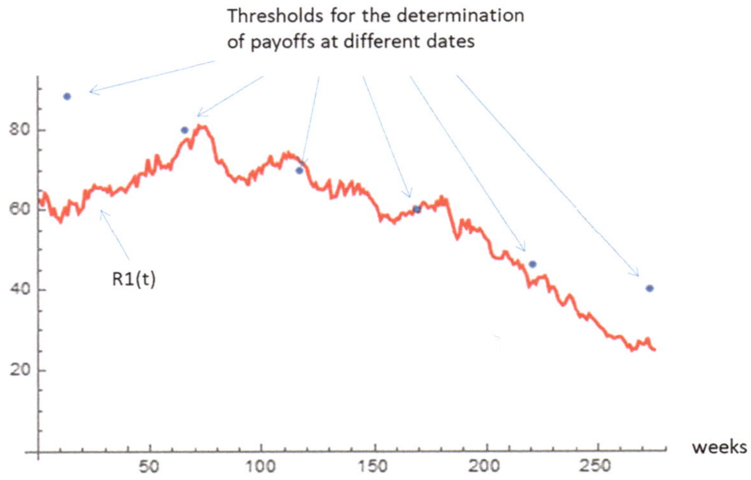

Fig. 3.85 Path of $R1(t)$ in the above scenario and payoff thresholds

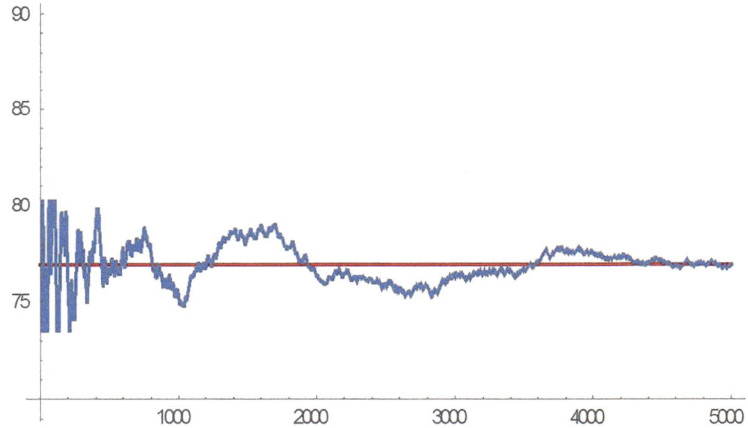

Fig. 3.86 Convergence behaviour of the MC simulation for determining the value of $P1$

To determine the value of Product $P2$ on 1 June 2008, we proceed in exactly the same way, we only adjust the relevant data. All of the stocks now start at the same normalized value of 100. The volatilities are higher on average, the investment period is 1 year longer, and the barriers and payout levels are adjusted accordingly. A further discount factor is used for the payout date on 31 August 2014, namely, 0.7649.

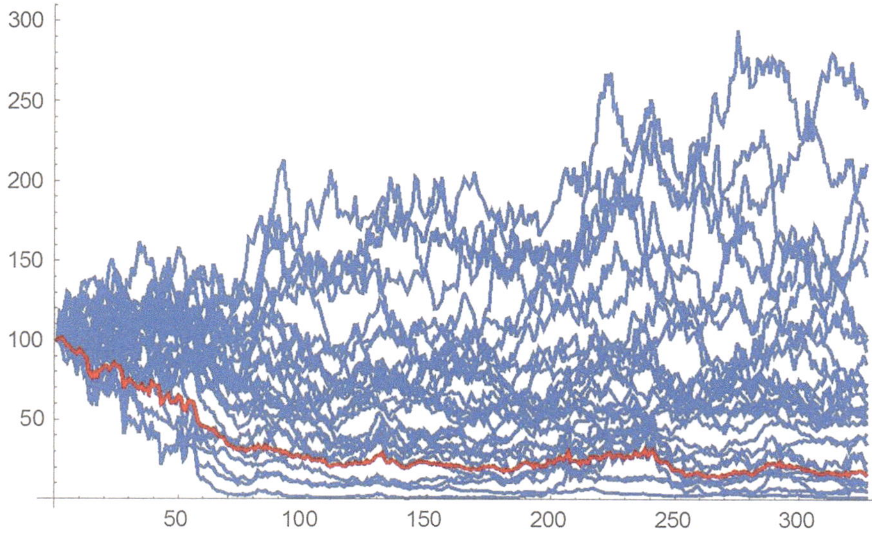

Fig. 3.87 Possible scenario of the performance of the 30 stocks (blue) in Stock Basket 2 from 1 June 2008 to 31 August 2014 (325 weeks) and performance of the reference value $R2(t)$ (red) of each week's ten worst performing stocks

In Fig. 3.87 we now see a typical scenario of the paths of 30 stocks and the corresponding reference value $R2(t)$ for Basket 2. We see that the stocks here all start out with a value of 100.

We see in Fig. 3.88 that in this scenario, the reference value $R2(t)$ remained below the relevant payoff thresholds throughout. Therefore, the payout on the product's expiration on 31 August would be the reference value, which in this scenario would have been only 15.89.

Finally, in Fig. 3.89 we see the convergence behaviour of this Monte Carlo simulation and get a relatively stable **value for Product $P2$ on 1 June 2008 of 40.04**.

Although certain inaccuracies were accepted in the calculations, the models, and the parameter choices, this result indicates a much lower value of Product $P2$ compared to Product $P1$ even when allowing a generous tolerance. We therefore conclude that the bank's representation—or implicit suggestion—that the exchange was a goodwill concession to the investor does not hold.

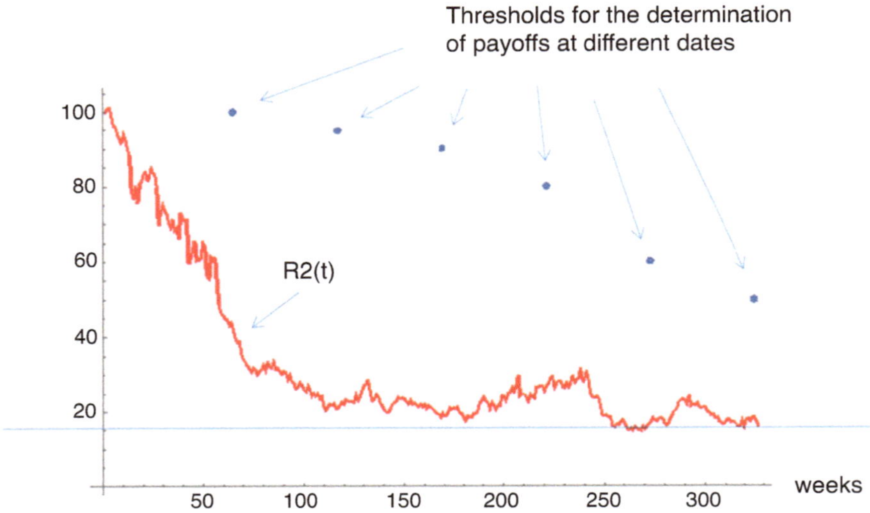

Fig. 3.88 Path of $R2(t)$ in the above scenario and payoff thresholds

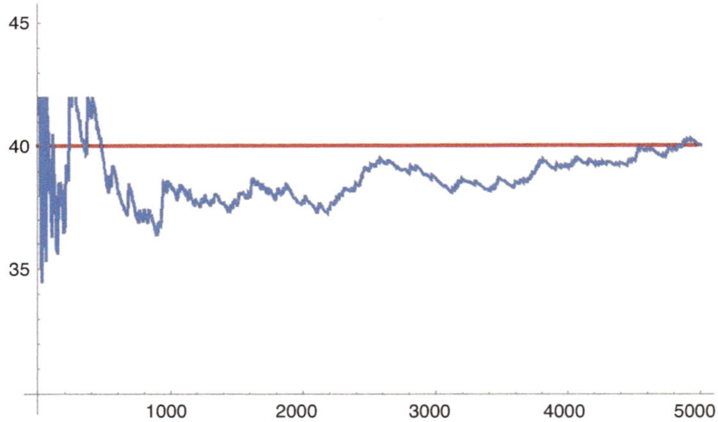

Fig. 3.89 Convergence behaviour of the Monte Carlo simulation of the fair price of Product $P2$

3.11 Case Study XI: Optimal Hedging with Future Contracts in Connection with Liquidity Risks and the "Metallgesellschaft" Case

The following case was motivated by heated debates in the mid- and late 1990s following the near bankruptcy of the German Metallgesellschaft in 1993. At the time, Metallgesellschaft AG (now part of the GEA group) was one of Germany's largest commodity trading companies. The turmoil was triggered by high losses in oil futures. A key component that led to the difficulties—and which we will address below—was the following:

While the oil futures in question were perfectly hedged in principle, the hedging instruments that were used required extremely high temporary liquidity so that they could be rolled on. The company did not have this liquidity, and the banks involved were not prepared to provide it. As a result, the hedges—allegedly at Deutsche Bank's insistence—had to be unwound at the worst possible time, resulting in immense losses (and mutual accusations of incompetence).

In the following, we will present this key component—in a simplified and abstracted form—formulate the problem, and solve it. The problem and its solution were not subsequently applied as such in a real case. However, this is an example of research work in pure mathematics undertaken as a result of a real-world case. Above all, this example is intended to demonstrate the following:

In terms of practical applications, more can often be achieved with "pure" mathematics than by simulation, numerics, and intensive computer use. This chapter is therefore (after the problem has been presented) the most challenging in mathematical terms in this book. The proof sketch of the problem's mathematical solution in Sect. 3.11.3 is intended for mathematically "open-minded" readers. But if you allow yourself to delve into the presentation of the proof ideas, you will likely be surprised at how accessible they are and will enjoy the beauty of the arguments and the solution.

3.11.1 Presentation of the Hedging Strategy and the Problem

A company, referred to as MG in the following, agrees to deliver q units of a certain commodity in certain regular (e.g. monthly) future intervals over a certain period of time $[0, T]$ (e.g. q times 1000 barrels of Brent crude oil) at a **fixed price of a EUR** per unit to Buyer A.

For simplicity of exposition, we denote the time interval by $[0, N]$, where N is an integer, and denote the times at which deliveries are to occur by $1, 2, \ldots, N$ (see Fig. 3.90).

We further assume, for simplicity, a risk-free interest rate $r = 0$, so that no discounting is needed in the following.

The **actual price of one unit of the commodity at time n is denoted by S_n**. For the changes of S_n over time, we adopt the following simple model:

$S_n = c + \sum_{i=1}^{n} X_i$ with mutually independent $N(0, \sigma^2)$-distributed random variables X_i. Thus, we are dealing with a discrete random walk with normally distributed increments. The expected value $E(S_n)$ is equal to c.

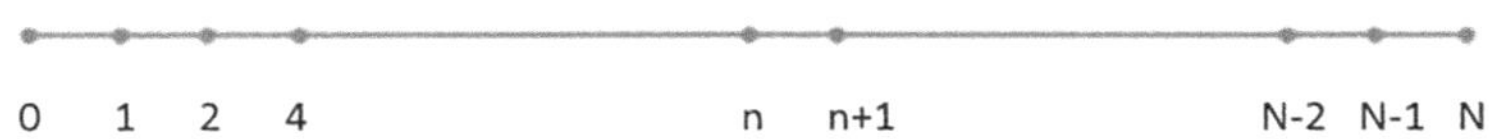

Fig. 3.90 Delivery times

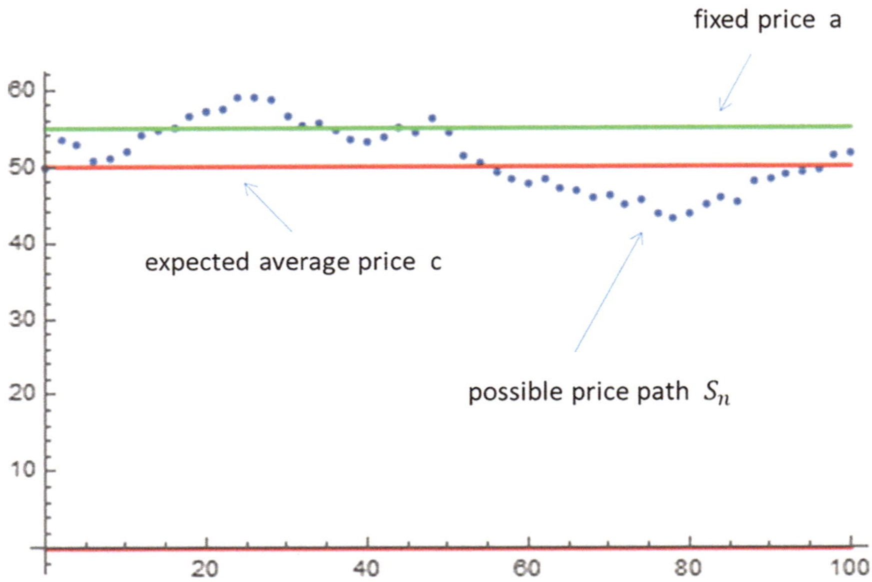

Fig. 3.91 Possible price path of one unit of the commodity ($c = 50$, $\sigma = 1$)

For this modelling to be useful, at least to some extent, especially so as to obtain a high probability that the prices S_n remain positive in this model, the standard deviation σ of the increments should be small compared to c.

Furthermore, for the business model to make sense, a must be (at least slightly) larger than c. Buyer A pays the difference $a - c$, i.e. the difference between the fixed price a and the average expected price c, as a premium for the assurance of a stable long-term price.

A possible path of the price S_n and of the business model is shown in Fig. 3.91.

The profit (or loss) that MG makes at time n is $q \cdot (a - S_n)$. The profit process C_k of the aggregate profits up to time k is therefore obtained by

$$C_k = q \cdot \sum_{n=1}^{k} (a - S_n) = q \cdot k \cdot (a - c) - q \cdot \sum_{n=1}^{k} (k - n + 1) \cdot X_n.$$

So, for the expected profit $E(C_k)$, we obviously get $E(C_k) = q \cdot k \cdot (a - c)$. The remaining term, the random variable $q \cdot \sum_{n=1}^{k} (k - n + 1) \cdot X_n$, has expected value 0 but may possibly also take large values: If k is large, i.e. converging to N, then most of the random variables X_n (those with smaller index n) have factors of order N. Of course, for the stability of the MG company, this is not a really desirable situation.

Figure 3.92 shows some possible simulation paths for the profit process C_k. For the sake of simplicity, as in everything that follows, we set $q = 1$. The expected profit trend $q \cdot k \cdot (a - c)$ is plotted in black.

For some paths, we see significant deviations from the expected profit trend.

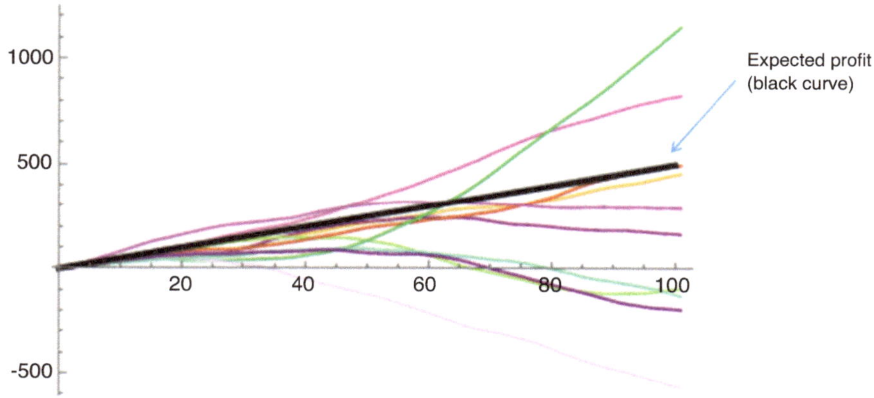

Fig. 3.92 Ten simulated profit paths versus expected profit paths (black)

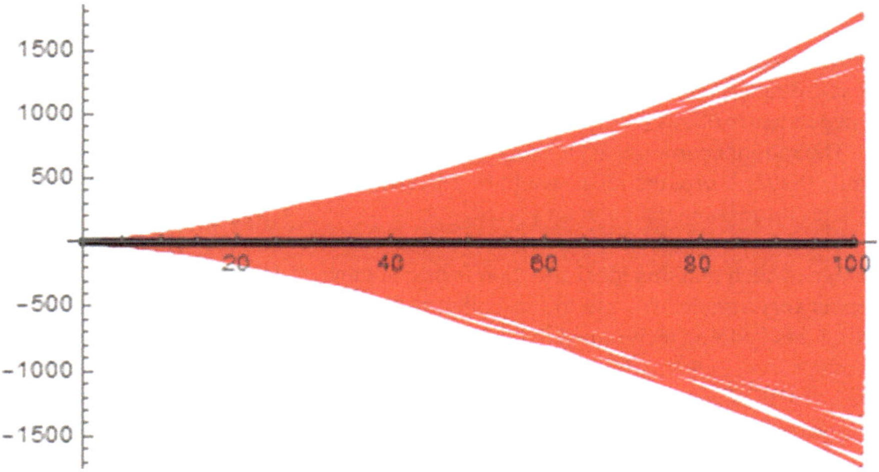

Fig. 3.93 Random component of the profit process—1000 possible paths

In Fig. 3.93 we only plotted, including with a view to facilitating later comparisons, the random residual component $R_k := \sum_{n=1}^{k} (k - n + 1) \cdot X_n$ of the profit process in 1000 possible paths (intentionally all-in-one colour). These paths no longer fluctuate around the expected profit process $k \cdot (a - c)$, but around 0.

So for the MG company, it would certainly make sense to contemplate certain hedges and hedging strategies for this investment. One possibility would be to hedge with short-dated futures: At every point in time n, MG buys from another market participant Z a certain number of future contracts (say, g_n contracts, assuming here that each contract refers to exactly one unit of the commodity) expiring at time $n+1$. This gives MG the right to buy exactly g_n units of the commodity at time n at the strike price F_n agreed at time n.

The price of the futures contract is 0.

The strike price that MG and Z will reasonably agree on is the commodity's price that is expected at time n to apply at time $n+1$ (recall that we set $r=0$). Hence:

$$F_n = E\left(S_{n+1}\,|S_n\right) = E\left(c + \sum_{i=1}^{n+1} X_i\,|S_n\right) = E\left(S_n + X_{n+1}\,|S_n\right) = S_n.$$

So, the profit that MG makes from this future trading at time $n+1$ is $g_n \cdot (S_{n+1} - Sn) = g_n \cdot X_{n+1}$. The cumulative profit process H_k for MG up to time k purely from the future transactions is thus $H_k = \sum_{n=1}^{k} g_{n-1} X_n$.

Therefore, the total profit process $\widetilde{C}_k$ from the agreement and the hedging (future trading) up to time k is

$$\widetilde{C}_k = k \cdot (a-c) - \sum_{n=1}^{k} (k-n+1-g_{n-1}) \cdot X_n.$$

Now, irrespective of the chosen hedging strategy, that is, of the specific choice of g_n, the expected value of the total profit process up to time k remains the same, hence, equal to $E(\widetilde{C}_k) = k \cdot (a-c)$. Here, we assume, of course, that g_n is chosen independently of x_{n+1}.

The random component $\widetilde{R}_k := \sum_{n=1}^{k} (k-n+1-g_{n-1}) \cdot X_n$ again has expected value 0. Of course, the ideal would be for fluctuations of $\widetilde{R}_k$ to stay within narrow bounds throughout the contract period.

The question that will arise below is as follows: How can the hedging strategy be designed such that this ideal situation is best achieved? The answer to this question seems very simple: We use the so-called **rolling-stack hedge**! The motivation for using this hedging technique is as follows: "In principle, it does not matter how the profit process behaves during the contract period. What is important and ideal is that we know in advance exactly how much profit will be made at the end of the contract period, that is, at time N".

How can we achieve that? Well, we simply need to eliminate the random component $\widetilde{R}_N = \sum_{n=1}^{N} (N-n+1-g_{n-1}) \cdot X_n$, and we achieve that by setting $g_{n-1} = N-n+1$, thus $g_n = N-n$ for all n. In this way, we get $\widetilde{C}_N = N \cdot (a-c)$ irrespective of the commodity's price dynamics. In other words, MG gets exactly what it hoped would happen by the end of the contract period.

For other times k, $\widetilde{R}_k$ has the form $\sum_{n=1}^{k} (k-N) \cdot X_n$, thus generally not 0, which, however, should be irrelevant. In Figs. 3.94 and 3.95, we now show (just like we did for C_k and R_k in the case without hedging) a selection of ten possible paths also for $\widetilde{C}_k$ and a selection of 1000 possible paths for $\widetilde{R}_k$ in comparison with the paths from the case without hedging.

We see that already at an early stage of the contract period, fluctuations may be considerably higher when using the rolling-stack strategy than in the case without hedging. Even if all uncertainties are reliably eliminated by the end of the contract period, this fact may well lead to liquidity problems while the strategy is being run.

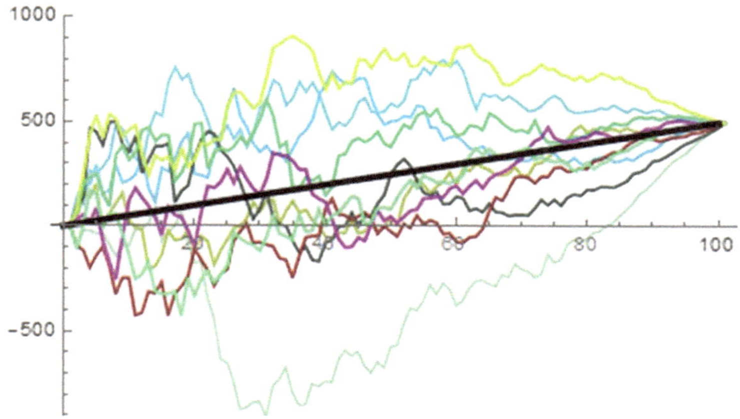

Fig. 3.94 Ten simulated payoff paths $\widetilde{C}_k$ versus expected payoff path (black) when using a "rolling-stack hedge"

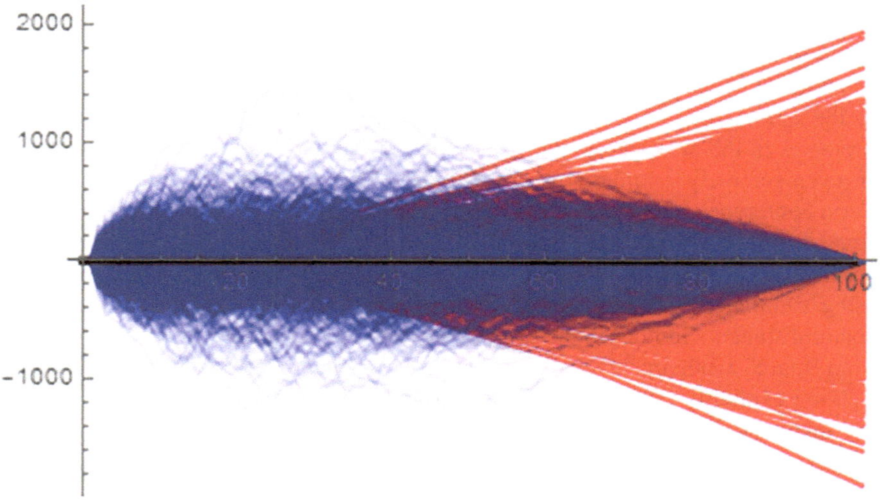

Fig. 3.95 Random component $\widetilde{R}_k$ of the payoff process—1000 possible paths for the case with "rolling-stack hedge" (blue) and without hedging (red)

There may not be enough cash available to continue this strategy in periods with strong temporary (!) losses.

In the light of this, what would be a more meaningful objective to aim for?

The random variable $\widetilde{R}_k$ cannot be definitively constrained over the entire period; in principle, it can attain arbitrarily high values. However, we could try to minimize the probability of $\widetilde{R}_k$ attaining high values at some point during the contract period through appropriate hedging.

In other words, for any values of x, we would like the probability W that $\widetilde{R}_k$ becomes larger than x at some point to remain small. That is, we want

$$W\left(\max_{k=1,2,\ldots,N} \widetilde{R}_k > x\right)$$

to be as small as possible. This probability can be determined relatively well by approximation. We have

$$W\left(\max_{k=1,2,\ldots,N} \widetilde{R}_k > x\right) \approx e^{-\frac{x^2}{\max_k \sigma_k^2}}.$$

Here, σ_k^2 denotes the variance of the random variable $\widetilde{R}_k$. So the probability that the error term stays below a value x throughout the entire contract period increases as $\max_k \sigma_k^2$ decreases.

Our task therefore is: Choose the hedging strategy such that the maximum of the variances of $\widetilde{R}_k$ becomes as small as possible.

When this issue was brought to my attention in 2001 by Paul Glasserman (who, by the way, is the author of the highly recommended book Monte Carlo Methods in Financial Engineering [7]), he had already proposed two better strategies—in the above sense—than the rolling stack (we will briefly present these two strategies in the next subsection) [8]. However, he had not been able to determine an even remotely optimal strategy. On the occasion of the international Monte Carlo conference MCM back then, we discussed ways of approximating an optimal strategy through Monte Carlo or quasi-Monte Carlo simulations. However, it soon became apparent that the problem was too complex for an approximation approach using simulation methods. Yet I subsequently succeeded, together with my then doctoral student Gunther Leobacher, in solving the problem exactly using methods from "pure" mathematics and stating the very interesting and beautiful optimal solution explicitly (see [9]). In the following we will present this optimal solution and give a **vague** proof sketch for the optimality of this solution, as well as an idea of how we managed to discover the relatively complex solution.

3.11.2 Translation of the Problem into the Continuous Case and Glasserman's Strategies

As the first step towards solving the problem, we translate the discrete problem at hand into a problem of continuous analysis (for this purpose, we will briefly revisit a few basic concepts of stochastic analysis). This is a frequently used trick, since the field of continuous analysis provides some tools that are considerably more powerful than are available for discrete problems. The procedure is roughly as follows:

We compress the time range $[0, N]$ into the time range $[0, 1]$. The time points $n = 0, 1, 2, \ldots, N$ thus become $0, \Delta t, 2 \cdot \Delta t, \ldots, N \cdot \Delta t = 1$. We then need to adjust the random variables X_n accordingly: The $N(0, \sigma^2)$-distribution for X_n becomes an $N(0, \Delta t \cdot \sigma^2)$-distribution. In the following we will limit the analysis to $\sigma = 1$. Thus, X_n is $N(0, \Delta t)$-distributed.

If we choose large N (i.e. very small Δt), we can denote the closely spaced time points $k \cdot \Delta t$ by the continuous time parameter t. Then $\widetilde{R}_k$ becomes $\widetilde{R}_t$, and the random variable X_k becomes $X(t)$. We can also take this $N(0, \Delta t)$-distributed $X(t)$ as $dW(t)$ for a standard Brownian motion $W(t)$. Thus, $\widetilde{R}_k = \sum_{n=1}^{k}(k - n + 1 - g_{n-1}) \cdot X_n$ then becomes $\widetilde{R}(k \cdot \Delta t) = \sum_{n=1}^{k}(k \cdot \Delta t - (n-1) \cdot \Delta t - (g((n-1) \cdot \Delta t)) \cdot dW((n-1) \cdot \Delta t)$, which becomes the Ito integral $\widetilde{R}(t) = \int_0^t (t - s - g(s))dW(s)$ when $\Delta t \to 0$. Here, $g(s)$ denotes the number of futures held at time s for our hedging purposes.

This means that, instead of working with the variance σ_k^2 of the random variable $\widetilde{R}_k$, we can now use the variance σ_t^2 of the random variable (of the Ito integral) $\widetilde{R}_t = \int_0^t (t - s - g(s))dW(s)$. And this variance σ_t^2 can be calculated using Ito isometry. We recall that

$$\sigma_t^2 = \int_0^t (t - s - g(s))^2 ds.$$

Our task therefore is: Determine a function $h : [0, 1] \to \mathbb{R}$ such that for $g \equiv h$ the expression

$$F_g := \max_{0 \le t \le 1} F_g(t) := \max_{0 \le t \le 1} \int_0^t (t - s - g(s))^2 ds$$

becomes minimal. Before we proceed to solve this problem in the next subsection, let us revisit—for purposes of this new continuous setting—the two "no hedge" and "rolling-stack hedge" strategies discussed above as well as the two strategies proposed by Glasserman.

"No hedge" means $g(s) = 0$ for all s. Thus we get

$$\sigma_t^2 = F_g(t) = \int_0^t (t - s - g(s))^2 ds = \int_0^t (t - s)^2 ds = \frac{t^3}{3}.$$

We see the evolution of σ_t^2 in Fig. 3.96, and the value that we are ultimately most interested in—$\max_t \sigma_t^2 = \max_t \frac{t^3}{3}$—is assumed for $t = 1$ and has the value $\frac{1}{3} = 0.333$.

"Rolling-stack hedge" means $g(s) = 1 - s$ for all s. (We recall that in the discrete case we had $g_n = N - n$, N became 1, and n is the free parameter which now became s.) We therefore have

$$\sigma_t^2 = F_g(t) = \int_0^t (t - 1)^2 ds = t \cdot (t - 1)^2.$$

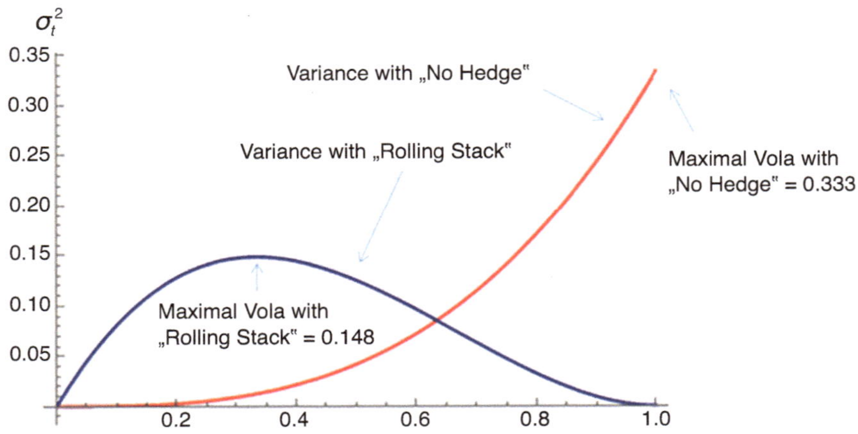

Fig. 3.96 Variance of the total profit process using "no hedge" (red) and "rolling-stack hedge" (blue), including times and values of maximum variance

We also see the evolution of σ_t^2 in Fig. 3.96, and the value that we are ultimately most interested in—$\max_t \sigma_t^2 = \max_t (t \cdot (t-1)^2)$—is assumed for $t = \frac{1}{3}$ and has the value $\frac{4}{27} = 0.148$. Again, we clearly see—as we already know—the standard deviation 0 at the end of the contract period at time 1.

Glasserman (see [8]) suggests an "optimal horizon" and an "optimal fraction" hedging strategy.

"Optimal horizon" is a rolling-stack hedging strategy, yet not for the contract's terminal date 1 but an earlier date τ. Under this strategy, one calculates the maximum variance and then determines τ such that this maximum variance becomes minimal.

First of all, this means $g(s) = \tau - s$ for all $s \leq \tau$ and $g(s) = 0$ for $s > \tau$. Then, if $t \leq \tau$, we get

$$\sigma_t^2 = F_g(t) = \int_0^t (t - s - g(s))^2 ds = \int_0^t (t-\tau)^2 ds = t \cdot (t-\tau)^2$$

and if $t > \tau$

$$\sigma_t^2 = F_g(t) = \int_0^t (t - s - g(s))^2 ds = \int_0^\tau (t-\tau)^2 ds + \int_\tau^t (t-s)^2 ds =$$
$$= \tau \cdot (t-\tau)^2 + \frac{(t-\tau)^3}{3}.$$

Now τ has to be determined such that the maximum of this function is minimal. An elementary curve discussion shows that this is the case for $\tau \approx 0.733$, and for this choice of τ, we obtain for F_g the value 0.0583. The evolution of σ_t^2 is shown in Fig. 3.97.

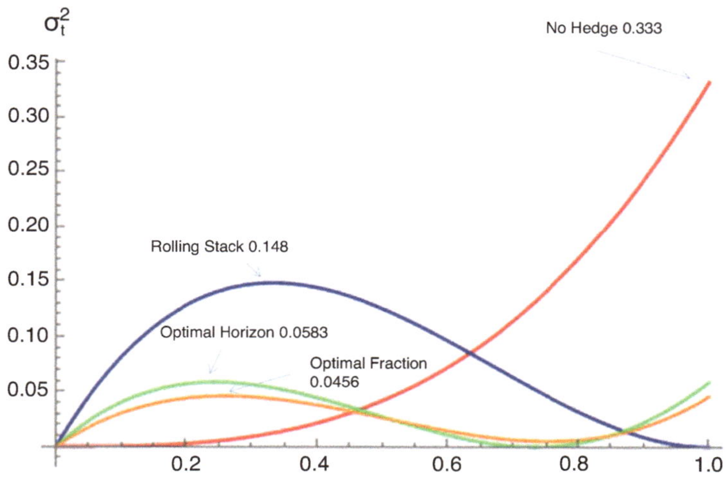

Fig. 3.97 Variance of the total profit process using "no hedge" (red), "rolling-stack hedge" (blue), "optimal horizon" (green), and "optimal fraction" (orange), as well as times and values of maximum variance

In the **"optimal fraction"** case, rolling-stack hedging is run for the terminal date, but not for the entire agreed delivery of goods but only for a portion γ thereof. Under this strategy, one calculates the maximum variance and then determines γ such that this maximum variance becomes minimal.

First of all, this means that $g(s) = \gamma \cdot (1 - s)$ for all s. Thus we get

$$\sigma_t^2 = F_g(t) = \int_0^t (t - s - \gamma \cdot (1 - s))^2 ds = \frac{(\gamma - t + t(1-\gamma))^3 - (\gamma - t)^3}{3 \cdot (1 - \gamma)}.$$

An elementary analysis of the last expression gives us the optimal value for γ if $\gamma \approx 0.62996$, and for this choice of γ, we obtain for F_g the value 0.0456.

The behaviour of σ_t^2 is also shown in Fig. 3.97.

Thus, "optimal horizon" and "optimal fraction" already show significantly better results than "rolling stack". But what is the optimal strategy h?

3.11.3 The Solution to the Optimal Hedging Problem

The optimal solution h looks like this:

$$h(s) = \begin{cases} 3t_0 - s & \text{for } s \in [0, t_0) \\ e^{-\frac{y}{2}} \cdot \cos \frac{\sqrt{3}y}{2} - s & \text{for } s \in [t_0, \frac{1}{2}) \\ 1 - s & \text{for } s \in [\frac{1}{2}, 1) \end{cases}$$

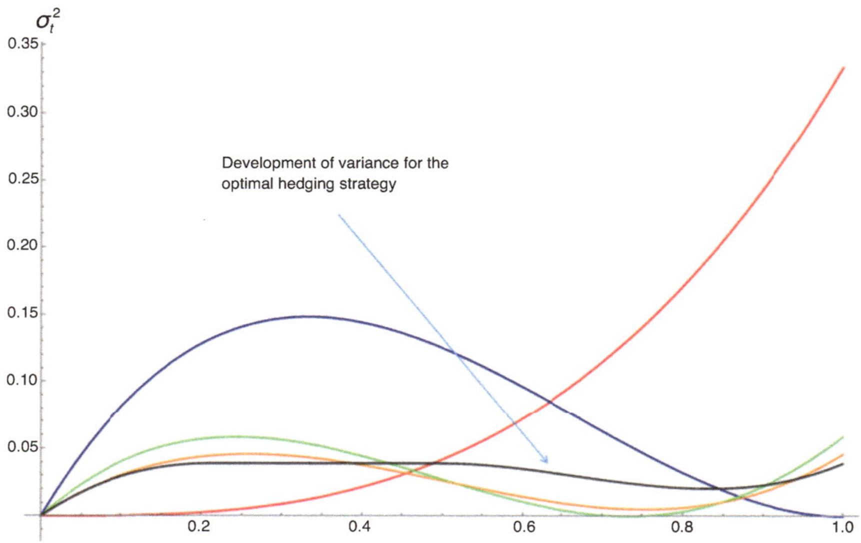

Fig. 3.98 Variance of the total profit process under otimal hedging strategy (black) compared to the variance using "no hedge" (red), "rolling-stack hedge" (blue), "optimal horizon" (green), and "optimal fraction" (orange)

where y for given s is defined by the equation $s = -\frac{1}{6}e^{-\frac{y}{2}}\left(\sqrt{3}\sin\frac{\sqrt{3}y}{2} - \right.$
$\left. -3\cos\frac{\sqrt{3}y}{2}\right)$ and $t_0 = \frac{e^{-\frac{\pi}{6\sqrt{3}}}}{2\sqrt{3}}$. The optimal value F_h, that is, the smallest possible maximum variance, is $F_h = \frac{e^{-\frac{\pi}{6\sqrt{3}}}}{6\sqrt{3}} = 0.0388532\ldots$.

In Fig. 3.98 we now also see how the variance of the total profit process under this optimal strategy compares with the variance under the other hedging strategies. For better visibility, it is shown again in Fig. 3.99 as a separate graph. Maximum variance under this strategy is assumed to occur at the terminal date 1 and in a period between the times t_0 and $\frac{1}{2}$.

The representation clearly shows that the result of Glasserman's optimal fraction strategy is already relatively close to the result of the optimal strategy.

Figure 3.100 shows the four hedging strategies again, i.e. the functions $g(s)$ for rolling stack, optimal horizon, and optimal fraction as well as the function $h(s)$ for the optimal strategy.

3.11.4 The Proof Sketch

We intend to and can only give a rough sketch of the proof here. This will give the reader insights into the proof ideas. The mathematically exact proof that such

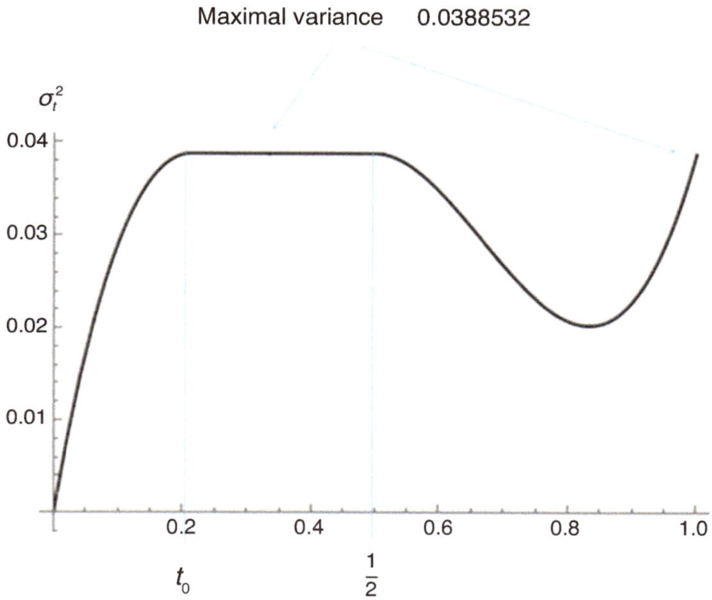

Fig. 3.99 Variance of the total profit process under optimal hedging strategy and times and values of maximum variance

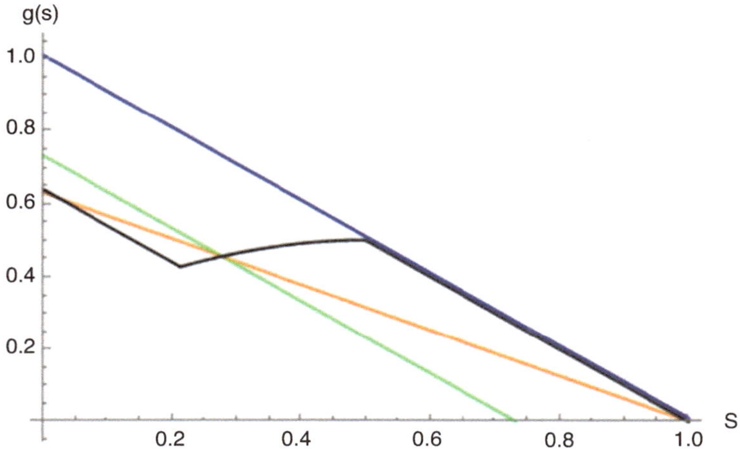

Fig. 3.100 The hedging strategies $g(s)$ for rolling stack (blue), optimal horizon (green), optimal fraction (orange), and the optimal strategy $h(s)$ (black)

an optimal strategy even exists is quite delicate and challenging and requires tools from various areas of mathematics, for example, from functional analysis.

So, what we have to do is determine the function g such that $\max_{0\le t\le 1}\int_0^t (t - s - g(s))^2 ds$ becomes minimal. We reformulate the problem slightly by combining s and $g(s)$ from the above expression into a function $h(s) := s + g(s)$ and try to find a function h such that $\max_{0\le t\le 1} F_h(t) := \max_{0\le t\le 1}\int_0^t (t - h(s))^2 ds$ becomes minimal.

For this we proceed in several steps, which we will merely outline below. The detailed proof is given in article [9] and can be read up on there.

After having shown, in the first step, using techniques of functional analysis, that a unique optimal strategy h must exist and that this optimal strategy h must be continuous, we proceed with:

Step 1

Here we show that the optimal h can only attain values between 0 and 1.

This is relatively obvious. After all, if h were to attain values smaller than 0 or larger than 1 at any points, then one could replace h, as indicated in the example in Fig. 3.101, by a function h_1 that only attains values between 0 and 1, for which we would then have $(t - h_1(s))^2 \le (t - h(s))^2$ for all s and $(t - h_1(s))^2 < (t - h(s))^2$ for at least one s and therefore $F_{h_1}(t) \le F_h(t)$ for all t, which would contradict the optimality of h or at least the uniqueness of the optimal solution h.

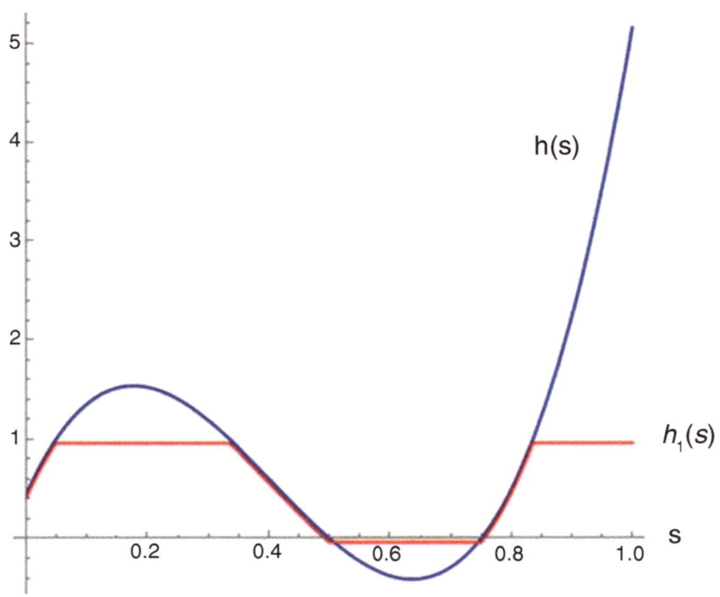

Fig. 3.101 Examples of functions h and h_1 in Step 1 of the proof

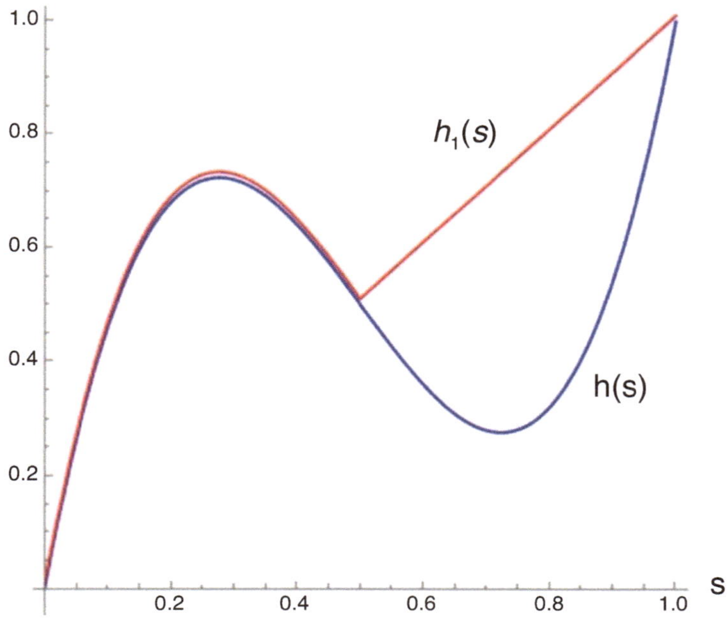

Fig. 3.102 Examples of functions h and h_1 in Step 2 of the proof

Step 2
Here we show that the optimal h satisfies $h(s) \geq s$ everywhere.

The proof of this step follows the same principle as that of the first step (which, by the way, goes for most of the following steps). After all, if $h(s)$ were to attain values smaller than s at any points, then one could replace h, as indicated in the example in Fig. 3.102, by a function h_1 for which $h_1(s)$ always attains values greater than or equal to s. Since in $\int_0^t (t - h(s))^2 ds$, the integration parameter s is always less than or equal to t, we would again have $(t - h_1(s))^2 < (t - h(s))^2$ for all s and $(t - h_1(s))^2 < (t - h(s))^2$ for at least one s and therefore $F_{h_1}(t) \leq F_h(t)$ for all t, which would contradict the optimality of h or at least the uniqueness of the optimal solution h.

Step 3
Here we show that the optimal h is monotonically increasing.

The beginning is the same again: If the optimal h were not monotonically increasing, then on sections where h is not monotonically increasing, we could change h as sketched in Fig. 3.103 into a function h_1. On the section where h was changed into h_1, h_1 has a constant value, namely, precisely the average value of h on the changed section. With a somewhat greater effort than in the first and in the

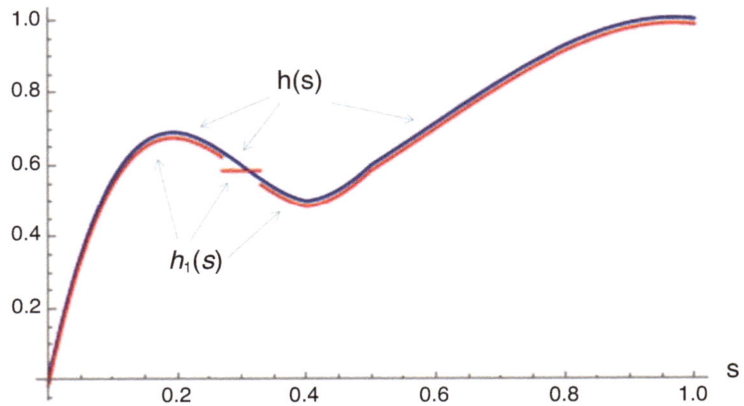

Fig. 3.103 Examples of functions h (blue) and h_1 (red) in Step 3 of the proof

second step, it can again be shown that $F_{h_1}(t) \leq F_h(t)$ holds for all t, which would contradict the optimality of h or at least the uniqueness of the optimal solution h.

Step 4

Let h again be the optimal function we are looking for.

By H we denote the set of values t in $[0, 1]$ in which F_h attains the maximum. This set H might, for example, be as shown in Fig. 3.104.

In Step 4 we will show that the value $t = 1$ is an element of the set H.

Interestingly, this step is the most difficult one in the proof. The procedure is again identical in principle to that of the above proof steps: Suppose that 1 is not an element of H, then h could be modified just slightly to a function h_1 so that $F_{h_1}(t) \leq F_h(t)$ then holds for all t, which would contradict the optimality of h or at least the uniqueness of the optimal solution h. We cannot go into the details of the further procedure here.

Step 5

Using the same approach in principle, we show that H must contain at least two elements.

Step 6

Here we prove yet another—highly essential—property of the set H: If the complement of the set H contains contiguous intervals (see, e.g. Fig. 3.104, where the green (open) intervals on the time axis represent the complement of H), then the optimal h is constant on these intervals.

Again, the process is as follows: If h were not constant on any one of these intervals, then one could replace h by a function h_1 ....

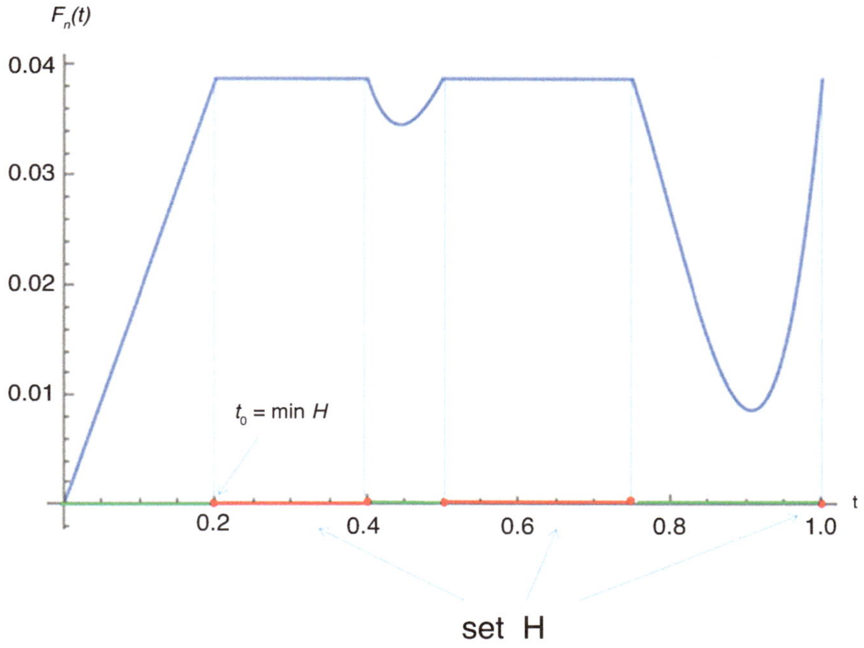

Fig. 3.104 Example of a set H as defined in Step 4

Step 7

In this step, we prove the (left-hand/right-hand) differentiability of the optimal function h and the associated function F_h, which is needed for the following final steps.

Step 8

Let t_0 be the minimum of the set H, i.e. the time at which the maximum variance is assumed for the first time (see example in Fig. 3.104).

In this step, we show that $h(s) = 3t_0$ for all $s \in [0, t_0]$.

We already know from Step 6 that h must be constant on $[0, t_0]$. Because of the continuity of h, it follows that h must also be constant on the closed interval $[0, t_0]$. Thus it suffices to show that $h(t_0) = 3t_0$.

F_h has a maximum in t_0. Therefore, $(F_h(t_0))' = 0$. Thus we differentiate $F_h(t) = \int_0^t (t - h(s))^2 ds$ with respect to t and set $t = t_0$.

For this differentiation we need the following differentiation rule for functions represented by an integral

$$\frac{d}{dt} \int_0^t G(s,t)ds = G(t,t) + \int_0^t \frac{dG(s,t)}{dt} ds$$

and we obtain $(F_h(t))' = (h(t)-t)^2 + 2\cdot\int_0^t (t-h(s))ds$, thus (since $h(s)$ on $[0, t_0]$ is always equal to $h(t_0)$):

$$0 = (F_h(t_0))' =$$

$$= (h(t_0)-t_0)^2 + 2\cdot\int_0^{t_0}(t_0 - h(t_0))ds = (h(t_0)-t_0)^2 + 2t_0\cdot(t_0 - h(t_0)).$$

It is then easily understood that $(t_0 - h(t_0))$ cannot be 0, and so it follows from the above equation that $h(t_0) = 3t_0$.

Step 9
In the same way, we show that for the optimal h we have $h(t) = 1$ for all $t \in [\frac{1}{2}, 1]$.

Final Step 10
What we know so far is that the optimal h must be constantly equal to $3t_0$ on $[0, t_0]$ and constantly equal to 1 on $[\frac{1}{2}, 1]$. What we do not yet know is the form of h between t_0 and $\frac{1}{2}$. This is what we will determine now:
In the first step, we show that F_h must be constant on that interval $[t_0, \frac{1}{2}]$. So $(F_h(t))'' = 0$ there. Using the above rule for differentiating functions given by integrals, we determine the second derivative $(F_h(t))''$ and obtain $0 = (F_h(t))'' = 2\cdot(t-h(t))\cdot(1-h'(t)) + 4t - 2\cdot h(t)$, which leads to the differential equation $h'(t) = \frac{3t-2h(t)}{t-h(t)}$.

This differential equation can be solved explicitly, and we get a solution for h which still contains the unknown value t_0. This value can be determined based on the fact that h must be continuous and must come together in $t = t_0$ resp. in $t = \frac{1}{2}$ with the already known parts of the function h. This then completes the proof.

Although the above presentation of the proof idea is a far cry from the actual complete proof, it has hopefully provided at least a glimpse into the complexity, creativity, beauty, and depth of certain mathematical thought processes and proofs.

3.12 Case Study XII: Computing the Payout Ratios for a Financial Market Game

3.12.1 Description of the Game

The following question (among others) arose in the context of a planned cooperation with a developer of a financial market game:

As part of the game, players were to be given the opportunity to bet on the future performance of stock prices.

In the following, we will explain (in slightly simplified terms) the mechanism of the game in its basic version (!), using the example of bets on the performance of the SPX.

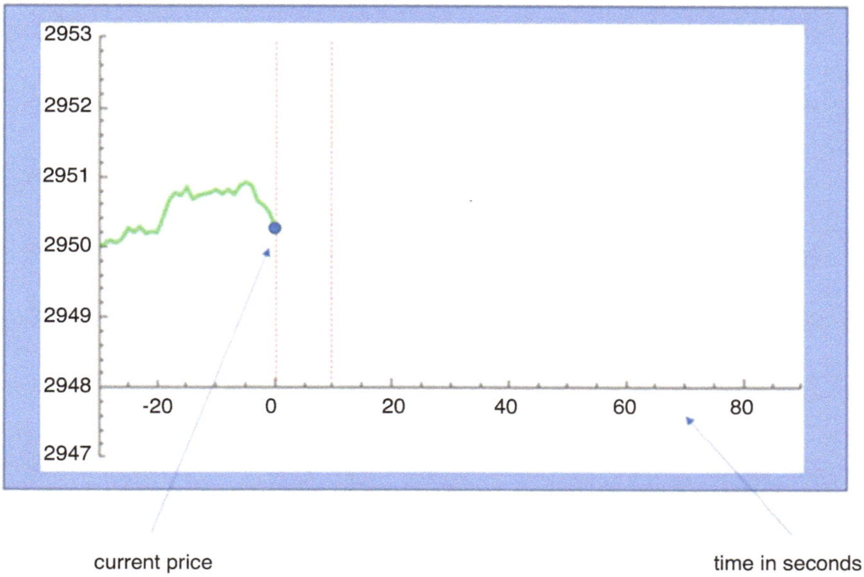

Fig. 3.105 Start of the game

Once signed up, a player gets an account that is initially filled with a (virtual) amount of 10,000 dollars.

After pressing a button labelled "New game", the screen will display the current movements of the SPX in real time based on tick data of the last half minute.

The current moment in time is 0 and stays fixed (on the horizontal x-axis). The last—thus current—price always remains fixed at time zero, and so the price curve "creeps" slowly towards the left side of the graph (see Fig. 3.105). On request, the player can also view the performance history.

Using a mouse or touch screen, the player can now open an axis-parallel rectangle B for a period from 10 to 90 seconds into the future (i.e. to the right of the right vertical line in Fig. 3.105).

Upon opening such a rectangle, a field appears with a "payout ratio", for example, "payout ratio = 1:1.8".

This payout ratio means: If one bets at this moment that within the next 90 seconds, the index price will touch this rectangle, then one would receive a payout of 1.80 dollars for the stake of 1 dollar (thus pocketing a profit of 0.80 dollars). If the index price does not touch the rectangle, then the invested dollar is lost.

Of course, the ratio changes continuously as the current price of the SPX changes, and it also changes whenever the rectangle is changed.

The player can now define the wager (e.g. 500 dollars).

As soon as the player then presses the button "Place bet", the bet is activated, and the SPX now moves in real time from second 0 to 90.

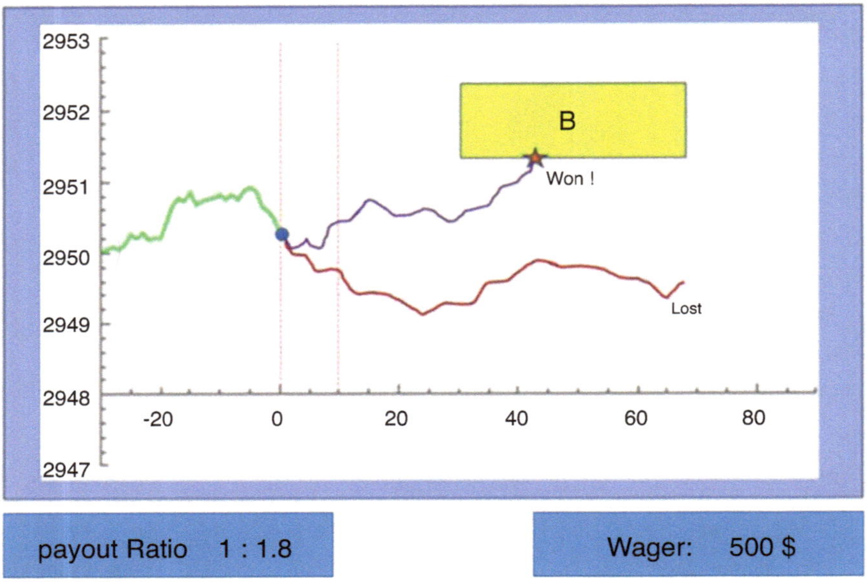

Fig. 3.106 End of the game

As soon as the price curve touches the rectangle for the first time, or has passed by the time range covered by the rectangle without touching the rectangle, the betting result is displayed (see Fig. 3.106).

In the case of the purple curve, the player pockets a profit of 400 dollars (for the investment of 500 dollars, the payout is $1.8 \times 500 = 900$ dollars). In the case of red curve, the player loses the invested amount of 500 dollars. The player's account is settled and a new game can be started.

The question that we were asked at the time was how to calculate the ratio correctly (and quickly).

3.12.2 Computation of the Payout Ratio

The starting point for our analysis was the assumption that the SPX follows a (continuous) geometric Brownian motion. (The assumption of continuity is to be questioned and we will do so later!) Thus, if we denote the price move of the SPX by $S(t)$, we get $S(t) = S(0) \cdot e^{\mu \cdot t + \sigma \cdot W_t}$ with a Brownian motion W_t.

The next assumption was that for the value σ, we would simply use the current value of the VIX. Any attempt to reliably estimate a trend μ for the SPX price move over the next 90 seconds will in all likelihood fail, so a risk-neutral move was assumed, i.e. a trend μ of the form $\mu = r - \frac{\sigma^2}{2}$ with a daily short-term risk-free dollar interest rate r.

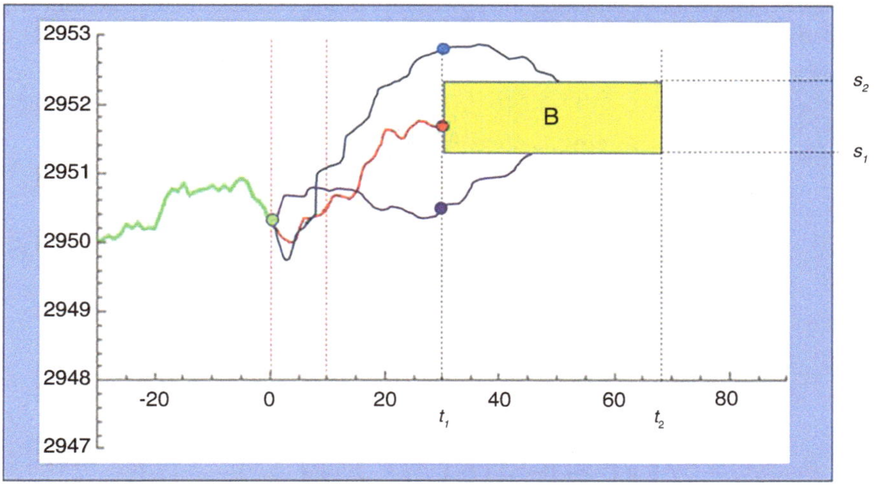

Fig. 3.107 Three cases of touching the rectangle

So, to determine the payout ratio, we had to compute the probability P with which the price $S(t)$ would touch the rectangle.

If this probability is known, then the ratio 1:x is to be determined such that the game is a fair game for both parties, i.e. that the expected profit is zero. This gives the equation $P \cdot (x-1) + (1-P) \cdot (-1) = 0$ for the ratio, and rearranging, we get $x = \frac{1}{P}$.

We now proceed to calculate the probability P.

The SPX price can touch the rectangle B in three different ways:

Case 1: The price touches the rectangle on its left side (red curve in Fig. 3.107).

Case 2: The price touches the rectangle at its upper side for the first time (blue curve in Fig. 3.107).

Case 3: The price touches the rectangle at its lower side for the first time (purple curve in Fig. 3.107).

We denote the probability of case i occurring by P_i. Then obviously (since the three cases are mutually exclusive) $P = P_1 + P_2 + P_3$.

3.12.3 Computation of Probability P_1

Probability P_1 is very easy to compute. Using the notations t_1, t_2, s_1, and s_2 illustrated in Fig. 3.107, we have (with a standard normally distributed random variable w and with W denoting the probability of the event defined in the following parenthesis):

$$\begin{aligned}
P_1 &= W\left(s_1 \leq S(t_1) \leq s_2\right) = \\
&= W\left(s_1 \leq S(0) \cdot e^{\mu \cdot t_1 + \sigma \cdot W_{t_1}} \leq s_2\right) = \\
&= W\left(s_1 \leq S(0) \cdot e^{\mu \cdot t_1 + \sigma \cdot \sqrt{t_1} \cdot w} \leq s_2\right) = \\
&= W\left(\log\left(\frac{s_1}{S(0)}\right) \leq \mu \cdot t_1 + \sigma \cdot \sqrt{t_1} \cdot w \leq \log\left(\frac{s_2}{S(0)}\right)\right) = \\
&= W\left(\frac{\log\left(\frac{s_1}{S(0)}\right) - \mu \cdot t_1}{\sigma \cdot \sqrt{t_1}} \leq w \leq \frac{\log\left(\frac{s_2}{S(0)}\right) - \mu \cdot t_1}{\sigma \cdot \sqrt{t_1}}\right) = \\
&= \mathcal{N}\left(\frac{\log\left(\frac{s_2}{S(0)}\right) - \mu \cdot t_1}{\sigma \cdot \sqrt{t_1}}\right) - \mathcal{N}\left(\frac{\log\left(\frac{s_1}{S(0)}\right) - \mu \cdot t_1}{\sigma \cdot \sqrt{t_1}}\right),
\end{aligned}$$

with $\mathcal{N}$ again denoting the distribution function of the standard normal distribution (and setting $\mu = r - \frac{\sigma^2}{2}$).

3.12.4 Computation of Probabilities P_2 and P_3

To calculate the probabilities P_1 and P_2, we will use a result that will provide us with information on the distribution of the minimum and the maximum of a Brownian motion with drift, as formulated in Chap. 1, as follows:

Let $B_{\mu,\sigma}(t) := \mu \cdot t + \sigma \cdot B_t$ *for* $0 \leq t \leq T$ *denote a Brownian motion with drift on* $[0, T]$ *and* $M_{\mu,\sigma}(T) := \max(B_{\mu,\sigma}(t)\,|0 \leq t \leq T)$ *resp.* $m_{\mu,\sigma}(T) := \min(B_{\mu,\sigma}(t)\,|0 \leq t \leq T)$.

(a) Let $a > 0$. *Then we get*

$$W\left(M_{\mu,\sigma}(T) \leq a\right) = \mathcal{N}\left(\frac{a - \mu \cdot T}{\sigma\sqrt{T}}\right) - e^{2 \cdot \frac{\mu \cdot a}{\sigma^2}} \cdot \mathcal{N}\left(-\frac{a + \mu \cdot T}{\sigma\sqrt{T}}\right)$$

(b) Let $a < 0$. *Then we get*

$$W\left(m_{\mu,\sigma}(T) \leq a\right) = \mathcal{N}\left(\frac{a - \mu \cdot T}{\sigma\sqrt{T}}\right) + e^{2 \cdot \frac{\mu \cdot a}{\sigma^2}} \cdot \mathcal{N}\left(\frac{a + \mu \cdot T}{\sigma\sqrt{T}}\right)$$

We start by determining P_2 (P_3 is then determined in analogy to that). In order for case 2 to occur:

1. $S(t_1)$ needs to have a value z greater than s_2.
2. $\min_{t_1<t<t_2} S(t)$ needs to be less than or equal to s_2.

Since $S(t_1) = S(0) \cdot e^{\mu \cdot t_1 + \sigma \cdot \sqrt{t_1} \cdot w}$, $S(t_1) = z$ is equivalent to $w = \frac{\log\left(\frac{z}{S(0)}\right) - \mu \cdot t_1}{\sigma \cdot \sqrt{t_1}}$ and $S(t_1) > s_2$ is equivalent to $w > \frac{\log\left(\frac{s_2}{S(0)}\right) - \mu \cdot t_1}{\sigma \cdot \sqrt{t_1}}$.
Therefore,

$$P_2 = \int_L^{\infty} \frac{1}{\sqrt{2\pi}} \cdot e^{-\frac{w^2}{2}}.$$

$$\cdot W\left(\min_{t_1<t<t_2} S(t) < s_2 \,\middle|\, S(t_1) = S(0) \cdot e^{\mu \cdot t_1 + \sigma \cdot \sqrt{t_1} \cdot w}\right) dw,$$

where $L = \frac{\log\left(\frac{s_2}{S(0)}\right) - \mu \cdot t_1}{\sigma \cdot \sqrt{t_1}}$.

To illustrate what we did here in simpler terms: We "totalled up" all possible values of w, weighting the respective value w by its "probability" (its density!) and multiplying this "probability of the occurrence of w" by the probability that, if w does occur, we would indeed have $\min_{t_1<t<t_2} S(t) < s_2$.

All that is left for us to do now is determine the probability

$$W\left(\min_{t_1<t<t_2} S(t) < s_2 \,\middle|\, S(t_1) = S(0) \cdot e^{\mu \cdot t_1 + \sigma \cdot \sqrt{t_1} \cdot w}\right).$$

Since for a t between t_1 and t_2 (with a Brownian motion B), we have $S(t) = S(t_1) \cdot e^{\mu(t-t_1)+\sigma \cdot B_{t-t_1}}$, it follows that $\min_{t_1<t<t_2} S(t) < s_2$ is equivalent to

$\mu \cdot (t - t_1) + \sigma \cdot B_{t-t_1} < \log\left(\frac{s_2}{S(t_1)}\right)$ for a t between t_1 and t_2, thus

$\mu \cdot \tau + \sigma \cdot B_\tau < \log\left(\frac{s_2}{S(t_1)}\right)$ for a τ between 0 and $t_2 - t_1$.

The last relation, in turn, is equivalent to $m_{\mu,\sigma}(t_2 - t_1) < \log\left(\frac{s_2}{S(t_1)}\right)$. And this gives us

$$W\left(\min_{t_1<t<t_2} S(t) < s_2 \,\middle|\, S(t_1) = S(0) \cdot e^{\mu \cdot t_1 + \sigma \cdot \sqrt{t_1} \cdot w}\right) =$$

$$= W\left(m_{\mu,\sigma}(t_2 - t_1) < \log\left(\frac{s_2}{S(t_1)}\right)\right).$$

We calculate the last probability using the above result on the distribution of minima of a Brownian motion with drift. We get

$$W\left(m_{\mu,\sigma\cdot\sqrt{\tau}}(t_2-t_1) < \log\left(\frac{s_2}{S(t_1)}\right)\right) =$$

$$= \mathcal{N}\left(\frac{\log\left(\frac{s_2}{S(t_1)}\right) - \mu\cdot(t_2-t_1)}{\sigma\sqrt{t_2-t_1}}\right) +$$

$$+\ e^{2\cdot\frac{\mu\cdot\log\left(\frac{s_2}{S(t_1)}\right)}{\sigma^2}} \cdot \mathcal{N}\left(\frac{\log\left(\frac{s_2}{S(t_1)}\right) + \mu\cdot(t_2-t_1)}{\sigma\sqrt{t_2-t_1}}\right).$$

For $S(t_1)$ we now have to set the relation $S(t_1) = S(0)\cdot e^{\mu\cdot t_1+\sigma\cdot\sqrt{t_1}\cdot w} := z(w)$, and we get

$$W\left(m_{\mu,\sigma\cdot\sqrt{\tau}}(t_2-t_1) < \log\left(\frac{s_2}{S(t_1)}\right)\right) =$$

$$= \mathcal{N}\left(\frac{\log\left(\frac{s_2}{z(w)}\right) - \mu\cdot(t_2-t_1)}{\sigma\sqrt{t_2-t_1}}\right) +$$

$$+\ e^{2\cdot\frac{\mu\cdot\log\left(\frac{s_2}{z(w)}\right)}{\sigma^2}} \cdot \mathcal{N}\left(\frac{\log\left(\frac{s_2}{z(w)}\right) + \mu\cdot(t_2-t_1)}{\sigma\cdot\sqrt{t_2-t_1}}\right).$$

Substituting in the above integral representation for P_2, we obtain

$$P_2 = \int_L^{\infty} \frac{1}{\sqrt{2\pi}}\cdot e^{-\frac{w^2}{2}}\cdot$$

$$\cdot\left(\mathcal{N}\left(\frac{\log\left(\frac{s_2}{z(w)}\right) - \mu\cdot(t_2-t_1)}{\sigma\sqrt{t_2-t_1}}\right) +\right.$$

$$\left. +\, e^{2\cdot\frac{\mu\cdot\log\left(\frac{s_2}{z(w)}\right)}{\sigma^2}} \cdot \mathcal{N}\left(\frac{\log\left(\frac{s_2}{z(w)}\right) + \mu\cdot(t_2-t_1)}{\sigma\sqrt{t_2-t_1}}\right)\right) dw,$$

with $z(w) = S(0)\cdot e^{\mu\cdot t_1+\sigma\cdot\sqrt{t_1}\cdot w}$ and $\mu = r - \frac{\sigma^2}{2}$.

For actual application purposes, we do not attempt to bring this integral into a more convenient form; instead we let mathematical software solve the integral approximately by means of numerical methods. This is easily achieved in "real time" with sufficient precision.

Proceeding analogously and using the formula for the distribution of the maximum of a Brownian motion with drift, it follows for the third probability

$$P_3 = \int_{-\infty}^{\frac{\log\left(\frac{s_2}{S(0)}\right)-\mu\cdot t_1}{\sigma\cdot\sqrt{t_1}}} \frac{1}{\sqrt{2\pi}} \cdot e^{-\frac{w^2}{2}} \cdot$$

$$\cdot\left(1-\mathcal{N}\left(\frac{\log\left(\frac{s_1}{z(w)}\right)-\mu\cdot(t_2-t_1)}{\sigma\sqrt{t_2-t_1}}\right)+\right.$$

$$\left.+\, e^{2\cdot\frac{\mu\cdot\log\left(\frac{s_1}{z(w)}\right)}{\sigma^2}}\cdot\mathcal{N}\left(-\frac{\log\left(\frac{s_1}{z(w)}\right)+\mu\cdot(t_2-t_1)}{\sigma\sqrt{t_2-t_1}}\right)\right)dw.$$

3.12.5 Testing the Results Using Monte Carlo Simulation and Concluding Remarks

To confirm the results, we performed a test using Monte Carlo simulation. For the example illustrated in Fig. 3.108, the calculated probability of touching the rectangle was $P = 0.4865$, and the payout ratio was therefore 1:2.055.

This probability P resulted from the sum of the values $P_1 = 0.1077$, $P_2 = 0.0967$, and $P_3 = 0.2821$.

For testing purposes, we simulated 10,000 continuations of the initial path on the basis of the geometric Brownian motion with the parameters selected above and counted the number of "hits" or, more precisely, the number of paths for which

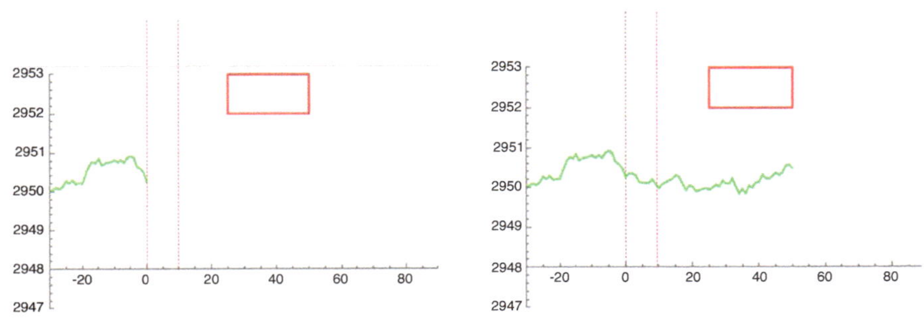

Fig. 3.108 Calculation and test example

case 1, case 2, or case 3 had occurred. We thus obtained empirical approximations $\widetilde{P}_1$, $\widetilde{P}_2$, and $\widetilde{P}_3$ for P_1, P_2, and P_3.

It turned out that: If the simulation steps were chosen with a length of 1 second, very good approximations were obtained with $\widetilde{P}_1$ for P_1, while the values $\widetilde{P}_2$ and $\widetilde{P}_3$ were noticeably lower than the values P_2 and P_3, respectively.

Only when choosing the simulation steps in tenths of seconds (or shorter) did $\widetilde{P}_2$ and $\widetilde{P}_3$ also give very good approximations for P_2 and P_3.

The reason for the discrepancy in the case of larger simulation steps is that the actual discrete simulated paths then deviate too much from the continuous paths for which the theoretical calculations were performed.

Yet for tick data in SPX, the resulting paths should be sufficiently close to the continuous paths so that ratios can be computed reliably on the basis of the theoretical results. (Otherwise, the theoretical results would have to be adapted accordingly.)

3.13 Case Study XIII: Trading Strategies Based on Hedging

Hedging provides a way to protect against losses when selling derivatives. However, as we will see in this section, it can also be used to develop strategies with which profits can be made when the financial market deviates from the frictionless market.

The following exposition is based on the article [10]. The analyses and the report were written by Bernhard Heinzelreiter.

3.13.1 Background

As we saw already in Volume II Section 1.17, the implied volatility of the S&P500 generally differs from the actually realized volatility. Analyses show that the implied volatility, denoted here by σ^i, is usually greater than the realized volatility, which we denote by σ in the following.

Furthermore, in Volume I Section 4.22, we discussed the dependence of call option prices on volatility. It was found that the fair price of a call option increases monotonically with the volatility used for calculation.

From that also follows that a higher volatility implies a higher price of the option.

With these two facts—that the implied volatility is frequently too high on the one hand and the connection between option prices and volatility on the other hand—we can conclude that the prices of call options on the S&P500 are usually too high, hence overpriced.

Thus, we have found that, in terms of arbitrage theory, call options often represent overpriced financial products, that is, they are not traded at their fair price.

Now we know that, if a product is not fairly priced, this creates arbitrage opportunities. This of course begs the question whether there might be a possibility to make a long-term systematic profit from overpriced call options.

3.13.2 An Initial Idea of a Trading Strategy

Let us consider a specific call option that is overpriced in the above sense. And let us also assume that we know the volatility σ that was actually realized during the option's life. In the following we denote the fair price of the option by C if the realized volatility is used for the volatility, and by C^i we denote the fair price calculated with the implied volatility. The price C^i is also the price at which the option is offered on the market.

3.13.2.1 The Strategy

An intuitive trading strategy would be to sell the overpriced product, i.e. to short the call option. Thus we get C^i, which, in terms of arbitrage theory, is too much for our product. However, this purchase does not yet represent a safe profit, since the short call position also can cause high losses.

So the question remains as to how we should proceed to secure a profit. Hedging could be the way to go.

We now hold a short call position. From Volume I Section 4.18, we know that a call option can be hedged. Since the realized volatility is σ and not σ^i, an initial capital of C is sufficient for hedging purposes, and we create and trade a replicating portfolio of the option.

It should be noted here that for hedging and especially for computing the delta, the realized volatility is used. For our entire portfolio, this then results in the following:

The initial value of the portfolio is 0 and is composed of:

- Short call option $-C^i$
- Replicating portfolio C
- Cash $C^i - C > 0$,

where the expression on the right represents the value at the current time. Upon expiry of the option, at a price of the underlying asset of S_T, the components of the portfolio will then have the following value:

- Short call option $-\max(S_T - K, 0)$
- Replicating portfolio $\max(S_T - K, 0)$
- Cash $e^{rT}(C^i - C)$.

In total, this results in a terminal value of $e^{rT}(C^i - C) \geq 0$. So, in theory, this strategy yields a safe profit at the option's expiration. The approach in this strategy can also be briefly summarized as follows:

1. Short the overpriced call option.
2. Hedge the short call option until expiration based on actual volatility.

3.13.2.2 Implementation in Reality

In reality, this strategy can only be implemented with certain limitations. First, hedging comes with transaction costs due to continued trading, and second, when trading the replicating portfolio, certain products may be tradeable only in contracts rather than individually. These limitations of the strategy represent deviations from the theory but become less relevant as the investment increases.

On the other hand, in reality there is a limit to how much capital is invested, and it is also in the investor's interest to keep it low. The strategy entails a certain capital commitment at all times. This capital commitment is due to the short call option position and the hedging.

Holding short call positions requires a certain margin. This margin is composed of a certain percentage of the strike price, and we assume 15% here and the current price of the option.

Now, if hedging is done with a replicating portfolio based on a direct investment in the S&P500, then that hedging requires a high capital investment. So, all in all, relative to the achievable profit, the capital requirement is very high. To keep the amount of tied-up capital low, hedging in this case is done with futures on the S&P500. This has two advantages: Firstly, futures have the price 0 and we only need a margin of about 5% of the strike price, and secondly, futures are settled daily, which allows us to own the value of the replicating portfolio in cash every day. Thus, the price of the short call option and the value of the replicating portfolio largely offset each other.

And so we can reduce the minimum investment of the strategy to about a quarter. The required investment I_t of the strategy at time t is thus essentially composed of

$$I_t = \underbrace{0.15 \cdot K + C_t^i}_{\text{margin of the option}} + \underbrace{0.05 \cdot S_t - C_t}_{\text{replicating portfolio}},$$

where the subindex t indicates the respective value at each time t.

Specifically, S_t represents the value of S&P500 at time t.

Thus, to execute the strategy, we need capital in the amount of

$$I = \max_{t \in [0,T]} I_t .$$

All of the above, real-trade constraints and conditions were taken into account in the following experiments, thus ensuring that the analyses are based on real-world conditions.

3.13.2.3 Estimating Realized Volatility

Of course, at time 0 of the strategy's execution, it is not yet certain what value the actual realized volatility will take in the future period in which the strategy is to be executed. To test the strategy, we may only use data that are known at that time, in particular historical price values. Here, we use the historical volatility at time 0 to estimate the realized future volatility. In terms of the best historical time span to choose for this strategy, the 3-month historical volatility calculated on a daily basis has proved to be a good choice. Shorter periods mean too high fluctuations in volatility and do not provide useful values for the future.

3.13.2.4 Executing the Strategy on an Example

To explain how the strategy can be executed in reality, we use an example to illustrate the procedure.

We choose a call option on 1 September 2017 with expiration 3 November 2017, that is, the option has a life of about 2 months until it expires. For the option's strike price, we base it on the value of the S&P500, which had a closing price of 2476.95 on 1 September 2017. The closest possible strike to this closing price is 2475. The option is offered at a price of 42.38 USD.

We estimate future realized volatility using historical data from the past 3 months. This estimate gives us 7.51%. The VIX at that time was 10.59. So there is a considerable difference. What is important, however, is the difference between the option price offered and the fair price estimated and calculated by us. Using the estimated volatility, the price is 26.48 USD, which represents a difference of 15.90 USD. This option is therefore overpriced if the actually realized volatility turns out to be 7.51%.

Once all the conditions for executing the strategy are in place, our first step is to short the option. However, as noted above, we need several option contracts to ensure sufficiently good hedging. Since we hedge with futures and their smallest tradeable contract size is 50, we execute the strategy with $50 \cdot 100 = 5000$ units of our option, i.e. with 50 contracts.

This gives us a total of $42.38 \cdot 5000 = 211{,}900$ USD. These options require a margin as a percentage of the strike and the current price, that is, approximately $(0.15 \cdot 2475 + 42.38) \cdot 5000 = 2{,}068{,}150$ USD.

Now, to assemble our hedging portfolio, we put up 211,900 USD and buy delta units of futures on the S&P500 per option. Since our hedging is based on estimated volatility, we need to use this estimated volatility to calculate the delta. Thus, the delta at this time is 0.51. This gives a total of $0.51 \cdot 5000 = 2550$ units of the futures. Since futures can only be traded in contracts, we have to round this number to a multiple of 50. Yet this is not necessary in this specific case. The strike price of the futures is approximately equal to the current price of the S&P500, so the required margin for the futures is approximately $0.05 \cdot 2476.55 \cdot 2550 \approx 315.760$ USD.

This gives a total margin of $2{,}068{,}150 + 315{,}760 = 2{,}383{,}910$ USD. We have 211,900 USD in cash, which means that we need additional capital of at least 2,172,010 USD.

The purchased futures are then settled on the second trading day. For exact hedging, we need to adjust the number of futures held. To do this, we calculate the new delta on that day. Depending on whether this delta is larger or smaller than that of the previous trading day, we have to either buy new futures or sell them again to obtain the desired number. The new margin is computed in exactly the same way as before, except that we have to use the current values of the options and the futures. The maximum margin that would have been required during the trading month would have been 2,557,630 USD.

This procedure is repeated until the option expires. Upon expiration, the replicating portfolio has a value of $109.85 \cdot 5000$ =549,250 USD. The SPX closed at 2587.84 on 3 November 2017. The call option therefore yielded a payoff of $112.84 \cdot 5000 = 564{,}200$ USD. Since we shorted the option, we have to pay that payoff. Furthermore, initially we received the difference of 15.90 per option, for a total of $15.90 \cdot 5000 = 79{,}500$ USD. Accrued interest is negligible over such a short period of time. The strategy's final value is therefore $549{,}250 + 79{,}500 - 564{,}200 = 64{,}550$ USD. To put this profit into perspective, we need to put it in relation to the maximum margin and thus get a return on investment of $64{,}550/2{,}557{,}630 \approx 2.52\%$. This corresponds to an annual return of 16.13%.

3.13.2.5 Analysing the Strategy Through Backtesting

We are now going to backtest the strategy described above. We choose the period from January 2017 to February 2019 for that purpose. The strategy is evaluated on one trading day per month. More specifically, the first trading day of each month is used for this purpose.

Actual volatility is then estimated using historical data from 3 months before that trading day up until that trading day, and implied volatility is calculated from the VIX. Once the trading day has been defined, we need to choose the options' strike price and expiration date.

For the strike prices of the call options, we select certain values within the range of the underlying asset's current value. The current price of the S&P500 is rounded to 25 for this purpose. This then is the base strike price K_0. Based on K_0 we determine all strike prices with $K = K_0 + dK$ with $dK \in \Delta_K = \{-100, -50, -25, 0, 25, 50, 100\}$. The contract periods of the options are selected such that they are as close as possible to 1, 2, or 3 months. This gives 21 experiments per month. The results are depicted in Fig. 3.109. The chart shows the VIX and the above-described 3-month historical volatility at the respective point in time. Each point represents an execution of the strategy with one option type. The corresponding y-value is the annualized return of the strategy relative to the minimum capital investment I. The various vertically stacked values each represent one configuration per month, i.e. different contract periods and strikes. The different colours represent contract periods of 1, 2, or 3 months.

The chart shows that the strategy essentially generates profits—at times substantially so—but occasionally also results in losses. In January 2018, the heaviest losses amount to about 70% of the investment. Most of the strategy's annualized returns up until 2018 are roughly between minus 5 to plus 10%. After that, the strategy mostly generates heavy losses. The minimum investment required is about 2 million USD.

3.13.2.6 Extension of the Strategy

In the previous section, we saw that the strategy does not always generate profits. So in reality *particularly* good conditions are needed to execute the strategy. In number terms, this means that C^i should deviate strongly from C, i.e. that the difference

Fig. 3.109 Analysis of the first strategy in the period from January 2017 to February 2019 without extension

$C^i - C$ should exceed a certain threshold. Thus, as an extension of the strategy, one could introduce a minimum threshold δ for this difference in order to decide whether to execute the strategy. That is, the strategy is executed if and only if $C^i - C \geq \delta$. For purposes of this article, we choose δ as 10% of the option price C^i. In addition, we take into account the maximum loss at every trading moment. If that loss falls below the value of 90% of the total investment, the strategy is aborted to reduce the maximum loss. Historically, the strategy has performed better with this extension. As can be seen in Fig. 3.110, some cases in which heavy losses would have been made are avoided due to this strategy. Yet there are still some cases with heavy losses, however.

It has been shown that with about 3 months to expiration, the strategy generates fewer losses and, above all, less extreme losses. This is illustrated in Fig. 3.111. Yet even in this case, there are occasional losses of about 10% of the investment. We now also look at various strike price ranges to see if there is a certain correlation between the profit/loss and the chosen strike price. For this, we look at strike prices that are close to the money and strike prices that are 50 points above and 50 points below the current price of the S&P500. These graphs are depicted in Figs. 3.111, 3.112, 3.113 and 3.114. We see that losses in these three cases essentially occur when the

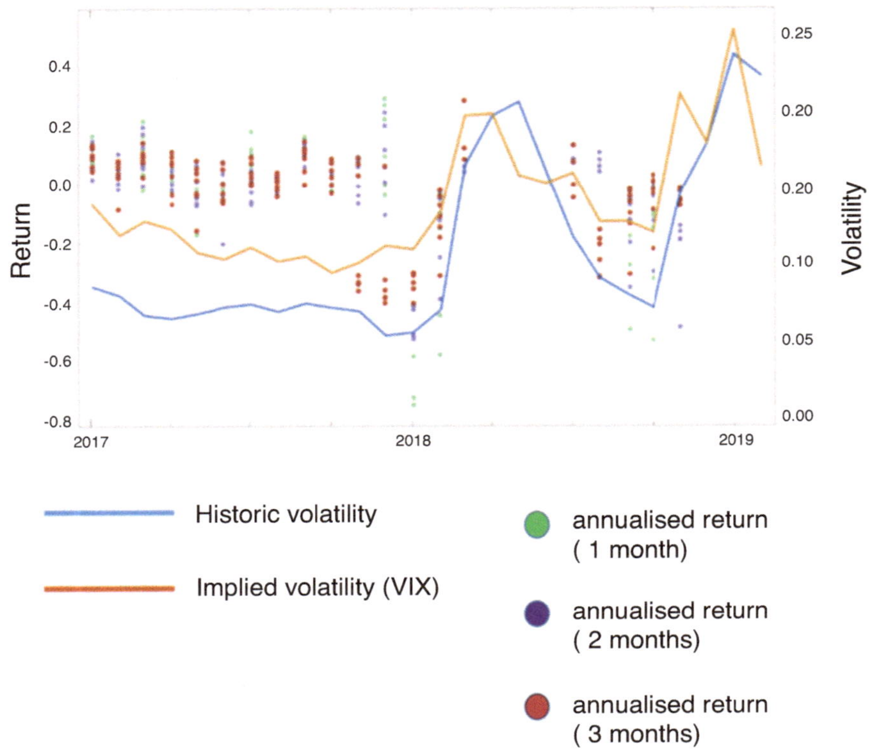

Fig. 3.110 Analysis of the first strategy in the period from January 2017 to February 2019 with extension

chosen strike price is about 50 points above the current price of the S&P500. For the other two cases, the annual returns are mostly between 3 and 10%. The exact annual returns are listed in Table 3.27.

3.13.3 A Second Variant of the Strategy

A major drawback of the strategy presented above is that it is necessary to estimate in each case how high the actually realized volatility is. A possible estimate of this volatility has been presented and used, but this approach may not be satisfactory in certain cases. This problem is eliminated in a modified second strategy, which is exactly the same as the first one, except that hedging is not done with the realized volatility but with the implied volatility. It can be shown that in theory, the profit at expiration is still always non-negative if the implied volatility is greater than the actually realized volatility. For a detailed profit analysis, we refer the reader to [10].

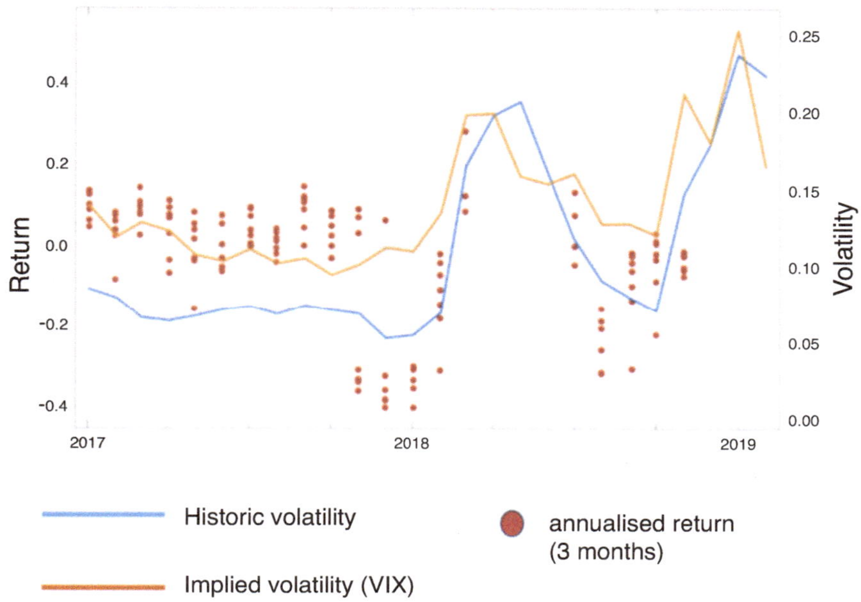

Fig. 3.111 Analysis of the first strategy in the period from January 2017 to February 2019 with about 3 months to expiration, with extension

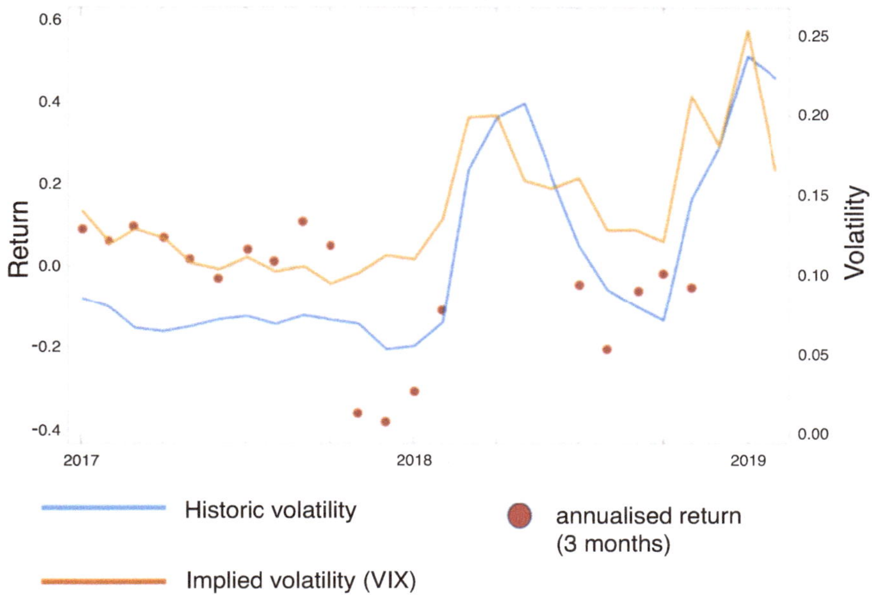

Fig. 3.112 Analysis of the first strategy in the period from January 2017 to February 2019 with about 3 months to expiration and strike price close to the S&P500 plus 50 points, with extension

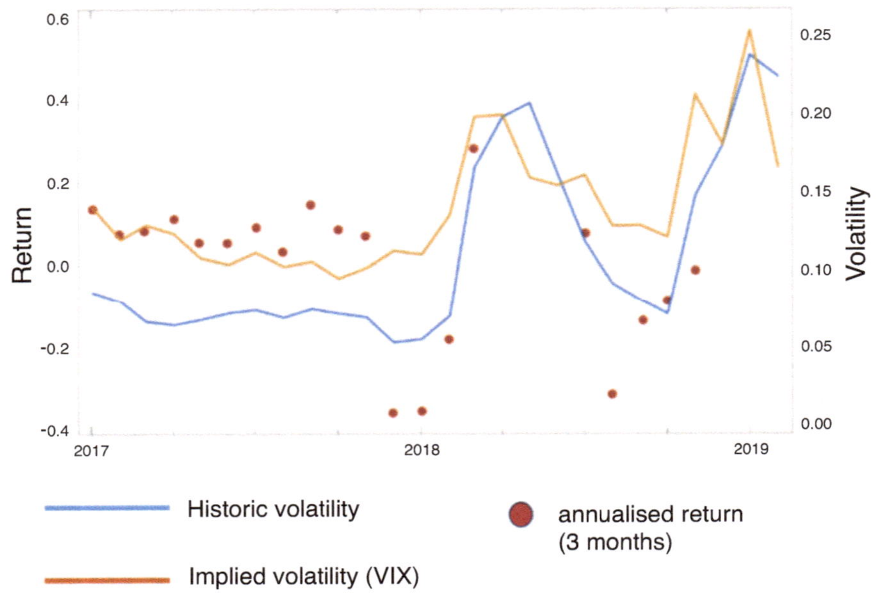

Fig. 3.113 Analysis of the first strategy in the period from January 2017 to February 2019 with about 3 months to expiration and strike price close to the S&P500 minus 50 points, with extension

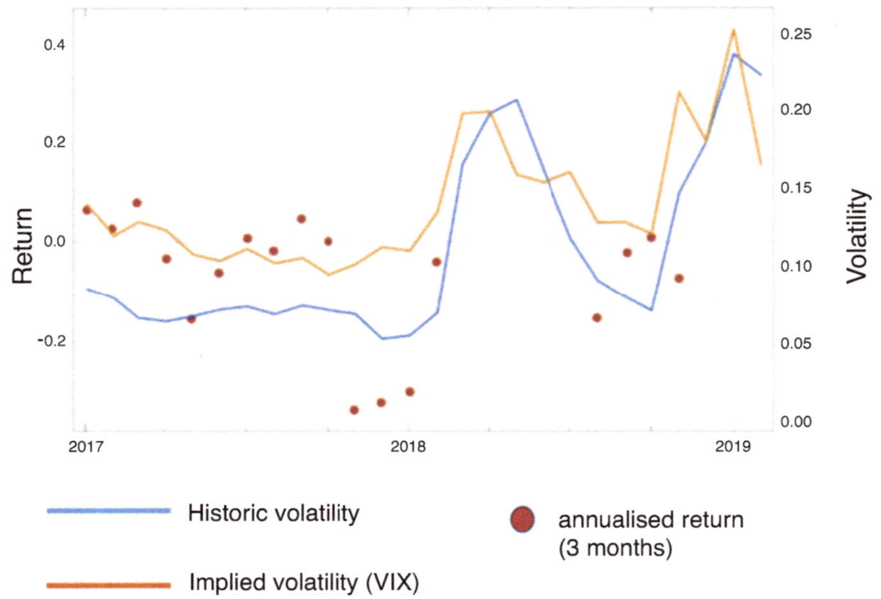

Fig. 3.114 Analysis of the first strategy in the period from January 2017 to February 2019 with about 3 months to expiration and strike price close to the S&P500, with extension

Table 3.27 Annual returns of the first strategy in the period from January 2017 to February 2019 with extension for various strike prices and about 3 months to expiration

Month	Strike near SPX price	Strike near SPX price + 50	Strike near SPX price − 50
2017			
January	9.13%	6.49%	13.77%
February	6.25%	2.68%	7.69%
March	9.82%	7.93%	8.43%
April	7.05%	−3.35%	11.38%
May	1.85%	−15.52%	5.64%
June	−2.86%	−6.19%	5.55%
July	4.14%	0.72%	9.32%
August	1.30%	−1.80%	3.47%
September	10.91%	4.65%	14.76%
October	5.07%	0.13%	8.77%
November	−36.02%	−33.74%	7.21%
December	−38.13%	−32.28%	−35.79%
2018			
January	−30.72%	−30.11%	−35.36%
February	−10.79%	−4.01%	−17.98%
March	–	–	28.36%
July	−4.47%	–	7.77%
August	−20.40%	−15.38%	−31.34%
September	−5.95%	−2.22%	−13.44%
October	−1.72%	0.87%	−8.55%
November	−5.07%	−7.29%	−1.18%

3.13.3.1 Analysing the Strategy Through Backtesting

The same tests were performed on this strategy as on the first strategy, except that the extension is already included. The results are shown in Fig. 3.115. We see that the results of the second strategy are structurally similar to the first one. While prior to 2018, the higher returns are higher for this second strategy, the losses are in a similar range to the first strategy. It is noteworthy that in 2018, the first strategy generates almost exclusively profits, while the second strategy presents almost exclusively losses.

Again, like with the first strategy, we used different expiration periods and strike prices. And again it pays to choose longer expiration periods, as can be seen in Fig. 3.115. Figures 3.116, 3.117 and 3.118 again show the result for different strike prices. The respective annual returns are listed in Table 3.28.

For further experiments and analysis of the strategy please, refer to https://app.lsqf.org/book/option-hedging.

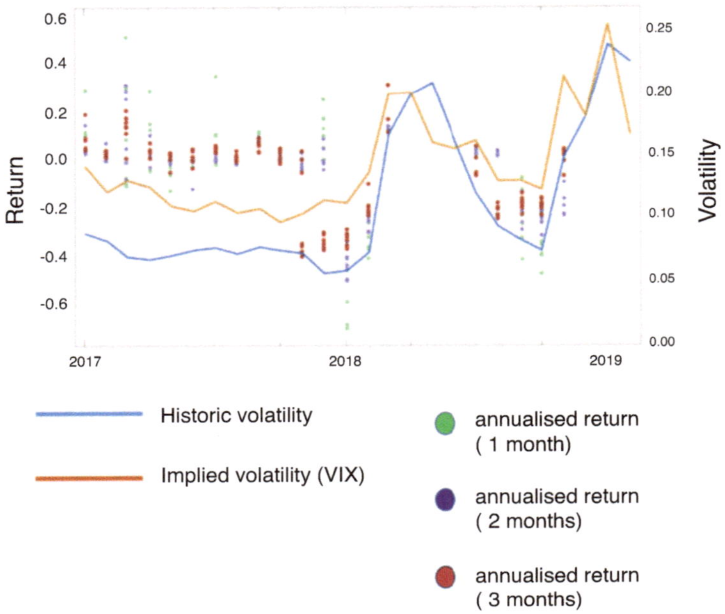

Fig. 3.115 Analysis of the second strategy in the period from January 2017 to February 2019 with extension

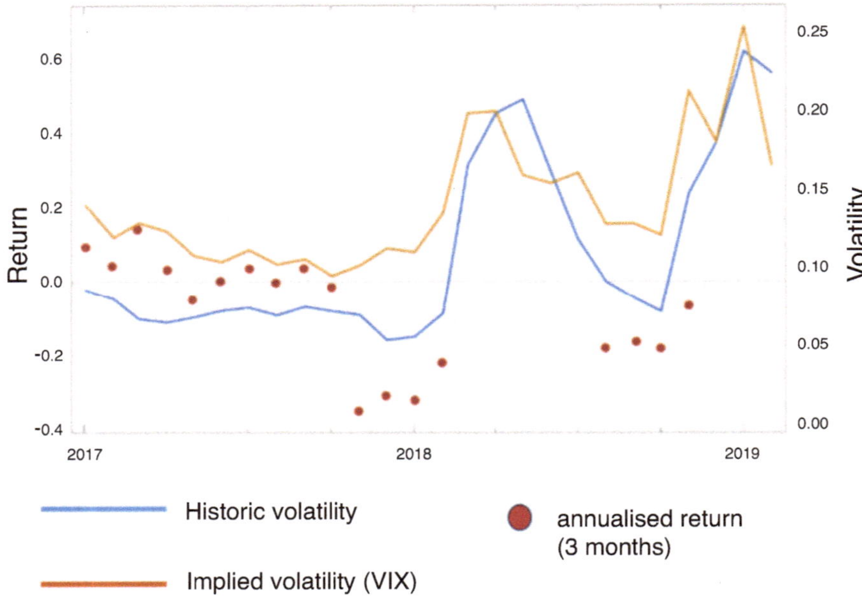

Fig. 3.116 Analysis of the second strategy in the period from January 2017 to February 2019 with about 3 months to expiration and strike price close to the S&P500 plus 50 points, with extension

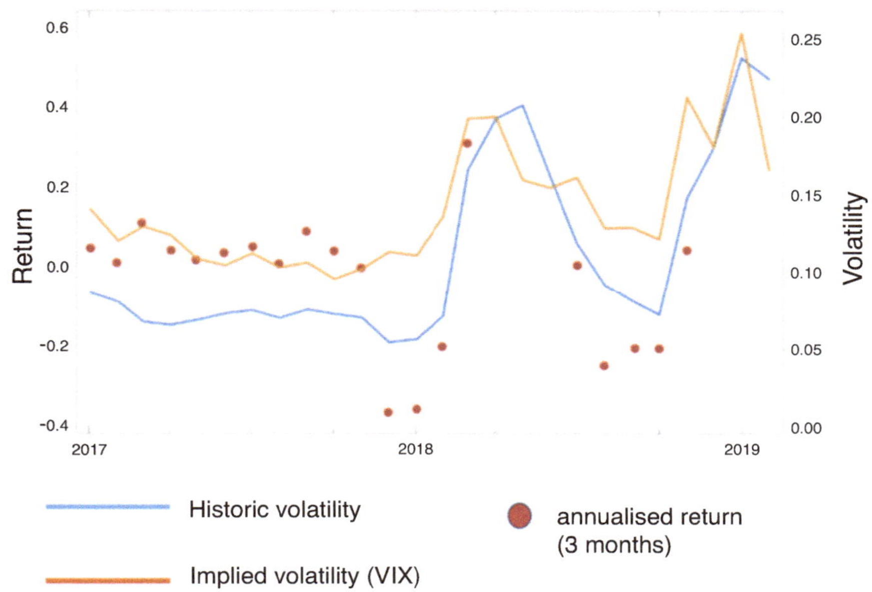

Fig. 3.117 Analysis of the second strategy in the period from January 2017 to February 2019 with about 3 months to expiration and strike price close to the S&P500 minus 50 points, with extension

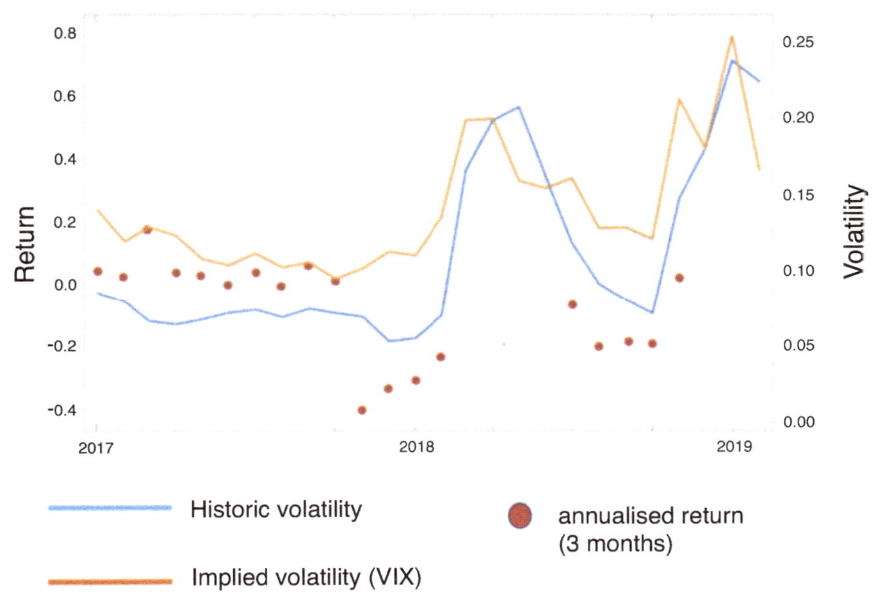

Fig. 3.118 Analysis of the second strategy in the period from January 2017 to February 2019 with about 3 months to expiration and strike price close to the S&P500, with extension

Table 3.28 Annual returns of the second strategy in the period from January 2017 to February 2019 with extension for various strike prices and about 3 months to expiration

Month	Strike near SPX price	Strike near SPX price + 50	Strike near SPX price − 50
2017			
January	4.49%	9.44%	5.06%
February	2.60%	4.28%	1.42%
March	17.71%	14.18%	11.39%
April	3.90%	3.22%	4.53%
May	2.93%	−4.99%	2.08%
June	−0.07%	0.17%	3.90%
July	3.89%	3.54%	5.40%
August	−0.51%	−0.29%	1.06%
September	6.09%	3.60%	9.28%
October	1.22%	−1.49%	4.34%
November	−40.44%	−35.07%	0.07%
December	−33.62%	−30.91	−36.63%
2018			
January	−30.92%	−32.11%	−35.80%
February	−23.37%	−22.09%	−20.03%
March	–	–	31.36%
July	−6.62%	–	0.77%
August	−20.00%	−18.10%	−24.79%
September	−18.49%	−16.40%	−20.41%
October	−19.10%	−18.23%	−20.48%
November	2.17%	−6.72%	4.60%

3.14 Case Study XIV: Analysis of the Derivative Trading Strategy "Lambda +"

3.14.1 Introduction

In this chapter we will again analyse a derivative trading strategy and its variants, which had been commissioned by an asset management company. We are again looking at a trading strategy based on SPX options. However, this time, futures (specifically mini futures) on the SPX will also be used as a component of the strategy—at least in some variants thereof.

To enable an analysis of the strategy and its variants, another extensive analysis software based on historical option prices was developed and has been made available to the readers of this book.

See https://app.lsqf.org/Lambdastrategie.

Therefore, we will not run any detailed test evaluations in this chapter. Instead we will present and motivate the basic strategy including its variants, explain the use of the analysis software, and provide just a few strategy evaluations. Beyond that, readers are encouraged to independently review the performance of various strategy variants in the past with the help of the analysis software. The software was developed and programmed by Lukas Larcher.

3.14.2 Definition of the Strategy and Its Variants

The strategy derives its name from its basic structure (compare Fig. 3.119), which is a fixed part of the strategy in every variant, namely: At all times throughout the strategy's life, we hold a combination of a short call position and a short put position on the SPX with the same expiration and the same strike price K.

In most variants, especially in the basic variant, the option combination's strike price K is almost exactly at the money when entered into, i.e. very close to the current SPX price.

The basic strategy component is therefore always a formation in the profit function that looks like an uppercase Greek lambda, hence the name "Lambda +" for this group of strategies (Lambda formation plus variants).

Upon opening, the options usually have a lifetime of about 1 week to 1 month until expiration.

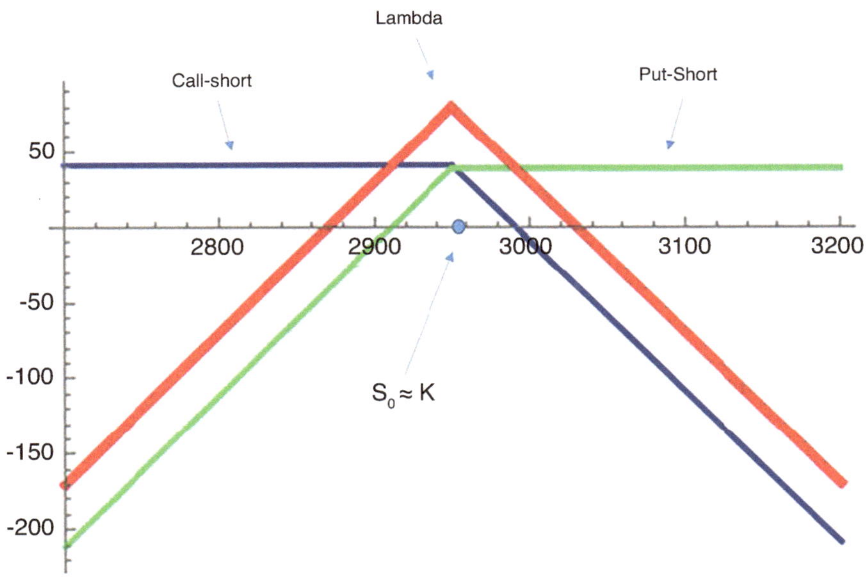

Fig. 3.119 Basic structure of Lambda +

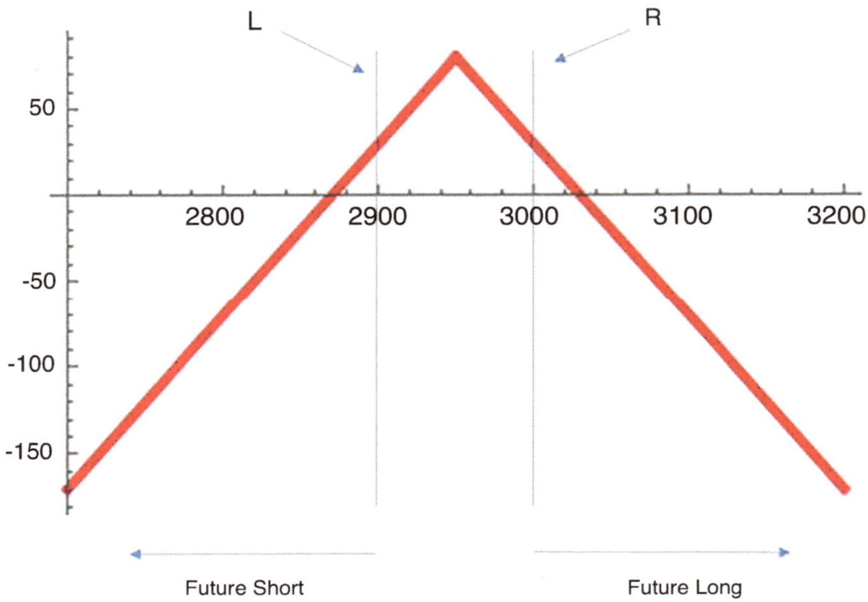

Fig. 3.120 Variant 1

Different variants are now used in attempts to hedge against possible losses:

Variant 1

In Variant 1, two values L and R are defined to the left and right of the strike price in such a way that both values are still within the profit range of the basic strategy (in Fig. 3.120, for example, these values are $L = 2900$ and $R = 3000$ to the left and right of strike 2950). Now, if and as soon as the SPX falls below L, futures are gone short (and closed again as soon as the SPX moves back above L), while if and as soon as the SPX rises above R, futures are gone long (and closed again as soon as the SPX falls below R). In all cases, the number of futures is equal to the number of individual options. Thus, for each contract (100 units) of the short put/short call combination, there are two contracts of mini futures (2×50 units).

If the SPX moves up beyond R, the resulting profits from the long future position offset the losses from the short call position. If the SPX continues to fall below L, the resulting profits from the short future position offset the losses from the short put position.

So, if the SPX price moves between L and R over the strategy's life, or if it exhibits a continuous upward movement beyond R or a continuous downward movement below L, the strategy should always generate a profit (viz. part of the premium initially received from the short option positions).

Losses may only be incurred if the thresholds at L or R are exceeded multiple times, that is, if future contracts are entered and exited multiple times, causing reaction time losses in each case (it is generally not possible to execute a trade exactly at the planned threshold and thus so-called slippage losses are incurred).

In principle, the limits L and R in this case (and in the following variants) can be time-variable (and also variable with regard to the current volatility of the underlying, i.e. as a function of the current option prices).

Variant 2

In Variant 2, one leg of the strategy is hedged by a long option position, and only fcr the second leg do we use futures as set out above. Thus (Case 1) a long put position is bought with a strike M lower than K. The strike M is chosen such that even if the SPX is below M at expiration, the strategy will generate a profit (or at most a small loss) (Fig. 3.121). Upward hedging is done with the future strategy described above. (This approach is used above all when there are clear indications that the SPX will fall.)

Or (Case 2): A long call position is bought with a strike N greater than K. Here, the strike N is chosen such that even if the SPX is above N at expiration, the strategy will generate a profit (see Fig. 3.122). Downward hedging is done with the future strategy described above. (This approach is used above all when there are clear indications that the SPX will move up.)

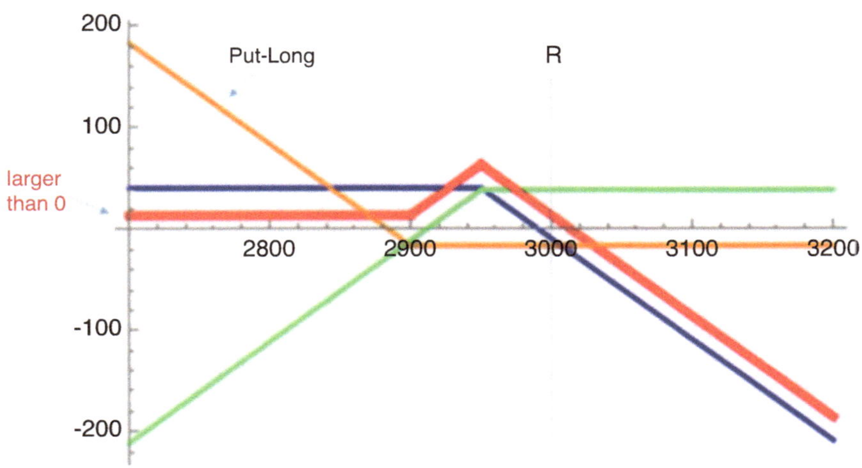

Fig. 3.121 Variant 2, downward hedging with long put

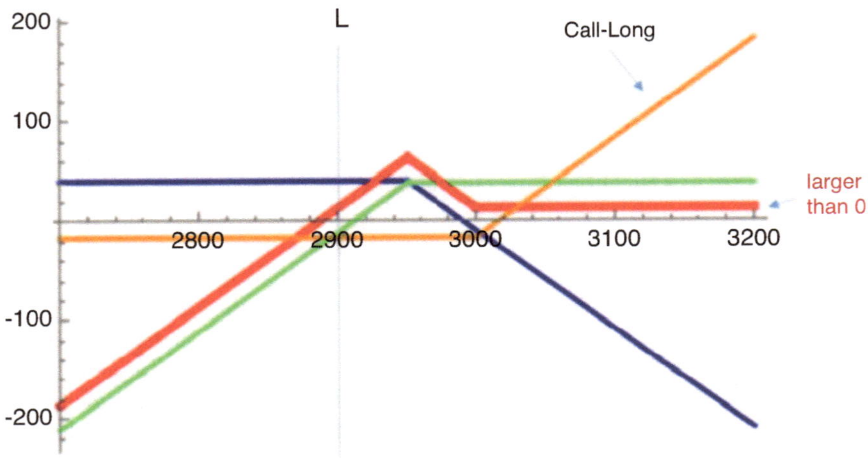

Fig. 3.122 Variant 2, upward hedging with long call

Variant 3
Variant 3 is like Variant 1, except that we do not use futures if the (time and volatility dependent) variable bounds L or R are reached; instead, if that happens, all option positions are closed.

Variant 4
Variant 4 is like Variant 2, except that we do not use futures if the respective (time and volatility dependent) variable bounds R or L, to be hedged by futures, are reached; instead, if that happens, all option positions are closed.

Variant 5
Variant 5 is like Variant 2, except that we do not use futures if the respective (time and volatility dependent) variable bounds R or L, to be hedged by futures, are reached; instead, if that happens, a further short call and short put position is entered with a strike directly at the new level of the SPX (i.e. with strike at R or at L) and with the same remaining time to expiration as the other options. Note that this variant requires provision of sufficient margin. The resulting typical structure of the profit functions (in black) is illustrated in Figs. 3.123 and 3.124.

As can be seen from the graphs of the (black) profit functions, the additional short combination to the right of R and to the left of L increases the profit range.

Entering into another short combination, as was done here, usually makes sense only when the options have a short period left to expiration.

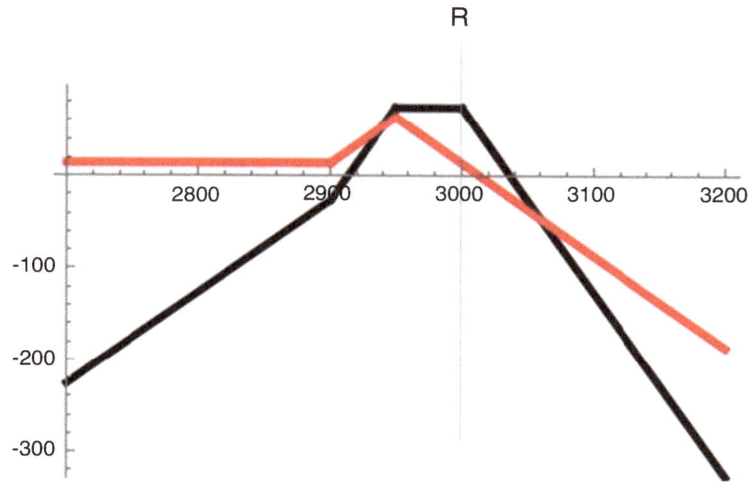

Fig. 3.123 Variant 5 with downward hedging through put position, original version (red), with second short combination with strike at R (black)

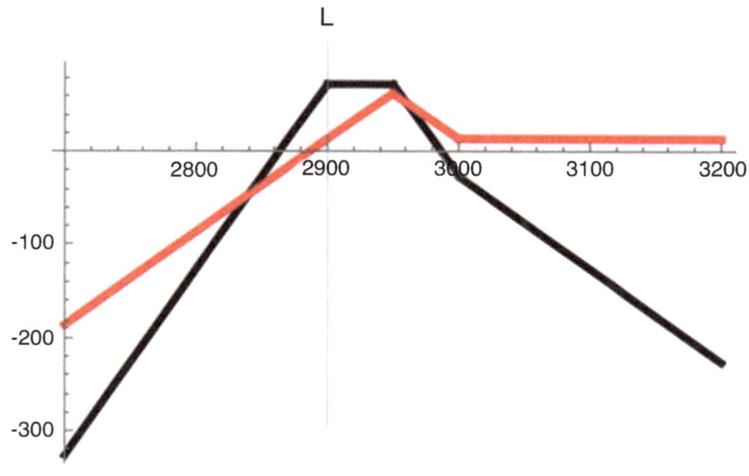

Fig. 3.124 Variant 5 with upward hedging through call position, original version (red), with second short combination with strike at L (black)

Variant 6

In Variant 6 (as well as in Variant 7), one starts from the situation described in Variant 1 after entering a relevant future position.

For Variant 6, this means: When opening the strategy, a combination of short call and short put is entered into with a strike K "far" below the current SPX level S_0, and at the same time a long futures position is opened on the SPX. The

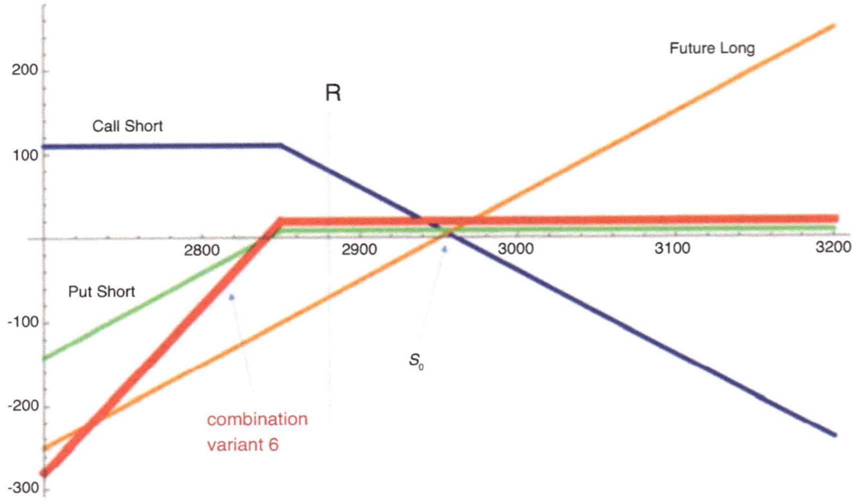

Fig. 3.125 Variant 6

corresponding profit function (identifying the future position with the SPX) can be seen in Fig. 3.125.

When a level R to the right of K is reached, all positions are closed.

Variant 7

When opening the strategy, a combination of short call and short put is entered into with a strike K "far" above the current SPX level S_0, and at the same time, a short future position is opened on the SPX. The corresponding profit function (identifying the future position with the SPX) is shown in Fig. 3.126.

When a level L to the left of K is reached, all positions are closed.

Of course, with these two variants, additional directives need to be given as to how far above or below S_0 that strike price K should be chosen.

3.14.3 The Task and the Analyses

The task given by the client was:

- To reliably test the performance of the individual variants on the basis of historical data and under real trading conditions
- To test the performance of certain indications that determine which variant is chosen in each case

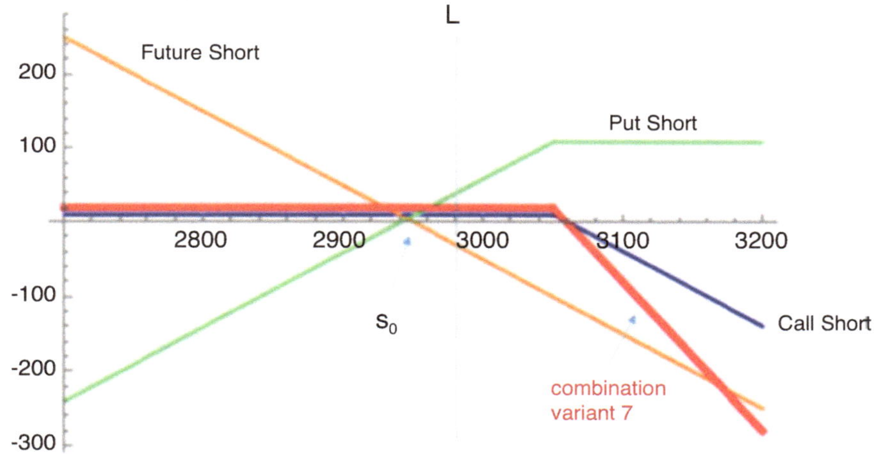

Fig. 3.126 Variant 7

Subsequently, we implemented the strategy's hedging component, which, depending on the variant, consisted of either closing all positions or entering and exiting future positions, into the Interactive Brokers trading system such that it automatically executed the hedging component when the strategy was opened. Of course, in implementing this automated future trading, various measures had to be taken to prevent oscillating future trading (i.e. frequently entering and exiting future contracts with oscillating SPX prices above the L or R bounds) as well as large reaction time losses.

This is not the place, however, to address the comprehensive analyses and implementations in detail.

Still, as part of our software offering, we want to provide readers with a program that can be used to test some of the variants (at least the basic variants) very easily and thoroughly.

In the following we will only give **one** example of an analysis of a simple variant, accompanied by an explanation of the constraints.

In the following **numerical analysis**, we test **the basic Variant 2 with consistent downward hedging by means of put options** and **closing all positions as soon as the SPX moves above a certain level *R*** (i.e. Variant 4).

The parameters that can be chosen freely in this variant are the strike M of the long put position and the exit level R.

The initial investment capital was assumed to be 100,000 USD.

"Essentially" **one contract** of the option combination was traded for each existing investment of 100,000 USD.

In the relevant time periods, the maximum margin required to execute the strategy (the SPX stayed below 3000 points throughout) was essentially $\max(100 \cdot (K - M), 100 \cdot (R - K)) + 15\%$ of $100 \cdot K$ USD.

Since in all situations, both $K - M$ and $R - K$ were less than 100, the maximum margin required was less than 55,000 USD.

Here is an example of how such a trade would be executed:

- Choose the long put strike such that if the SPX falls, a profit of at least 1% of the investment sum is made.
- Exit the strategy at a level R that is exactly at the break-even on the right boundary of the strategy.
- Re-enter the strategy on the next third Friday.

With the option prices on 18 October 2019 (3rd Friday of October; see Fig. 3.127) and the SPX at 2986.39 points and assuming an initial cash investment of exactly 100,000 USD, the basic constellation in this example would have been as follows:
Sale of a put contract on the SPX with expiration 14 November 2019, strike 2985 at the price of 3840 dollars.
Sale of a call contract on the SPX with expiration 14 November 2019, strike 2985 at the price of 4190 dollars.
Purchase of a put contract on the SPX with expiration 14 November 2019, strike 2940 at the price of 2510 dollars.

The total premium received is 5520 USD. (We ignore the negligible transaction costs of about 6 USD here.)

If the SPX were to fall below 2940, a payoff of $100 \cdot (2985 - 2940) = 4500$ USD would have to be paid at expiration for the short put/long put combination, resulting in a profit of 1020 USD (i.e. slightly more than 1% of the investment amount). Losses would be incurred if the SPX were to move up to a level above $2985 + 55.20 = 3040.20$. Upon the SPX reaching the threshold of 3040 points, all positions are closed.

The profit situation **at expiration** is shown in Fig. 3.128.

It is not possible to make a definitive advance estimate as to how high the loss would be if all positions were closed upon the SPX rising to 3040 points. The loss depends both on the relevant remaining life of the strategy and the implied volatility of the SPX at that moment. That said, it is generally to be expected that the implied volatility will have fallen due to the strong increase in the SPX and that the positions can be closed at comparatively favourable prices.

As a rough estimate, the expected loss can be assumed to be in the range of 0% to 5%, depending on when the level R is reached. The maximum profit that can be obtained with this approach is 5520 USD, i.e. about 5.52% of the investment.

Below we run the strategy with a **different choice of parameters** over a time period of almost a year and analyse the performance using our software on our website:

- The initial investment is 100,000 USD.
- The period under review is 1 January 2018 to 20 November 2018.
- The traded options have a lifetime of essentially 2 weeks (each with an expiration date as close as possible to the 1st of a month and the 15th of a month).

Calls		Puts
Bid x Ask	Strikes	Bid x Ask
116.00 x 120.70	2885	14.60 x 15.10
111.90 x 116.30	2890	15.30 x 15.80
107.60 x 112.00	2895	16.10 x 16.60
103.50 x 107.70	2900	16.90 x 17.40
99.40 x 103.50	2905	17.70 x 18.30
95.40 x 99.30	2910	18.60 x 19.10
92.30 x 94.20	2915	19.50 x 20.10
88.20 x 90.20	2920	20.40 x 21.60
84.20 x 86.20	2925	21,40 x 22.00
80.30 x 82.20	2930	22.50 x 23.10
75.40 x 78.30	2935	23.60 x 24.20
72.60 x 74.50	2940	24.70 x 25.40
68.80 x 70.70	2945	26.00 x 26.60
65.00 x 66.90	2950	27.20 x 27.90
61.40 x 63.30	2955	28.50 x 29.20
57.80 x 59.60	2960	29.90 x 30.70
54.80 x 55.70	2965	31.40 x 31.10
51.40 x 52.20	2970	32.90 x 33.70
48.00 x 48.90	2975	34.50 x 35.30
44.80 x 45.60	2980	36.30 x 37.00
41.60 x 42.40	2985	38.10 x 38.90
38.50 x 39.20	2990	39.90 x 40.70
35.50 x 36.20	2995	41.90 x 42.80

Fig. 3.127 Option prices on 18 October 2019

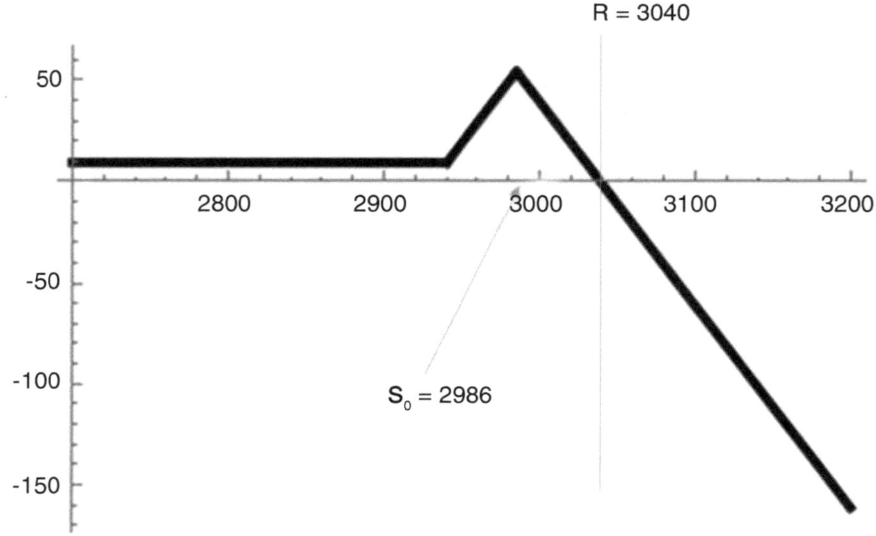

Fig. 3.128 Profit situation at expiration in the above example

- We choose the smallest possible strike for the long put position so that, if the SPX falls below this strike at expiration, the maximum loss would be 500 USD.
- In the case of a sharp rise in the SPX, all positions will be closed if the SPX reaches a level of more than 100 points above the short strike.

We deliberately present only one singular test example over a very short period of time, since we want to "push" our readers to visit our website, use our software, and run their own tests on a wide range of variants of this strategy. See https://app.lsqf.org/Lambdastrategie.

The following Fig. 3.129 lists the payoffs in this strategy with all the details as they are shown in our software.

Our software then graphs the profit curve as shown in Fig. 3.130:

Finally, in Fig. 3.131 the profit curve of this strategy (blue) is compared with the profit achieved when investing 100,000 dollars directly in the SPX (red).

Now, these analyses obviously cannot take into account individual strategy modifications and adaptations to various situations; they can only be based on strictly predefined strategy instructions. Whether using strictly predefined strategies is the optimal choice for successful performance is another question altogether, of course. Yet whatever choices are made, one thing is certain: When trading on behalf of third-party investors (i.e. when not investing one's own money), the strategies to be executed must be clearly defined and communicated in advance, including for one's own protection. It is the only way to ensure one has a solid defence against

return: 2.88 %
return p.a.: 3.246 %

net worth	date of trade	under-lying	maturity	short-strike	long-strike	premium short-op.	costs long-op.	number contracts	win	result	cumul. result
100 000,00	02.01.2018	2 695,81	16.01.2018	2 695	2 675	2 500,00	760,00	3	✗	-6 417,00	-6 417,00
93 583,00	16.01.2018	2 776,42	31.01.2018	2 775	2 740	4 210,00	800,00	3	✗	-1 486,00	-7 903,00
92 097,00	31.01.2018	2 823,81	14.02.2018	2 825	2 785	5 150,00	1 335,00	3	✗	-200,00	-8 103,00
91 897,00	14.02.2018	2 698,63	28.02.2018	2 700	2 650	6 425,00	1 765,00	3	✓	3 262,00	-4 841,00
95 159,00	28.02.2018	2 713,83	14.03.2018	2 715	2 670	6 220,00	1 855,00	3	✓	902,00	-3 939,00
96 061,00	14.03.2018	2 749,48	28.03.2018	2 750	2 710	5 340,00	1 465,00	3	✗	-140,00	-4 079,00
95 921,00	28.03.2018	2 605,00	11.04.2018	2 605	2 540	8 435,00	2 115,00	3	✓	2 586,00	-1 493,00
98 507,00	11.04.2018	2 642,19	25.04.2018	2 640	2 580	7 305,00	1 730,00	3	✓	5 500,00	4 007,00
104 007,00	25.04.2018	2 639,40	09.05.2018	2 640	2 585	6 710,00	1 490,00	3	✗	-574,00	3 433,00
103 433,00	09.05.2018	2 697,79	23.05.2018	2 700	2 665	4 420,00	1 215,00	3	✗	-139,00	3 294,00
103 294,00	23.05.2018	2 733,29	06.06.2018	2 735	2 705	3 555,00	1 000,00	3	✗	-1 195,00	2 099,00
102 099,00	06.06.2018	2 772,35	20.06.2018	2 770	2 740	3 710,00	905,00	3	✓	2 522,00	4 621,00
104 621,00	20.06.2018	2 767,32	06.07.2018	2 765	2 730	4 255,00	1 050,00	3	✓	2 672,00	7 293,00
107 293,00	06.07.2018	2 759,82	20.07.2018	2 760	2 725	4 125,00	1 030,00	3	✗	-1 103,00	6 190,00
106 190,00	20.07.2018	2 801,83	03.08.2018	2 800	2 770	3 880,00	1 060,00	3	✗	-1 230,00	4 960,00
104 960,00	03.08.2018	2 840,35	17.08.2018	2 840	2 815	3 170,00	935,00	3	✓	1 207,00	6 167,00
106 167,00	17.08.2018	2 850,13	31.08.2018	2 850	2 820	3 890,00	1 075,00	3	✗	-2 352,00	3 815,00
103 815,00	31.08.2018	2 901,52	14.09.2018	2 900	2 870	3 760,00	1 005,00	3	✓	2 242,00	6 057,00
106 057,00	14.09.2018	2 904,98	28.09.2018	2 905	2 875	3 490,00	880,00	3	✓	1 697,00	7 754,00
107 754,00	28.09.2018	2 913,98	12.10.2018	2 915	2 885	3 730,00	1 085,00	3	✗	-370,00	7 384,00
107 384,00	12.10.2018	2 767,13	26.10.2018	2 765	2 700	8 445,00	2 290,00	3	✗	-360,00	7 024,00
107 024,00	26.10.2018	2 658,69	09.11.2018	2 660	2 575	10 770,00	2 490,00	3	✗	-3 836,00	3 188,00
103 188,00	09.11.2018	2 781,01	23.11.2018	2 780	2 735	5 825,00	1 615,00	3	✗	-305,00	2 883,00

Fig. 3.129 Payoffs of the chosen strategy

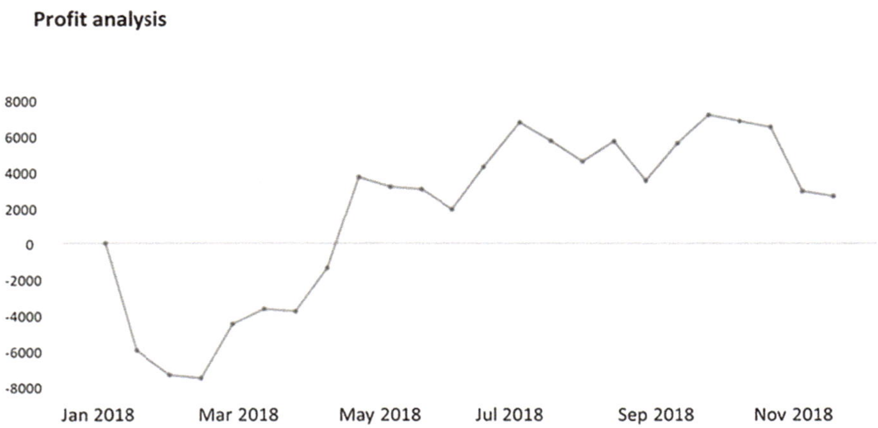

Fig. 3.130 Profit curve of our test example

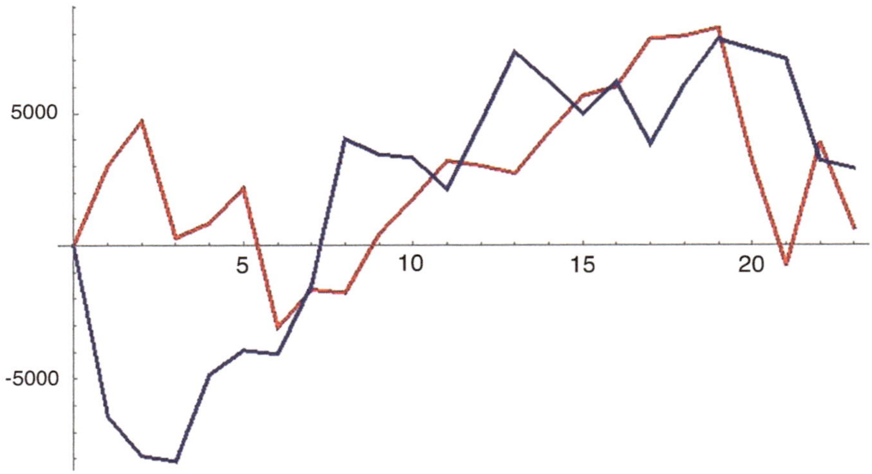

Fig. 3.131 Profit curve of our test example (blue) in comparison with the profit curve when investing directly in the SPX (red)

claims by investors that losses, in case the strategy was not successful, were caused by allegedly " faulty decisions".

References

1. Jason Ungar and Matthew Moran. "The Cash-Secured Put Write Strategy and Performance of Related Benchmark Indexes". In: The *Journal of Alternative Investments* 11.4 (2009), pp. 43–56.
2. Theodore Day and Craig Lewis. "Stock market volatility and the information content of stock index options". In: *Journal of Econometrics* 52.1–2 (1992), pp. 267–287.
3. Pedro Santa-Clara and Alessio Saretto. "Option strategies: Good deals and margin calls". In: *SSRN Electronic Journal* 12.3 (2009), pp. 391–417.
4. Lucia Del Chicca, Gerhard Larcher, and Michaela Szölgyenyi. "Modeling and Performance of Certain Put-Write-Strategies". In: *The Journal of Alternative Investments* 15.4 (2013), pp. 74–86.
5. Alexander Brunhuemer, Gerhard Larcher, and Lukas Larcher. "Analysis of Option Trading Strategies Based on the Relation of Implied and Realized S&P500 Volatilities". In: *ACRN Journal of Finance and Risk Perspectives* 10, Special Issue 18th FRAP Confer. (2021), pp. 166–203.
6. Lucia Del Chicca and Gerhard Larcher. "A Comparison of Different Families of Put-Write Option Strategies". In: *ACRN Journal of Finance and Risk Perspectives* 1.1 (2012), pp. 1–14.
7. Paul Glasserman. *Monte Carlo Methods in Financial Engineering*. Springer, 2004.
8. Paul Glasserman. "Shortfall Risk in Long-Term Hedging with Short-Term Futures Contracts". In: *Option pricing, interest rates and risk management, eds. J. Cvitanic, E. Jouini und M. Musiela, Cambridge University Press* (2001), pp. 477–508.
9. Gerhard Larcher and Gunther Leobacher. "An optimal strategy for hedging with short-term futures contracts". In: *Mathematical Finance* 13.2 (2003), pp. 331–344.
10. Riaz Ahmad and Paul Wilmott. "Which Free Lunch Would You Like Today, Sir?: Delta Hedging, Volatility Arbitrage and Optimal Portfolios". In: *Wilmott Magazine* (Jan. 2005).